THE
NORTON
TRILOGY

THE NORTON TRILOGY

JACK L. AUGUST JR.

TCU
PRESS

Fort Worth, Texas

All photographs, unless otherwise identified, are from the photo collection of John R. Norton III.

Library of Congress Cataloging-in-Publication Data

August, Jack L., Jr., 1954-
 The Norton trilogy / Jack August.
  p. cm.
 Includes bibliographical references and index.
 ISBN 978-0-87565-547-5 (hardcover : alk. paper)
1. Norton, John R., 1854-1923. 2. Norton, John R., Jr., 1901-1987. 3. Norton, John R., III, 1929- 4. J.R. Norton Company—History. 5. John Norton Farms—History. 6. Agricultural industries—Southwest, New—History—19th century. 7. Agricultural industries—Southwest, New—History—20th century. 8. Water resources development—Southwest, New—History—19th century. 9. Water resources development—Southwest, New—History—20th century. 10. Water in agriculture—Southwest, New—History—19th century. 11. Water in agriculture—Southwest, New—History—20th century. 12. Arizona—History. 13. Southwest, New—History—1848- I. Title.
 HD9010.N67A94 2013
 338.7'630922791—dc23
          2012038200

TCU Press
P.O. Box 298300
Fort Worth, Texas 76129
817.257.7822
www.prs.tcu.edu

To order books: 1.800.826.8911

Designed by Bill Brammer
www.fusion29.com

CONTENTS

FOREWORD

THE HONORABLE
JON KYL, UNITED STATES SENATOR

I HAVE KNOWN ARIZONA NATIVE JOHN R. NORTON III FOR MOST OF MY PROFESSIONAL LIFE, and first met him in the late 1960s when I was a young lawyer commencing my career at the Phoenix firm of Jennings, Strouss & Salmon. By that time he was one of Arizona's agribusiness leaders, though his headquarters were then located in Blythe, California, on the west side of the Colorado River, where his land and water rights were perfected and secure. His regional agricultural and ranching operations extended from New Mexico through Arizona to California, and he had already begun his parallel service to the community, whether through board affiliations, philanthropy, or political engagement. I recall also in 1980, as my legal practice moved increasingly into the areas of water and natural resources, that John, a licensed pilot since the early 1950s, was kind enough to fly me in his personal aircraft to Denver, where we attended a quarterly meeting of the Mountain States Legal Foundation. He sat on the board of directors and I served on the board of litigation, and our mutual interests in land and water issues were evident. His service to his industry and his record of achievement in it, as well as his distinguished record in the military during the Korean War, led ultimately to his nomination and confirmation as US Deputy Secretary of Agriculture in the administration of President Ronald Reagan in 1985.

During my time in elective office, which began in the House of Representatives in 1987 and then moved to the US Senate in 1995, John has been a supporter, friend, and champion of Arizona and the nation. His support of the University of Arizona, from which we both graduated, has been manifested in numerous acts of philanthropy. The John and Doris Norton School of Consumer and Behavioral Sciences in the College of Agriculture and Life Sciences at UA, for example, stands as concrete testament to his support for higher education. Over the past two decades, moreover, John and his wife, Doris, have made charitable donations to the arts, health care, and countless other causes that improve the quality of life in Arizona and the Greater Southwest.

Yet Dr. Jack L. August Jr.'s *The Norton Trilogy* is about more than John's remarkable life and career. It extends back two generations, tracing and interpreting the lives of John's grandfather, John R. Norton, and father, John Jr., whose careers served as metaphors for the growth and development of the region. From a sparsely populated and primitive territory struggling with environmental limitations and excesses—the vicissitudes of flood and drought—the eldest Norton, along with his pioneer brethren, forged a rough-hewn existence in an unforgiving arid environment. When John R. Norton first arrived in Arizona in 1881 to work on the grading of the new railroad track bed across the northern part of the territory, scarcely forty thousand people lived within its borders. From that initial foray, Norton, by 1884, had found his way to the embryonic agricultural outpost of Phoenix, where he became foreman of the crew that built the 40.5-mile Arizona Canal, which opened up Salt River Valley lands to a new economic world of irrigated agriculture. This altered dramatically the arc of agricultural development in this emerging crossroads of commerce. In 1889, Norton was one of three members of the congressionally supported Breakenridge Survey, which located the future site of Roosevelt Dam, the enduring symbol of President Theodore Roosevelt's federal conservation policy. The dam, among the first built under the Newlands Reclamation Act of 1902, became the cornerstone of the Salt River Project, one of the first and most successful federal reclamation projects in American history. Appropriately, John R. Norton attended the dedication ceremonies on March 18, 1911, at which then-former President Roosevelt spoke to an enthusiastic crowd of territorial citizens. John R. Norton lived through the first two decades of the twentieth century, and his agricultural pursuits, livestock interests, and civic engagement made him a well-respected and even beloved figure in the emerging oasis in the desert. His life and career laid the foundation for his distinguished descendants in the twentieth and twenty-first centuries.

John Jr. was born in 1901, and his lifetime saw dramatic agricultural cycles that shifted from the extremes of two national depressions to halcyon days brought about by the market demands of World War I and World War II. He matured during the early years of federal reclamation in the Salt River Valley and witnessed the inexorable urbanization that took place during the middle decades of the twentieth century. When the dust cleared in the late 1940s and 1950s, John Jr. had fashioned a distinguished career in produce and livestock that was the envy of many of his contemporaries. By midcentury he looked out upon a region of growth, modernization, and progress thanks in large part to the development and wise use of the Southwest's natural resources. He was one of the "Produce Gang" of mid-twentieth-century Arizona and indulged his fondness for horses and cattle ranching throughout his lifetime. Unlike John R. Norton, an old-time states' rights southern Democrat, John Jr. found his political views consistent with the so-called "Pinto Democrats" who by the second half of the twentieth century registered as Republicans. Not as politically active as his father, John Jr., who in 1940 tired of President Franklin Roosevelt's seemingly endless New Deal, took the time to serve as state chairman of "Democrats for Wilkie," the surprise Republican nominee for president in that year. In the long term John Jr. served as a business exemplar for his son, John III, who later built upon and learned from the agricultural and ranching enterprises his father established beginning in the aftermath of the Great Depression.

Though he refers to himself as a "farmer," John III proved to be an astute and insightful businessman who throughout his career demonstrated wise use and stewardship of the Southwest's natural resources. As a water rights attorney prior to seeking elected office, I often sought John III's counsel on specific issues that affected irrigated agriculture. He foresaw the need to change and update technology in agricultural production when necessary and withstood the economic cycles endemic to agribusiness. He saw the critical role that water rights played in the produce, cotton, and livestock businesses and experienced firsthand the roiling legal and political controversies that shaped the history of Colorado River water rights. As Dr. August clearly points out, Norton III knew that the Blythe area had gained prior rights early in the hydropolitics struggle that characterized regional development during the middle decades of the twentieth century. With the securing of land that held prior rights, John III's fortunes rose. What the third John Norton foresaw—and doubtless he owes his father and grandfather a debt of gratitude on this score—was that water resource certainty drove the regional economy. Early in his professional career, John III sounded the alarm about the serious challenges that lay ahead for the states in the Colorado River Basin. He always focused on practical solutions and emphasized the need to study how future agriculture and energy production could keep pace as water supplies tightened. He advocated a measured approach to water issues and their impact on the economies of Arizona and other western states. And when I ran for office I could always count on John's support, whether it was on the campaign trail or supporting a policy that impacted the region's agribusiness sector. Finally, *The Norton Trilogy* addresses important themes that not only tie together these three fascinating lives but also provides historical perspectives on water and land use that will inform and shape our future. It adds much to our understanding of the mutually reinforcing roles of agriculture, water resource development, and the irrigation of arid lands, in the overall historical tapestry that comprises Arizona and the Southwest. Pioneer families like the Nortons, Haydens, Goldwaters, and Babbitts, among others, provide valuable insight into our historical and regional identity. Also, the roles of philanthropy and public service—giving back to the community—maintain a central place in this lively narrative, and these elements too are essential components to a region's maturity. Personal, political, and professional relationships, moreover, stand out as key factors in public stewardship, the kind demonstrated by my friend, John R. Norton III, and by extension, his father and grandfather. The Nortons, who never suffered from the lack of a work ethic, have made Arizona and the nation a better place. This book is as much an American story as it is an Arizona one.

ACKNOWLEDGMENTS

MY INTEREST IN THE HISTORY AND CULTURE OF THE AMERICAN SOUTHWEST BEGAN IN EARNEST DURING MY YEARS AS AN UNDERGRADUATE STUDENT AT YALE UNIVERSITY IN THE 1970s. This undiminished hunger for the history of this unique region of the country was rooted in the research and teaching of Howard Roberts Lamar, Sterling Professor of History and President Emeritus at Yale. Many students from western states who attended Yale gravitated toward Dr. Lamar, not only because he was pleasant, approachable, and the nation's leading scholar on the history of the American West, but also because he had a dog named Tucson. He often invited these wayward students from the West to what we called, "Rancho Lamar," where he served Mexican food, much to our delight. He tried to make us feel at home in an unfamiliar environment. Of course, I took every course he offered, and in my sophomore year he assigned Robert Utley's *Last Days of the Sioux Nation*. One evening, early in the semester, I picked up Utley's book, intending to thumb through the introduction. I did not put it down until I had read the last sentence at 6:00 a.m. the next morning. Twelve years later, after I had completed and defended my doctoral dissertation on water resource development in the Southwest, Robert Utley cooked me dinner at his home in Santa Fe, New Mexico.

The Norton Trilogy is a direct result of these early and potent influences on my life and career. It is a natural extension of my previous books on the American Southwest, and I owe a debt of gratitude to the various scholarly mentors and writers in a host of disciplines who shaped my views on the critical roles that federal reclamation, law, politics, and individual initiative played in the settlement and growth of this often unforgiving region of the country. In short, this work is a rumination on the history of water and agribusiness in the American Southwest through the lives of three generations of John Ruddle Nortons.

I owe much to John Ruddle Norton III and his wife, Doris, whose willingness to work with me in terms of securing appropriate sources as well as in providing guidance and wisdom throughout the process enabled me to complete the research and writing of a three-generation story that spans the nineteenth to the twenty-first century. John III"s business acumen, generosity of spirit, and public stewardship have touched many people and institutions during his lifetime. Some, like former Senator Dennis DeConcini (D-Arizona), former Governor and Secretary of the Interior Bruce Babbitt (D-Arizona), and Senator Jeff Flake (R-Arizona) were kind enough to offer quotes about John III and *The Norton Trilogy*. And I was honored that one of John III's good friends, former Senator Jon Kyl, wrote an insightful foreword to the book. I appreciate the endorsement of these distinguished Arizonans from both sides of the political aisle.

In addition to those individuals noted above, Megan Doyle, intern, Kathy Walton, editor, and Melinda Esco, production guru at TCU Press, whose patience and professionalism are evidenced in this work, deserve much credit for the final product. Their years on the editorial front lines have garnered them enviable reputations nationwide. The various archives in the universities of Arizona, California, New Mexico, and the National Archives in Washington, DC, deserve special mention. As most scholars involved in serious historical research know, these repositories of our collective memory maintain the most highly qualified staffs. Archivists' guidance into the innumerable collections under their charge is crucial for the historian and other scholars. To all of these people and my family and friends who understand and accept my passion for this work, I extend my gratitude.

INTRODUCTION

◇◇◇

THE PRIMARY REASON TO WRITE A BOOK IS TO TELL A STO-
RY THAT HAS NOT BEEN TOLD. Surprisingly, no one has addressed
the history of the John Ruddle Norton family and its three generations of agricultural enterprise in one of the most arid regions of the United States. Historians have not ignored the general topic of water and irrigated agriculture.
There exists a variety of accounts addressing the political, legal, and economic
implications of putting water to beneficial use in Arizona in particular, and
the Colorado River Basin states of the American West in general. This work
centers on the arid reaches of Arizona and southern California, and although
students of the region know of the century-long struggle over the rights to
Colorado River resources which divided these two great states, this story em-
braces both sides of that legendary interstate stream, which remains the lifeblood of the Pacific Southwest.

Several themes run through *The Norton Trilogy*—the lives of John R. Norton (1854-1923), John R. Norton Jr. (1901-1987), and John R. Norton III
(1929-present). The most important is the interplay between human values
and the waterscape. Technology, of course, played a monumental role in this
drama—technology that leaped from the early Native American digging stick
to the dynamite, bulldozers, and reinforced concrete that ultimately shaped
the agricultural economy. Another theme is the central role played by government—local, state, regional, and national—in shaping water policy. Finally,
the biographical profiles of each John Norton addressed in this work reveal
much about the history of Arizona and the central role that the quest for water
has played in the growth and development of the region.

Before the Europeans arrived in what is now Arizona and southern California, Indians manipulated water resources for beneficial use, but always
within the context of maintaining their relationship with the natural environment. Spanish and Mexican settlers in the region, from the seventeenth
through nineteenth centuries, viewed natural resources, and water especially,
as goods to be exploited for the community. Significantly, the common good
weighed more heavily than self-interest. With the discovery of gold in the

1

mid-nineteenth century, however, attitudes toward water changed. Thousands poured into California, thence to Arizona, and this unprecedented rush to the Pacific Southwest—the coastal plains of California and the interior regions to the east of them—reflected a frontier individualism and appetite for profit that elevated the exploitation of nature to new levels. This population established the framework for private ownership not just of land, but of moving water—a system that persisted into contemporary times. The population influx and its consequences transformed legal and political institutions. The region's emergence as a collection of water seekers—like the Nortons—is central to this larger story.

In the era just prior to the first John R. Norton's arrival in Arizona Territory, Spanish and Mexican officials established guidelines for local authorities on the haciendas and the few centers of settlement in Tubac and Tucson. Following the American takeover in the late 1840s and early 1850s, the federal system evolved in complex ways that have been reflected in every water project in the region, with the growing power of Washington increasing at the expense of local prerogatives. After 1902, with the passage of the Newlands Reclamation Act, Arizonans and Californians—and westerners in general—looked toward the federal treasury when it came to developing water for beneficial use.

Every society extracts matter and energy from the natural world, relying on its climate, soils, plants, animals, minerals, and sources of water. As the Nortons realized, dry farming was impossible over most of Arizona and southern California. If they wanted to raise crops they needed to provide more water than rainfall alone, so irrigation and water control were mandatory. But the sheer magnitude of water control and conveyance, the types of crops grown and their ultimate value, depended often on factors outside the local economy. Arizona farmers cultivated little cotton until World War I, when the British embargoed long-staple cotton at a time when the US military needed it for tires and airplane fabric. In response, the United States turned to Arizona, with its long growing season, for the government's cotton needs. By this time the Salt River Project was managing the valley's water, and the labor supply south of the border was sufficient to quickly alleviate the emergency. Thus both Norton and Norton Jr. witnessed how a global disruption created a national demand that intersected with Arizona's climatic factors, river systems, irrigation works, and labor availability to produce Arizona's cotton boom, which redounded to the Nortons benefit. The Nortons, however, farmed more, much more, than cotton.

Another theme reflected in the three generations of Nortons was the intimate relationship between private and government interests and goals. Under the Spanish and Mexican systems, a sustainable hydraulic system was built with Indian labor. With US sovereignty, private interests and government bureaucracies aided each other in their quest for power over the resource, for financing and for policy formation. As Robert Kelly argued in his compelling *Battling the Inland Sea: American Political Culture, Public Policy, and the Sacramento Valley, 1850-1986*, "local experience gains its full meaning for us only when we see it within the framework of national experience." This declaration applies directly to each Norton generation.[1]

When assessed from local, regional, and national perspectives, Arizona's water and agricultural achievements, as viewed through the Norton trilogy prism, placed it as a major participant in a booming economy based upon great water projects. These endeavors included irrigation, flood control, and reclamation programs—both private and public—in the late nineteenth through the twentieth centuries. Without the Salt River Project and the Boulder Canyon Project Act, the Nortons would not have been able to make their respective marks on the region's agricultural economy.

As the great historian Walter Prescott Webb noted in 1957, one cannot understand the American West unless we recognize the fact that the desert dominates it.[2] Much of the land is irrigated and green, but the desert, as the Nortons realized, did not disappear. Technology and human ingenuity modified it, but it refused to fade. The literary critic Edmund Wilson noted on a visit to the region that visitors from the East felt a strange sense of unreality toward the West that made human existence appear hollow. Everything seemed out in the open, not underground as in older, more densely populated areas. Wilson mused that the climate and landscape caused such feelings. The sun, rare yet torrential rains, dry mountains, and the vast and void landscape created a uniformity that impacted emotional and psychological well-being in what he concluded was the "Dry West."[3]

During the twentieth century technology mitigated the environment—the Dry West—to a significant degree, whether by windmill, cloud seeding efforts, groundwater drilling, and, in some instances, huge hydroelectric generating dams. Through these means westerners like the Nortons utilized new technologies to bring water to arid areas. In effect, what developed from the early labors of John R. Norton in the late nineteenth and early twentieth century was a kind of "oasis civilization" that clustered almost exclusively in towns

3

and cities. The American West of the twentieth century became an urban civilization, and though the Nortons maintained their agricultural activities in the hinterlands of these emerging population centers—Los Angeles, Phoenix, Tucson, San Diego, Salt Lake City, and Las Vegas—they saw a major portion of the fruits of their labors flow to these exemplars of the New West.

In the course of the Nortons' three lifetimes, the development of the region passed through at least two distinct stages. Between 1890 and 1940 the relationship of the West to the East was like that of a colony to its mother country. This arrangement did not cease entirely with the advent of World War II, but the conflict transformed the region in profound ways that in the succeeding half century, during the better part of Norton III's lifetime, it became a pacesetter for the nation, a blueprint for a new society—socially, culturally, economically, and politically.

During much of the first half of the twentieth century, Arizona and southern California were largely dependent on eastern goods and services for virtually every aspect of the region's development. The East was where the majority of potential new settlers lived, where western farmers and livestock men like the Nortons found markets, and where westerners, in turn, could secure manufactured goods necessary to exploit their natural resources. The region's citizens looked eastward also for cultural values and institutions that could be transplanted, and where political power could be harnessed to facilitate growth and development. Westerners of the late nineteenth and early decades of the twentieth century, like Norton and Norton Jr., were aware of their dependence on the East; they resented it and sought to assert independence in a variety of ways. But World War II wrought changes that began to reverse the situation. Instead of imitating eastern culture, economic practices, and social values, the region began showing signs of innovation that other regions—the Eastern Seaboard in particular—began to emulate. In Norton III's lifetime, the Pacific Southwest became the America of tomorrow.

The Norton Trilogy illustrates the evolution and development of agriculture, from subsistence to corporate, in an environment of water scarcity. Agricultural endeavor as an economic enterprise in the far Southwest and inland California was essential for sustaining a taxpaying civilization. This remained so through the better part of the twentieth century, as evidenced by the lives and careers of the first two generations of Nortons. John R. Norton III, however, whose agribusiness career reached heights that his father and grandfather could never have imagined, effected a transition away from irrigated agriculture that serves as a signpost for the future. The era of corporate irrigated agriculture in Arizona, especially, reached its apex in the late twentieth century, and based on water use and availability, this economic activity and lifestyle appears to be in an inexorable decline. Elected officials and policy makers, responding to calls for the highest and best use of our finite resource, have singled out irrigated agriculture as one of the first areas where water will be sacrificed for other uses. John R. Norton III, the last of the Norton agribusiness line, understands this environmental and economic reality. His life and career, as do those of his forbears, reflect vividly the significant role that water and irrigated agriculture have played in the growth and development of this unique area of the United States for the last one hundred fifty years.

·1·

ON THE EDGE
OF A DESERT EMPIRE

"*The territory we have selected for our home is unlike any other portion of the United States. . . . Attached as we are nominally to the Territory of New Mexico, and situated many hundred miles from its seat of government, the western portion of Arizona is a region without the shadow of anything that claims to be law. . . . So far as we know, no judge or justice, either Federal or Territorial, has ever visited this portion of the country.*"

EDWARD E. CROSS, IN THE *Weekly Arizonan* (TUBAC), MARCH 3, 1859

"*Arizona . . . is just like hell; all it lacks is water and good society.*"

SENATOR BENJAMIN F. WADE, 1863

O n October 3, 1888, as the dread days of summer in the desert Southwest passed, John Ruddle Norton, the superintendent of the Phoenix-based Arizona Improvement Company, celebrated mightily when he learned that the US Senate had appropriated $100,000 for the US Geological Survey (USGS) to determine the feasibility of reclaiming lands in the arid West.[1] Norton, who moved to Arizona from Gibson City, Illinois, seven years earlier, had spent the majority of his time in the federal territory locating, developing, and conveying the region's scarce water resources to agricultural lands. He learned quickly that attracting private capital and embarking on various bond efforts proved as difficult as the construction of canals and laterals to put water to beneficial use. The challenge, he knew also, was worth the effort, though the return on investment might be decades in

7

the future. He observed with great interest the inexorable stream of settlers moving into central Arizona searching for their new start—their version of the American dream. Like Norton, his neighbors sought a secure, regulated, and predictable water supply.[2] Yet the unpredictable flow of the Salt River tempered the hopes of the most optimistic and innovative pioneer farmers.

Phoenix was a small and insignificant settlement dependent on irrigated subsistence agriculture, which at any moment could dry up and blow away on the desert winds, or conversely, could be washed away by an unexpected spring flood. Perhaps the survey signaled that Congress might act to help this emerging, taxpaying civilization on the far southwestern edge of the American frontier. In 1888, Norton foresaw not only the broad outlines of a vibrant agricultural civilization, but also a sustainable economy that would grow like an expanding oasis in the desert.[3]

On February 14, 1889, the US Senate created a US Senate Select Committee on Irrigation and Reclamation of Arid Lands, and three weeks later the Phoenix Chamber of Commerce adopted a resolution inviting the committee to visit Phoenix.[4] Chamber President Henry K. Kemp contacted the Chairman of the Senate Select Committee, Senator William M. Stewart of Nevada, and the senator replied promptly that they planned to arrive in September.[5] The committee visit prompted a flurry of organizational activity. Significantly, Maricopa County Supervisor C. R. Hakes suggested that the county appropriate funds for a survey that would locate the best reservoir sites and estimate their capacities and costs. On July 12, 1889, the supervisors appropriated $500 for the survey over the protests of Hakes, who wanted a much more substantial sum.[6] The supervisors directed Maricopa County Surveyor William N. Breakenridge to conduct the survey. "I was instructed to make a thorough examination of the Salt and Verde Rivers, with a special view to the availability of points on these streams for catchment purposes." John R. Norton, representing the Arizona Improvement Company, along with Colonel James H. McClintock, newspaperman, writer, and director of the Tempe Normal School, were to assist the county surveyor in his public charge.[7] The board paid Norton $135 for the rental of his two pack mules—one named Brigham, Norton's favorite animal; the other named "Paddy," who, according to McClintock, had ears that would keep time on the road "in a most regular manner." On July 17, 1889, Norton traveled to Mesa to await his fellow surveyors.[8]

On July 18, 1889, when they met in Mesa to embark on their trip, Norton announced to his companions that there would be "no kicking" (complaining), then dolefully predicted that the weather and unforeseen hardships before them were almost insurmountable. On day two of the expedition, McClintock noted in his account (published in 1916), "Arose with Norton making all manner of fuss, yet lustily swearing that he was not kicking." They followed the base of Superstition Mountain, over the same "Apache Trail" used for centuries by that nomadic group. As Breakenridge later wrote:

> After satisfying ourselves about the depth to bedrock and the character of the rock in that vicinity, we went to the upper end of Tonto Creek, the east Verde, the main Verde, and the site of Horseshoe Dam; and then to old Fort McDowell, where we swung back up the river toward the location we had made at the head of the canyon on Salt River. This location was so much larger and better for a reservoir site that we put in most of our time there in trying to get a fair measurement of it. On our return, as the water in Salt River was low, we journeyed down through the narrow canyon and had to cross and recross the river every few hundred yards, as the water would hit first one side of the canyon and then be deflected back to the other side. But it was much better traveling than over the rough Apache Trail.

Along the way they bestowed a valley with the name "Norton Valley," after our "efficient major domo," according to McClintock. On July 22, they spent the afternoon bathing and fishing in the river, and Norton and Breakenridge caught a number of fish. Their survey equipment was limited; they measured distances either by stepping them or by counting McClintock's blue mare's strides. They utilized an aneroid to determine elevations, a small sight compass to locate directions, and a simple hand level to measure inclines. Incredibly, later technological developments in surveying affirmed that the Breakenridge Survey measurements and elevations were substantially correct.

They returned to Phoenix on the morning of August 10 and in a report to the supervisors and an article written by McClintock for the *Phoenix Daily Herald*, the survey crew announced their findings. Indeed, they located a number of sites suitable for dams, but the water storage basins they located in the beginning were too small. Then they spied the wide valleys of the Tonto

Creek and Salt River, above the head of a narrow canyon. The party agreed they found an ideal location for a storage dam.[9] The *Phoenix Daily Herald* on August 12, 1889 reported that the Breakenridge Survey endorsed, as the most promising dam site, "a box canyon with sheer rock walls several hundred feet high located on the Salt River below the mouth of Tonto Creek. . . . A quick, but accurate survey disclosed that the catchment basin behind the dam site would hold more than a million acre-feet of water." The writer optimistically predicted that although the planned dam would create an enormous reservoir, it could be "secured at a cost relatively trivial in comparison to the great benefit to be derived."[10]

In fact, Breakenridge estimated the cost of the dam and reservoir at not more than $1.5 million, and "it would supply water enough to irrigate all the land in the valley from the Arizona Canal dam to Yuma." The lake behind the dam, he mused, "would be the largest artificial reservoir in the world." McClintock agreed with this assessment and added, "Surely the general government can find at no other place a more eligible site for water storage than this presents, and it should now be made the aim of our representative citizens to see that Congress is properly informed and to push matters so that, either by government aid or assisted private enterprise, the necessary $1.5 million shall be forthcoming for the construction of the dam, and that the prosperity of Central Arizona shall be established as firmly as the rocks."[11]

Only later would the then thirty-four-year-old Norton, who had moved to the Salt River Valley in 1883, realize that his participation in the publicly funded Breakenridge Survey of 1889 placed him at the center of a series of private and public actions that triggered transformative growth and development of the arid Southwest. In reality, Phoenix, Arizona Territory in the late 1880s and early 1890s was a small island community in a remote portion of the American Republic. Americans east of the Mississippi oftentimes questioned the expenditure of public moneys on a desert land described by one federal official as "just like hell." Similarly, Norton and his contemporaries wondered why a president and congress located thousands of miles away in Washington, DC, should exert power over a region about which they knew very little. Nevertheless, Norton's Phoenix judged itself by eastern standards and looked eastward for approval. Phoenicians of the Gilded Age boasted that their town was not "wild and wooly" or under constant threat of Indian attack, but identical to an average eastern town of like size. But the lifeblood of this eastward-looking town was the Salt River, and Norton and his friends knew the river's many moods and uses. Erratic and unpredictable, it flooded suddenly, washing away dams and canal banks, or, conversely, ran dry for months, ruining those who depended upon it for a steady water supply. Norton witnessed the pioneer spirit of cooperation deteriorate into acrimonious lawsuits and, at times, violence when canal company partners or landholders fought over shares or water rights. Farmers and businessmen alike talked incessantly of spring runoffs, groundwater, watersheds, and precipitation, and by the time Norton participated in the survey for a dam site on the upper reaches of the Salt River, he realized that a constant, regulated water supply was the most crucial issue facing the Salt River Valley.[12]

John R. Norton's route to Arizona mirrored that of many other late nineteenth-century pioneer settlers. He followed an east-west migration from the Midwest with a few family tragedies and fortunate detours along the way. Born in Sulphur Wells, Metcalfe County, Kentucky, on February 24, 1854, John was the only son of Robert Hunter Norton and Annie Norton. Robert was a farmer and machinist who maintained a penchant for inventing new equipment for woolen mills. He built the mill at Nicholasville, Kentucky, which he operated until it was gutted by fire in 1858. Robert then moved his family to Lexington, Missouri, where he dealt in hemp until his death in 1860. Shortly thereafter, Annie passed away from tuberculosis, and John was orphaned, so he returned to Kentucky and lived with an uncle where he grew and was forced to mature quickly.

In his late childhood and early teenage years he worked on a farm near Centralia, Illinois, where he remained for five years. Then in 1876, at the age of twenty-two, he began farming on his own near Gibson City, in Ford County, Illinois. Unfortunately, the large corn crop he planted suffered from too much rain that season and he lost it to the elements. At this juncture he determined to head to the arid Southwest, in part to avoid his mother's fate.[13] He surmised that a more salubrious climate might not only preserve his health, but also change his luck. The dry climates of Arizona and New Mexico had already drawn thousands of health seekers to the region, and Norton looked for an opportunity to emigrate to the far Southwest.

Early in 1880, William J. Murphy, known to his friends and family as "W. J.," a grading contractor in Gibson City, befriended Norton. Murphy, a Civil War veteran from New Hartford, New York, had bounced around Tennessee and then Illinois after the war. In 1879, after a brief flirtation with farming, Murphy found himself equipped to enter a new business.[14] A debtor owed him a large sum, and Murphy was forced to take in grading machinery and large excavators in lieu of money. He quickly put these acquisitions to work, winning contracts for grading roadbeds in Chicago and central Nebraska. One contract took him to northern Colorado, near Julesburg, where he worked on a branch of the Union Pacific. Murphy continued west, taking a subcontract on the Atlantic & Pacific line at Blue Water, west of Albuquerque, New Mexico. He offered Norton a job as the corral boss on the grading crew, and the latter readily accepted. Norton headed to New Mexico followed by Murphy's wife, Laura, and their three children. The work crew graded the track bed as it led through western New Mexico, and everyone noted how the water holes grew increasingly scarce and brackish. By 1881 they had crossed into Arizona Territory and later, in mid-summer, searched for fresh water in the Turkey Tanks west of Flagstaff, an area now known as Fort Valley. Several searchers fell into one deep depression filled with water. Norton "came very near to drowning . . . but succeeded in swimming to the edge and scrambling out."

Meanwhile, Murphy rented a log building that had served as a warehouse to house his family. The other seven buildings in what comprised Flagstaff were saloons, and within a week the routine of drunken rowdies knocking on the Murphys' door drove him to homestead land fifteen miles northwest of the tiny town, in a small valley known as Leroux Prairie at the base of the San Francisco Peaks. The Murphys, Norton, and the rest of the work crew lived in tents on the homestead site. Remote and peaceful, Leroux Prairie suited Laura's conservative views on drinking, smoking, swearing, and gambling. Norton met the only other inhabitants in the area, John W. Young and W. G. Young, sons of the Latter Day Saints leader Brigham Young. They and their large families supplied ties for the railroad, which they whipsawed by hand out of pine timber in the area.

In early September 1881, Norton accompanied Murphy and his son on a scouting expedition to land one hundred fifty miles to the west, along the proposed railroad route. Murphy planned to bid as the prime grading contractor on a stretch near Peach Springs, fifteen miles east of Hackberry in Hualapai Indian country, in what is today Mohave County, Arizona. Upon returning to Flagstaff to confer with Atlantic & Pacific officials regarding the specifics of the contract, Murphy learned he had to travel to Albuquerque for final approval.[15] Meanwhile, Murphy entrusted Norton with his family's safety, signaling growing confidence in his young friend from Illinois. Norton chaperoned a large-scale move that included Murphy's family and employees to the proposed worksite. On November 9, 1881 the mix of wagons, twenty-three men on foot, horses, oxen, and other animals left Leroux Prairie en route to Peach Springs.

The weeklong trip, which closely approximated the fabled Route 66 constructed during the mid-twentieth century, elicited vivid descriptions from the travelers. Laura Murphy noted that "We traveled through Truxton Canyon, in some places weird and wild, fit haunts for the Hualapai [Indians]; in other parts delightfully picturesque and beautiful." Shortly thereafter she wrote that the wagon train had "camped within a mile of Hackberry, a mining town high up the side of the mountain where a good spring had attracted the miners to locate. . . . a few miners' cabins and a number of Hualapai Indians form the town." "Mail," she sniffed, "is carried regularly from this point." Then, the intrepid crew learned it had to retrace its route. "Before daylight, [W. J.] was at the tent door," Laura recounted, "He had been compelled to go to Albuquerque to contract for work and expected us to go to Peach Springs fifteen miles further, but the officials would not 'close' for that work [and instead] gave him six miles on the west side of Aubrey Valley which we had passed the Wednesday before."[16]

As Laura noted, Murphy secured a contract for six miles of roadbed grading on the west side of Aubrey Valley. He sublet one section, and in February 1882 he contracted for another four sections, then in August for three more, thus securing the roadbed work from Seligman to Peach Springs. Murphy established a well-provisioned camp in Aubrey Valley to work on the railroad right-of-way and Norton, the corral boss, was in charge of what was described as "most excellent work stock: forty fine mules, thirey-one oxen, and a half dozen or so horses." The camp also included a commissary that maintained grain, flour, groceries, dried and canned fruit, blankets, boots, shoes, and other clothing. Two cooks presided over the cook tent, where everyone but the Murphy family ate meals. Four more tents comprised the men's sleeping quarters. The well-planned camp enabled Murphy and his men to effectively execute their contract with Atlantic & Pacific.

Norton enjoyed Christmas Day, 1881, ending the day with a holiday dinner and a cigar. He was joined by thirty-two other employees. Nothing remained of the roasted meats, potatoes, onions, tomatoes, two large milk pans full of pudding, pies, and cheese. Laura noted that her husband "pronounced it a fine success which in a great measure helped me overcome my fatigue resulting from preparing fourteen pies, twenty loaves of bread, and two big pans of pudding, which fatigue manifested itself in a light numb spell." And William and Laura, strong temperance advocates, were delighted that the sumptuous meal had the desired effect—none of the men visited the local saloon a mile away. [17]

Norton remained in northwestern Arizona with Murphy's crew through the end of 1882 as they completed the railroad grading contracts. Entrusted with more responsibility within Murphy's organization, Norton began hauling supplies to other contractors and picking up wood and other necessities on his trips back to camp. He oversaw a first-class outfit; top quality mules and equipment enabled Norton and the crew to progress from subcontracting to prime contracting. With the completion of the Aubrey Valley job, however, the possibility of further work with the Atlantic & Pacific ended. That line terminated at Needles, California, on the west bank of the Colorado River, and contracts to that point had already been let. By January 1883 Murphy established a family residence in Prescott, and Norton, who had proven a dedicated worker and leader among men, remained close to Murphy as they scouted the next economic frontier. The contractor, Murphy, and the employee, Norton, found it in a fast-growing agricultural crossroads in central Arizona, one hundred miles southeast of their new yet temporary Prescott home. Although sparsely populated at the time, the Salt River Valley, with the recently named Phoenix as its population center, had a long, tempestuous history of peoples, cultures, and traditions, with water serving as the fulcrum of power and survival.[18]

By this time Norton realized that water in the Southwest sustained a culture and tradition that reached back centuries. Soon after the outbreak of the war between the United States and Mexico in 1846, for example, Colonel Stephen Watts Kearny and his Army of the West encountered the Pima Indians farming along the Gila River. The invading Americans, en route to their military destination in California, marveled at the Pimas' agricultural abundance and extensive irrigation works, and realized immediately, as the Spaniards and Mexicans had before them, that irrigation along the river predated the Pimas. By the time Spanish explorers had discovered prehistoric ruins in central Arizona, their builders, the Hohokam, had been gone for nearly two hundred years. Archeologists continue to debate the nature and extent of more than three hundred and fifty miles of canals along the Salt River in the Phoenix area and additional canals in southern Arizona. It was the most extensive prehistoric irrigation system in North America, and the Hohokam, according to recent scholarship, irrigated approximately one hundred thousand acres of land in the Salt River Valley alone. One scholar has noted that the Hohokam are recognized as the premier desert irrigation specialists of the prehistoric era, and that they actually used many methods to control and use water. Besides an extensive canal system, they developed terracing, check dams, rock piles, and linear and grid borders.[19] Additionally, like other prehistoric cultures in Arizona, Hohokam water resource development fell into two categories: irrigation methods (canals and ditches) and indirect methods utilizing soil moisture conservation. These methods provided powerful technological precedents for the region's European successors.[20]

Hohokam irrigation works evidenced the earliest impact of water management in Arizona's traditions. Popular western imagery evokes a place inhabited by rugged individualists, but in an arid region one must share a "plumbing system" with one's neighbors in order to thrive or even survive. In fact, the earliest forms of government in the world may have arisen in the Middle East over the need to construct and share water distribution facilities. In the Southwest, the irrigation works of the Hohokam present similar early evidence that such cooperative endeavor is necessary for the growth of civilization.

As Spaniards colonized the Southwest in the sixteenth and seventeenth centuries, they consciously adopted Native American settlement patterns, often displacing the inhabitants of existing towns and villages where water could be easily conveyed for irrigation. They also reconfigured agriculture in the region. They brought in new crops and introduced livestock and various other domesticated animals. The Spanish also created a new language of water use, much of it dating from the Moorish occupation of Spain from 711 to 1492. The Spanish word *acequia*, or canal, is still used today, and water users in Arizona continue to refer to the *zanjero*, or ditch tender. The acequia districts of northern New Mexico are the oldest extant governmental units of European settlement in the United States.

The most significant Spanish influence in Arizona and the Southwest was in the law. The Spanish maintained a near obsessive interest in water regulation, reflected in volumes of land grant documents and government officials' diaries about interminable water rights disputes. The single most important Spanish water legacy is reflected in the Latin phrase *qui est in tempore, potior est in jure*—first in time, first in right. The legal doctrine of "prior appropriation" became the cornerstone of water law, not only in Arizona but also throughout the trans-Mississippi West. Though Spanish law recognized "first in time, first in right" in theory, Spanish settlers, in many cases, used a more pragmatic, communal approach to water that recognized the inherent value of a scarce resource in an arid region. The doctrine of prior appropriation remained in place during the Mexican period (1821-1848) and formed the legal framework regarding the use and distribution of water when Colonel Kearny promulgated his legendary code of 1846 when he took Santa Fe, New Mexico without a shot. That code resulted in continuing Spanish traditions of water and land use until the creation of the Arizona Territory in 1863.[21]

Spanish priests, soldiers, and civilian explorers of the seventeenth and eighteenth centuries took note of the inhospitable, arid landscape and inadequate water supplies of the Salt and Gila River systems. "With few major exceptions," according to the distinguished historian of Mexico, Michael Meyer, "the water sources (the Rio Grande, the Colorado, the Fuerte, the Yaqui, and the Gila being among the most notable) which the Spanish dignified with the word 'Rio' were scarcely rivers at all. Not even the largest, the Rio Grande, proved valuable for transportation or commerce either before or after conquest. Although scientific evidence suggests that they carried a larger flow than they do now, most rivers were not perennial; they ran only part of the year, trying their best to carry the excess from an exceptional winter snow cover in the surrounding mountains. The more common pattern was for the water that reached them to sink quickly into the sandy bed and, within a short distance, to disappear from human sight. On occasion, however, they ran partly above surface, then underground, protected from the evaporative powers of the environment, to be forced to the surface again by the geological structure of a given area."[22]

To place the concept of aridity in regional and historical context, with the exception of eastern Texas, the Mexican north, which the Spanish first encountered in the sixteenth century, was generally arid, semi-arid, and on occasion, extremely arid. The availability of water spelled the difference between desolation and abundance with countless variations between the two. This vast desert region had been occupied continuously for several thousand years, but, in the mid-sixteenth century, the population density was low, perhaps less than two people per square mile. Significantly, aridity increased as one moved west from Texas and Coahuila to New Mexico and Chihuahua, and then to Arizona and Sonora and southern California and Baja California. With the exception of the higher elevations, evaporation was high and humidity low. The topography and natural vegetation doubtless reminded the first Spaniards of southern Spain. They were not surprised that the sun could crack the soil and blister the land. They fully comprehended moisture deficiency and knew that the critical challenges of aridity encouraged the development of a special kind of human society. Like their successors, the nineteenth-century Anglo-American pioneers, they were not surprised to learn that the labor of controlling water and putting it to beneficial use could occupy much of the working day. Forging an existence in this land would be a continuous struggle. This vast region was much more varied and capricious than its counterparts in Andalusia and Castile. It had a wider range of altitudes, soils, animal life, and drought-resistant vegetation, and annual rainfall was more unpredictable. The mountains were more rugged and towering; the canyons virtually impenetrable. Erosion and sedimentation had created a physiography at once harsh and captivating—frightening and alluring, as Laura Murphy noted. The rainy season extended from July to September, but few areas of the desert received more than twelve or thirteen inches of precipitation per year. In drier parts, like central Arizona, years of less than seven inches were not uncommon. The mountains of this inhospitable land captured most of the moisture carried by prevailing Pacific or Gulf of Mexico winds and left the valley parched for most of the year. The winter snow cover in the mountains was almost always insufficient to provide lower elevations with a reliable source of water, except during the early spring thaw.[23]

If these rivers, like the Salt and Gila, did not always carry a significant amount of water, they nevertheless proved amazingly attractive, drawing the surrounding animal life and providing the modicum of moisture required for desert flora. It was along rivers like the Salt, arroyos, and other quixotic streams that most Indian populations (like the departed Hohokam) adapted to desert life. The alluvial plains, ranging in width from a few feet to several miles, were rich and a source, although an unreliable one, of water. Here, too, Spanish towns, missions, and presidios would cling to a precarious existence. And as these two groups—the Spaniards and the Indians—were forced by physical and historical circumstances into increasingly closer contact, precious water soon came to dominate their contests for power and survival.

Indeed, from the time of Father Eusebio Francisco Kino's extension of the "Rim of Christiandom" into the lower Santa Cruz and Gila Valleys in the 1690s, the Salt and Gila Rivers played prominent roles as transportation routes as the Spanish furthered their aims. Spanish diarists often noted the remnants of the Hohokam civilization that marked much of the lower reaches of the Gila, below its confluence with the Salt.[24] Sergeant Juan Bautista de Anza, on a reconnaissance of central Arizona in November 1697, took note of ruins on the north side of the "irregular" river: "On the 18th we continued west over an extensive plain, sterile, without pasture; and at the end of five miles, we discovered, on the other side of the river (the Gila), other houses and edifices. The sergeant . . . swam over with two companies to examine them; and they said the walls were two yards in thickness, like those of a fort; and that there were other ruins about, but all of an ancient date."[25]

Later, in 1775-1776, Don Juan Bautista de Anza led a colonizing expedition from Tucson to San Francisco. Father Pedro Font, who irritated Anza greatly, nevertheless kept the best diary of this historic expedition, which traversed central Arizona via the Santa Cruz River to the Gila, then down to its confluence with the Colorado. The Gila portion of the journey brought forth noteworthy observations of its flow. According to Font, there were Indian agricultural systems diverting water; there were dry stretches and occasional deep reaches that coursed slowly down the streambed. In effect, the Gila, in the fall of 1775, was intermittent and erratic, and in many reaches, dry.[26] References to the Salt and Gila from the period of the Mexican Revolution (1810-1821) through the Mexican period (1821-1848) vary little: they describe anemic flow with occasional destructive flooding and spring freshets.[27]

With the Treaty of Guadalupe Hidalgo (1848) and the subsequent Gadsden Purchase (1854) affirming American title to the land bisected by the Gila River, much changed in the region's legal, political, and social foundations as they pertained to land use and water resource development, though some traditions—like the legal doctrine of prior appropriation—carried over to the American period. Evolving concepts and public policies concerning central Arizona's natural resources emanated from Washington, DC, rather than from Madrid or Mexico City, and the outlines of these policies shaped the world of John R. Norton, who had yet to realize that his future lay in this unstable and inhospitable desert environment.

On November 11, 1867, Jack Swilling, a Confederate deserter, along with approximately twenty downtrodden miners from Wickenburg, unwittingly established the first permanent Anglo-European settlement along the thirty-eight-mile course of the Salt River that ran from the Verde Valley to the Salt's confluence with the Gila. Indeed the valley was a vast alluvial plain that stretched from the Superstition Mountains on the east to the Sierra Estrella on the west. All of Arizona's major rivers except the San Pedro and the Colorado flow together there, so even though the valley is low and hot desert land, the snowmelt and runoff from the Mogollon Rim and the White Mountains surge down the drainages on their way to the Gulf of California. Its rivers, therefore, made the Salt River Valley the greatest conjunction of arable land in the Southwest. The Hohokam had, in fact, built some of the largest cities there, but no Europeans settled near the junction of the Salt and Gila until 1865, when the army established Camp McDowell along the Verde River twenty miles to the northeast. In essence, the US military gave birth to Phoenix just as the Spanish military created Tucson nearly a century earlier.[28]

Swilling and his group of miners-turned-farmers, aware that Camp McDowell needed food and supplies, claimed six thousand miner's inches of water from the Salt River for irrigating purposes, and they recorded the filing with the Yavapai County Recorder in Prescott. A few days later, on November 16, the group formed the Swilling Irrigation and Canal Company. In December they began construction of an irrigation canal on the north side of the Salt River near Tempe Buttes. A hard formation of caliche, however, halted their efforts and they moved downstream and commenced work on a canal four miles west of their original attempt. They realized that the canal could not be completed in time for the planting season, so they cut a temporary ditch along a long-abandoned Hohokam Canal and planted several hundred acres of corn and beans. Work continued on the Swilling Ditch, later known as the Salt River Valley Canal or Town Ditch.[29]

These efforts proved successful and opened the Salt River Valley to agricultural enterprise and development. In 1868 Swilling and Thomas Barnum carved out a new ditch three-quarters of a mile upstream from the Swilling Ditch head. This "north extension ditch," organized under the name Phoenix Ditch Company, later became known as the Maricopa Canal. To the south of the Swilling Ditch, Jacob and Andrew Starer built the Dutch Ditch. Several pioneer farmers settled along the Dutch Ditch, and by 1870 a farming community of two hundred and fifty souls cultivated fifteen hundred acres of land.[30]

During the 1870s, the decade prior to Norton's arrival in the Salt River Valley, local residents relied on the Salt River, irrigation ditches, or wells for their domestic water supplies. In August 1870, the *Arizona Miner* reported that in Phoenix, "Digging wells has commenced. One of sixteen feet has a large supply of delicious water. At present the ditches supply nearly every demand of man, beast, and crop." While wealthier residents dug wells for personal use, most residents used water from the irrigation ditches both for irrigation and domestic purposes. One Phoenix resident wrote in an August 31, 1872, letter: "Recent rains have damaged the ditches—so town has no water for a week, except that which has been hauled from river or drawn from wells."[31] During this period town commissioners arranged to supply irrigation ditches with water from the Salt River Valley Canal, which ran along Van Buren Street on the northern edge of town. For community use, the commissioners contracted for a well that was dug in the town plaza. And, in 1879, Judge John T. Alsap erected the first windmill in Phoenix for pumping water.[32]

During this decade of community building, saloons and hotels took the lead in the installation of plumbing fixtures and distribution systems. In 1878 Phoenix hotel owner John Gardner stunned residents and travelers alike when he introduced shower baths. Toilets added the following year completed this innovative plumbing system and turned the hotel into a "resort" for the dust-covered traveler. Even more exciting, Gardner added a swimming pool the next year. The owner of the Capitol Saloon at 27 East Adams Street kept pace with Gardner and constructed a small water distribution system. Using a dependable well in the back of the saloon, the bartenders pumped water to an elevated tank and from there distributed it to five or six other businesses, including a Chinese laundry, two other saloons, and a bathhouse. One pioneer resident, John Rau, said that he first saw the system in 1879 and commented that "it was old then." In fact, the fountain from the well at the Capitol Saloon provided most of the residents of Phoenix with their drinking water. Besides the plaza well, which was at times contaminated, only three other wells in Phoenix were convenient for locals to obtain drinking water: one in Dr. Thibodo's backyard, a second at Monihon's Corral, and a third in the stage corral. The 1880s census recorded 1,708 residents in Phoenix, and that fall a group of residents raised money to place a pump in the plaza well. By the end of 1880, public-spirited Phoenicians completed the project, placing a cover over the old well to prevent debris and refuse from contaminating it, and installing a hand pump for the benefit of the citizenry.

Central Arizona in the mid-to-late nineteenth century was an arid land, but one that had supported successful development by prehistoric Hohokam, contemporary Indian groups, Spaniards, and Mexicans. For much of the period between 1850 and 1880, American water development followed the pattern of earlier cultures. While growth was slow and uneven, the seeds for future growth were planted. Furthermore, pioneering efforts in the early years of American political control in Arizona proved that the area had tremendous potential for economic development. Every successful venture eroded the general belief that Arizona was an inhospitable desert. Jack Swilling's success in transforming Hohokam irrigation canals to new agricultural uses demonstrated that a great civilization might someday rise again in central Arizona.

The Swilling Ditch, moreover, also served as the prototype for other joint-stock canal companies proliferating across the Salt River Valley. Unlike earlier Mormon irrigation projects, the canal companies were business propositions. Investors joined together to form the companies because they did not possess the capital to build dams and ditches on their own. The shareholders decided how to put their portions of water to beneficial use, and some actually farmed their own fields. Others leased their water and sold their land. Speculators insinuated themselves into the mix at every turn; they wanted profits.[33]

By 1872, farmers were cultivating over eight thousand acres of barley, wheat, beans, corn, sweet potatoes, grapes, and fruit trees. In an amazingly brief period the Salt River Valley developed into the most important agricultural region in Arizona. Area farmers supplied not only the military but also the mines, which would proliferate throughout the territory in the next two decades. The beneficent river, of course, could change radically and threaten the incipient community that depended upon it. In September 1868, heavy rains sent a huge flood roaring down the Salt. Six years later, in January 1874, the Salt flooded the valley for three days, destroying the Swilling headgates and wiping out William Parker's granary, which was filled with ten tons of wheat. Farmers dashed away from their crumbling adobe homes to seek refuge in the local schoolhouse and courthouse. For weeks religious services were held in a saloon. After these two signal events, the farmers knew they faced serious environmental challenges in harnessing the river's resources.

In the first two decades of Anglo-American settlement in central Arizona, many irrigation projects were farmer owned. The Swilling Canal was one of these, and the builders either purchased shares in the canal company or exchanged labor for shares. The shareholders thus considered the water transported from the Salt River through their canals to be appropriations attached to their land. If floods washed out the canal, damaged a canal head diversion dam or other appurtenance, the owners joined together and made the repairs themselves or voted levies on their stock to pay for repairs. Improvements were addressed in the same manner. The system, at first, worked well.

A different type of canal ownership emerged after Congress passed the Desert Land Act of March 3, 1877. An attempt to adapt land policy to the arid West, the new federal policy resulted in a more realistic approach to western settlement compared to its more famous predecessors—the Homestead Act of 1862 and the Desert Land Act of 1876. Under the revised Desert Land Act, a person could obtain 640 acres of land—a full section—in any of the eleven western states and territories for $1.25 per acre, if he or she agreed to irrigate it within three years of filing. The Desert Land Act of 1877 at least indicated a willingness to recognize the need for irrigation on western land, but Congress continued to think of farm development largely as an individual, Jeffersonian effort. Proponents of the legislation believed that once given a larger grant by the government, settlers would find some way to irrigate land. In short, Western farmers, with the aid of federal land grants, could subdue the West as earlier generations of farmers had subdued the East. The law, however, reflected the general American ignorance of irrigation. No settler could bring 640 acres into irrigation within three years. Instead of helping settlers, the law proved a boon to speculators.[34]

For nearly one hundred years, beginning with the Land Ordinance of 1785, federal land policy, in its various forms and incarnations, had encouraged the transfer of the public domain to private hands, thus facilitating the development of a tax-paying civilization. In the arid West in general, and in central Arizona in particular, the acquisition of land and the development of scarce yet valuable water resources in the last quarter of the nineteenth century created a dynamic situation that shaped the early development of Phoenix and Maricopa County and formed the nexus of its economic future. Private business interests dominated water resource development and service in Arizona from 1880 to around the turn of the century. These desert entrepreneurs ranged from the corporate owners of the Atlantic & Pacific Railroad in Flagstaff to Hispanics in Tucson who carried water on the backs of burros from house to house. Typically, they were businessmen like John Gardner in Phoenix, but not surprisingly corporate entities soon followed the early individual entrepreneurs. Water systems required amounts of capital beyond the reach of smaller business owners. In addition to hard money, owners of water systems often had to manipulate political capital as well, and larger corporate organizations often proved more adept at turning the political system to their advantage—a key asset for the success of latter-day corporate water utilities.

Venture capitalists, private enterprise and initiative, and federal policies encouraging settlement all influenced the economic and environmental worlds of the earliest settlers in Arizona Territory, including those of the youthful John R. Norton.[35]

Phoenix incorporated as a city in 1881, and its earliest officials soon wrestled with complaints over the provision of services. Much of the public outcry was over water quality issues, as well as availability. In June 1881 a reporter for the *Arizona Gazette* lamented the filthy condition of water supply ditches and that "fully one-half of the population" received its domestic supply from the irrigation ditches that ran along both sides of major streets in the city. The critique ended with the admonition that residents had received better service before incorporation. An *Arizona Gazette* subscriber later complained of a flock of ducks living at the head of the town ditch and polluting the water supply. On March 10, 1882, *Phoenix Herald* readers received a stiff warning from the editor: "We have noticed for some time in front of several saloons spittoons put into the ditches stay there to soak. This is a very bad habit and, according to the law, a nuisance. Many citizens use this water for drinking and culinary purposes. Desist, gentlemen, from this foul practice."[36]

In May 1882, Walter S. Logan, a powerful New York attorney who later served as President of the New York State Bar, visited Arizona. Logan asserted that he and his clients could bring some order, civilization, and, importantly, investment capital to the thirsty yet growing territory. Governor Frederick A. Tritle, along with then-Arizona Territory Attorney General Clark Churchill, later described by Logan as "the best lawyer within a thousand miles of Arizona," rode up and down the entire length of the Salt River, giving special attention to a much-discussed route for a new canal that would cut across the northern edges of the Salt River Valley. He later said that he "saw every now and then, under the small ditches built, their ranches and their gardens, their fields, and their stone houses, and all that Arizona even then could produce in such prodigal plenty." After the tour, Logan and Churchill "formulated the economic plan under which it became possible to build irrigating canals in Arizona with outside capital." Logan later took credit for organizing the Arizona Canal Company "to begin the work and it was a client of mine in the east and friends of W. J. Murphy in the West, who furnished the money to build it."[37] Logan's exaggerated rhetoric reflected the early, if ephemeral, optimism of the time, and he quickly inserted his brother, H. H. Logan, into emerging plans for a canal company that would help expand the agriculture developing in the Salt River Valley.

John R. Norton moved into this evolving world of water development, agricultural enterprise, and imperfect community-building in 1883, after his short respite in Prescott. The wheels had already been set in motion. With Phoenix growing at an unprecedented rate, and residents clamoring for sustainable amounts of potable and otherwise usable water, new canal companies were being organized under the broad guidelines of the Desert Land Act of 1877. On March 10, 1882, attorney William H. Hancock, merchant J.Y.T. Smith, rancher Wilson W. Jones, and banker Martin W. Kales filed for a canal location and 50,000 miner's inches of water. (A miner's inch was equivalent to a flow of 2,274 cubic feet within twenty-four hours.) On December 20, 1882, Hancock, Kales, and territorial adjutant general Clark Churchill incorporated the Arizona Canal Company with Churchill as president of the board of directors and Logan's brother, H.H., as secretary. Others would become associated with the company. First National Bank became institutional treasurer, and later this role would be assumed by the Valley Bank of Arizona. Frederick A. Tritle, governor of the territory, was named to the board of directors, as were Colonel F.C. Hatch and Moses Hazeltine Sherman. The latter, a former superintendent of schools, was made adjutant general of the Territory. Arthur Barry was the first engineer of the company, but was later replaced by Charles A. Marriner. This was an investor-owned corporation, which sold stocks, bonds, and water rights to finance construction and operation. Theoretically, the company maintained the canals, head works, and dams, and asserted ownership of water appropriated from the Salt River. To secure water delivery, farmers, ranchers, and other buyers of water rights signed contracts with such companies, in which they agreed to pay a set fee per acre of land. In some instances, companies rented water use.[38]

According to the articles of incorporation, the Arizona Canal began about three-quarters of a mile below the junction of the Salt and Verde Rivers and extended northwest, skirting the northern edge of the Salt River Valley, crossing Skunk Creek and New River, spanning the Agua Fria about eighteen miles north of its union with the Gila River, and continuing twelve miles to the White Tank Mountains before turning back south to the Gila. Two months after incorporation, Hancock, Smith, and Jones sold their water rights to the Arizona Canal Company for five dollars. In announcing the sale, the *Arizona Gazette* noted that the new canal would open "hundreds of thousands of acres of land" to cultivation. These grand assumptions might have been correct if the canal had extended west of the Agua Fria River as proposed, and if all of the land along the projected route could be claimed for agricultural use. As it turned out, more than half the acreage was off-limits to the canal company.[39]

The project was ambitious. It was, to one observer, "the largest, longest, had the highest point of diversion on the [Salt] river and served more acreage it was in a position to exert power . . . ."[40] Unlike the other major ditches, the Arizona Canal did not follow Hohokam precedents and instead tapped into the Salt more than forty miles upriver, along the northern edge of the Salt River Valley. By May 1883 the first twenty miles of the Arizona Canal had been surveyed and was ready for construction. Churchill prepared to hire a contractor. In Prescott, W. J. Murphy heard about the project and traveled to Phoenix to talk with Churchill, whom he had met, apparently, in Prescott.

At first, Murphy and Norton seemed to consider this project another in a growing list of contracts to burnish their respective resumes. This project, however, proved to be very special for both men. It involved them deeply in the development of the Salt River Valley, and both would spend the rest of their lives in the area. They had recently completed the job for the Atlantic & Pacific Railroad and had been in Arizona for two years. Norton and the crew were rested and ready to work. In late April 1883, Murphy signed a contract to build the 40.75-mile-long Arizona Canal. This contract called for the canal to end at Skunk Creek, east of the Agua Fria River, though it also outlined the construction of five bridges, the Arizona Dam to divert water from the Salt into the canal, lateral ditches, two sets of head gates, and waste weirs. Construction costs would be paid from the sale of company stock and $500,000 in 8 percent interest bonds with face value of $1,000.

Norton understood and grappled with the engineering complexity at hand. He learned that the canal was designed to be thirty-six feet wide at the top and thirty feet wide at the bottom for the first five or six miles, with a depth of about six feet. For the first three miles a uniform grade of one and three-quarters feet to the mile was to be held and for the rest two feet per mile. The slope of the bank was to be one and one-half to one foot for the looser material such as the dirt. The diversion dam, located three miles below the confluence of the Salt and Verde rivers, would consist of fascines, bundles of brush tied together and weighted by rocks secured inside. Engineers also planned to have some rock cribs—large wooden crates floated into place and then filled with rocks to make them sink to the bottom and remain stationary. The dam was to be approximately 1,000 feet in length with a fifty-foot base.[41] Another aspect of the project riveted John Norton's attention, and he questioned his friend and boss, Murphy, about the logistics and financing of the contract. It contained strict and precise language and provided that the canal should be completed to a certain point—the twenty-mile point—by March 1, 1884, less than a year from the commencement of operations. If not completed, the contractor was to forfeit the contract and also the unpaid balance of work already done. Furthermore, Norton questioned the idea of receiving payment solely in bonds and stock holdings in the Arizona Canal Company. This would require Murphy not only to administer the contract, but also to sell bonds in order to get money to operate. No cash need change hands—water rights and land served as two forms of payment. Whatever the understanding about the sale of these bonds—and Murphy grew disgruntled when officers of the company were slow, if not unavailable, to assist him in selling the bonds or borrowing money on them—it ultimately fell to Murphy to raise the money to pay his fees, the wages of his men, and the subcontractors. Given the burden on his boss, the twenty-nine-year-old Norton knew that he must shoulder increased responsibilities on the Salt River project.[42]

Norton moved to the Phoenix area on April 30, 1883 and readied his animals for the massive construction job. Murphy and Churchill, meanwhile, embarked on a range of land speculation activities. On April 25, 1883, two days before the construction contract was finalized, the two filed eighteen entries claiming parcels of land under the Desert Land Act at the US Land Office in Tucson. Significantly, seven of the entrants were Murphy family members or relatives of his wife, Laura Fulwiler Murphy. These included Murphy's father George, his sister Mary M. Culver, and Mary's husband, attorney Joseph F. Culver of Emporia, Kansas. Laura Murphy's brother, William Dunlap Fulwiler, a Hackberry, Arizona miner, and her sister, Julia L. Fulwiler, a music teacher in Prescott, were also listed. The eleven other entries also raised eyebrows. They included territorial governor and Arizona Canal Company Director Frederick A. Tritle, Churchill's wife Margaretha, company engineer Andrew Barry, company organizer Martin Kales, and civil engineer and Murphy's friend John D. Buckley of Greeley, Colorado. Significantly, entrants were required to sign an affidavit attesting that the filing was not made for the purpose of fraudulently obtaining title, but for the purpose of faithfully reclaiming a tract of desert land.[43] Most of the claimants had never seen the property, and the large number of family members, the identical handwriting, and language in all of the applications suggested that

John Ruddle Norton, Salt River Valley pioneer.

the filings were a coordinated effort by Murphy, Churchill, and the directors of the Arizona Canal Company to claim large tracts of land for irrigation by the proposed Arizona Canal.[44]

The questionable filings notwithstanding, Norton and the workforce organized themselves like a military operation, and its general, Murphy, traveled to San Francisco, Detroit, Chicago, New York, and other cities in the East seeking capital to construct the system. As it was in the early nineteenth century when there were no large pools of venture capital in the eastern United States, so there were none in the West in the latter nineteenth. As a result, Arizona and the Salt River Valley, like much of the West, "found itself under the sentence of economic colonialism," dependent on Eastern capital to finance its large and significant improvements. Investors in these cities, like many of those at home in Arizona, were primarily interested in speculation, not water distribution or quality of life issues in the economic colony. The Desert Land Act of 1877, which stipulated that settlers must irrigate 640 acres of land within three years, made profit-seeking investors inevitable. Thus the Arizona Canal Company experienced the trials and tribulations of speculation in western water and lands.[45]

Nevertheless, in early May 1883, Norton established his headquarters north of the old Fort McDowell Road. It was a huge undertaking. One local writer described the mobile community as a "little canvas town" with a boarding house, blacksmith shop, commissary, engineers' quarters, tents for three families, and a number of larger tents that housed sixty or seventy men each. Norton maintained his corrals, stacks of hay for the stock, piles of tools, wagons, and all of the accoutrements necessary for camping, clearing, excavating, and prosecuting the work in various conditions and varying circumstances. His 225 two-mule teams enabled the 450-man workforce to execute their jobs. His earlier experience as corral boss in northern Arizona served him well. As the excavation progressed westward, across the northern edge of the plat, he faced endless work hours and ever-greater challenges and responsibilities on the grandest man-made water delivery system in the history of the region: construction of the Arizona Canal.[46]

·2·

CORPORATE WATER

◇◇

"The agricultural advantages of Arizona are, I think, generally underestimated abroad. There is no more productive soil in America than is to be found in the valleys of Arizona, and it is believed that a greater variety of productions can be raised here than elsewhere in the United States, providing water can be had for irrigation. Not only does the soil produce fine crops of cereals, but fruits of all kinds, and vegetables of the finest quality."

HON. FREDERICK AUGUSTUS TRITLE, GOVERNOR OF ARIZONA, IN HIS "REPORT TO THE
US DEPARTMENT OF TREASURY FOR THE YEAR 1884."

"I arrived in Phoenix in 1884 and the Arizona Canal was then under construction. W.J. Murphy, whom I met in San Francisco, was its promoter and builder. ... The subsequent development of the Valley is due, in large measure, to his foresight and his faith in the future of the country."

HON. RICHARD E. SLOAN, LAWYER, JUDGE, AND LAST TERRITORIAL GOVERNOR OF
ARIZONA (1909-1912).

L ike other western areas not yet visited by the growing latticework of the transcontinental railroad system, the Salt River Valley, in 1883, remained an agricultural island of vast distances, rugged mountains, scorching deserts, and a cultural admixture of Hispanics, Anglos, and Indians. In spite of its eastern pretensions, Arizona was a land of rudimentary transportation, isolation, parochialism, and endemic violence. By 1883 John R. Norton knew that forging an existence in Arizona included violence as a condition of life. The raw frontier lured various types from all over the United States, even from throughout the world. They formed a civilization marked by youth, energy, ambition, daring, recklessness, greed, contempt for restraints, and a casual view of suffering and death. Local newspapers reported countless incidents of bloodshed and lawlessness. Norton knew, also, that three economic and social influences incited men to violence. Ambition—the scramble for quick money and the power that went with it—played a key role in the area's social psychology. Added to that was a toxic mix of liquor and guns. Nearly everyone went armed, and most of the young men in their teens and twenties drank constantly. The combination often proved deadly. Norton spent no small amount of time avoiding explosive situations in the Salt River Valley. He focused on the task at hand: building the Arizona Canal[1]

On one occasion Richard E. Sloan, a newly arrived attorney from San Francisco who would later become a territorial judge, governor, and, after statehood, Supreme Court Judge, visited the construction site with engineer Albert Barry. They located the burial site of John Coil, who had died in an accident during the construction process. According to Sloan, Norton approached Barry after the burial and said, "It's too bad that Coil's grave should remain out there in the desert without anything to mark it." The company engineer responded to Norton that he should craft a suitable elegy and place it on the head of the grave. The next afternoon, after Norton had worked on the project all day, he brought to Barry the result of his literary efforts: "Here lies John Coil; a son of toil; who died on Arizona soil; he was a man of considerable vim; but this here air was too hot for him." Sloan recounted that the headboard bearing Norton's inscription was erected and it remained there for several years.[2]

Norton's chief responsibility centered on the care and use of the 450 mules that were the primary engine of the project. They transported men, hauled supplies and water, and pulled excavators, plows, and scrapers. The animals consumed five tons of hay and 4,500 pounds of barley every day, and every day the logistics of providing energy for these animals consumed many man-hours of labor. Besides tending the stock, Norton scrutinized the various subcontractors, who charged varying prices per yard for excavation. The chief engineer, Barry, determined whether the work required the removal of soft dirt, loose rock, or hard rock. Most of the excavation was in loose dirt, and bucket chains conveyed the dirt over the bank. Norton had helped the crew perfect this dirt removal method while working on the Atlantic & Pacific project in northern Arizona and implemented it on the canal project.[3]

Daily life on the work crew grew monotonous. Norton knew, and the crew learned, that the Murphys ran a temperance camp, not selling liquor and not even allowing it on the premises, thus avoiding drunkenness, braggadocio, and violence. There was little in the way of recreation, though workers were encouraged to attend the Presbyterian Church in Phoenix. The summer heat and dust made life in the canvas tents almost unbearable. Plank floors served to tamp down the dust, though Norton observed that the heat made food preparation over a wood stove extremely difficult for the sweltering cooks, who had to take great care not to set the plank floors on fire. For the most part, life on the site settled into a routine of hot and dusty sixteen-hour days—a monotony occasionally broken when violent sandstorms blew up, followed by driving rains. In August of 1884 one contemporary wrote, "A great wind blew upon us about four o'clock and leveled the whole camp. Our quarters were greatly wrecked. The kitchen was so much so that we had to take the cook tent of the other camp. Our large tent was torn badly but before dark we had a shelter. I was so thankful that we escaped uninjured. We were all in the tent when it went over."[4]

During this period Norton was steadily earning the respect of the crew and the board of directors. Sometime early in 1884, Murphy rewarded him with a promotion to project foreman. In his new role he was in the field daily enduring the intense heat and frequent dust storms, and directing the rebuilding of canal works when they were washed out by flooding. Norton's sphere of influence broadened that same year, when county officials designated the camp as a precinct for the Democratic primary, and Norton was elected as a delegate. As Laura Murphy noted, "Today the Democratic Primary has completely upset work. . . . There was some voting in the morning and Norton and [Frank A.] Trott were elected delegates. At noon the camp stampeded to go to town to vote. Two four-horse teams went in and Norton said he would have to go in to see that the teams were taken care of and get the men started home right as no doubt there would be drinking." Thus Norton's responsibilities in both the private and public sectors grew incrementally, and at thirty years of age, he began to distinguish himself among his peers.

Shortly after his promotion, Norton responded to one of the subcontracting crew's calls that they had encountered a ridge of hard rock at a site south of Camelback Mountain. Instead of attempting to dislodge the rock ridge, Murphy and Norton conferred, and they decided to leave it in place, allowing the water to flow over the geologic formation. This created a twenty-foot-high waterfall that became known as Arizona Falls. It was an anomaly in canal construction that would, in time, become a favorite destination place for picnics and outings.[5]

The struggle to raise money for payment of wages, food, and supplies continued to weigh on Murphy. As early as the summer of 1883, the Phoenix bank Kales & Lewis refused to furnish funds for the construction, despite the fact that Martin W. Kales was one of the founders of the Arizona Canal Company. The bank had apparently abandoned active support of the canal project and turned its attention to land speculation, road construction, and the mining ventures that consumed latter-day prospectors and would-be magnates. Fortunately, Murphy convinced an old friend from his Atlantic & Pacific days, Colonel William Christy, to move to Phoenix and establish a new bank, and in September 1883 a charter was issued for the First National Bank of Phoenix, with authorized capital of $100,000, "$50,000 of which has been paid up," according to an account published in the *Arizona Gazette*. The same article described Christy as "a man of the highest standing and a banker of very large experience."[6] Christy, forty-two years old at the time, moved to Phoenix, purchased a 440-acre farm west of the city on a country thoroughfare that was soon being called Christy Road (now McDowell Road), and immediately made a contribution to the community. With zeal, thoroughness, and imagination, Christy studied the local cattle situation and concluded that Hereford would flourish better than the longhorn he found when he arrived in the valley. He imported the white-faced breed to prove his point. He also conducted water, soil, and weather tests and planted olive, citrus, and peach trees, some of which took root and others not.

But Christy and the new bank could not help Murphy in the short term, and the canal project needed cash immediately. Murphy had exhausted his own means, and only 3 percent of the canal work had been completed. Worse, Christy could not raise even $5,000, and Murphy, in frustration, wrote his wife, Laura: "The amazing questions that arise are how money shall be raised to start the bank and where shall I get money to keep the camp going till the bank opens? Something must be done before long. A carload of machines coming and nothing to pay freight with."[7] Murphy's application for a loan from an Albuquerque bank had been refused, and he had not heard back from the Bank of Arizona in Prescott, where he had applied for a similar, yet smaller loan. Prospects seemed bleak.[8]

In order to attract investors, dispose of the canal bonds, and raise money, Murphy drafted an Arizona Canal prospectus for distribution to prospective investors, something he considered "out of his line," but crucial for his fiscal survival. When Murphy visited one of his creditors, First National Bank in Los Angeles, which held his note for $16,000, the board of directors would not extend the loan. Murphy determined to try his hand in San Francisco. From the Bay Area he wrote Laura: "I had expected to have this business done by Christy. Did not anticipate so great difficulty in raising money on these bonds, but that turns out to be the hard business in the scheme, and so far nothing is done." Then J. W. Dodge of San Francisco apparently saved the day. Murphy's sojourn in San Francisco resulted in $42,000 in loans from Dodge and others, which paid off the Los Angeles bank debt and kept the canal project alive.

The First National Bank of Phoenix officially opened for business on November 22, 1883 in the newly constructed Ellis Building. Samuel J. Murphy, W. J. Murphy's brother, was appointed president; W. J., vice president; Christy was cashier, and E. J. Bennitt, Christy's brother-in-law, was his assistant. The newspaper notice allowed, "All the parties connected with the bank are men of sound financial standing and well-known integrity."

Murphy returned to Phoenix in mid-December. He vowed to continue his fundraising activities while the new First National Bank of Phoenix extended liberal credit to him. In short, the project remained fiscally afloat at year's end 1883.[9]

As Murphy addressed the fiscal challenges, Norton and the subcontractors sprinted to complete the first portion of the contract, which had to be finished on or before March 1, 1884. During this period the project faced its greatest crisis. Martin W. Kales, who supported the project in its earliest days, not only ceased financial backing in mid-1883, but also reversed course entirely and took several provocative actions to prevent its completion. Kales moved on Murphy and his partners while Murphy was in California in the summer and fall of 1883, working desperately to keep the canal project afloat. Kales knew that Murphy was unable to pay rental costs on equipment and mule teams at that point, and while Norton looked on in disbelief, Kales began urging canal subcontractors to file liens. According to contemporary accounts, Kales told one of his bank directors that Murphy and his partners were cutting into Kales & Lewis's banking business, and Kales sought to put Murphy into bankruptcy and cause the new First National Bank of Phoenix to go down with him. Kales intended to "bust Murphy."[10]

In early 1884 Murphy and his partners performed some fiscal sleight-of-hand in order to keep the project solvent. The unrelenting Kales pounced on the fact that the National Bank of Phoenix had extended more credit to Murphy than its national charter permitted, and notified banking officials. In an overnight reaction to Kales's complaint, the bank reorganized under territorial banking regulations and reopened as the Valley Bank of Phoenix. The officers of the new bank were the same as those of the old First National, with the exception of the president. Moses Hazeltine Sherman held that office. The local newspaper explained to its readers that the object of the change was to "untrammel the policy of the Bank and permit it to make larger loans."[11]

Murphy left Phoenix for San Francisco on another fundraising effort on February 20, 1884, the same day a huge storm swept into the Salt River Valley. The storm washed down the telegraph lines, cutting off the valley from the outside world. There was no communication of any kind in or out of Phoenix for two weeks, and reportedly the first message taken out was conveyed by a Pima Indian who swam across the Gila River. At this time, Kales induced all but one of the subcontractors to quit work and "attach Murphy's outfit with a view to preventing the completion of the contract stipulation."[12] The remaining subcontractor, J. T. Simms, resisted Kales's efforts, thanks to the entreaties of Will D. Fulwiler, Norton's good friend and Murphy's brother-in-law.

In the field Norton and Simms responded to environmental forces that threatened work which had already been completed. The heavy rains in February 1884 caused floodwaters to flow down the completed parts of the canal and spread over section seventeen, leaving that part so wet that Norton and the crews were forced to remove two miles south and work on that section until section seventeen dried out. Nonetheless, Simms completed the canal to the required point at 10:00 a. m. on March 1. The deadline was noon. Murphy thanked Norton and Fulwiler for their "rock firm" support in trying times.[13]

Unable or unwilling to reconcile his dual role of contractor and fundraiser, Murphy endeavored to have the company take over the sale of the bonds. The company attempted to oblige, and Churchill and Governor Tritle journeyed to Chicago and New York in an effort to assist in raising capital. They were unable to convince investors of the potential profitability of the canal venture, however, and another cash crunch ensued in September 1884 as Murphy traveled again to San Francisco to order materials for the Arizona Dam portion of the project. On September 9 he ordered a carload of powder, along with wagons and carts to haul lumber from Maricopa, and on October 2 he ordered a carload of iron for bridges. His $30,000 in notes in San Francisco were due, and another $13,000 past due had to be paid. Murphy had no money to repay these notes, much less to pay for these deliveries, and he hoped to secure funds before the goods reached Arizona. He returned briefly to Phoenix to oversee the beginning of the construction of the dam and then headed to New York in November to follow up on a lead developed by Governor Tritle and Churchill. He remained in the East working every angle to raise money in places like Hartford, New Haven, and Boston, and was able to borrow $5,000, putting up $25,000 in bonds as collateral.

Norton knew well and liked Governor Frederick Augustus Tritle, a tireless promoter of the Territory of Arizona in general, and of the Arizona Canal project, in which he maintained a pecuniary interest, in particular. A Republican appointee in a heavily Democratic federal territory, Tritle nevertheless developed longstanding relationships throughout the region. In his 1884 report for the Commerce and Navigation section of the US Department of Treasury, Tritle relied heavily on Murphy's Arizona Canal Company prospectus as he touted the virtues of the Salt River Valley and the canal project to his superiors in Washington, DC.[14]

In his prefatory comments to his report, Tritle wrote, "In reference to the benefits of the canal to the Territory, the following quotations from the prospectus of the canal company are instructive." He continued, "In the Salt River Valley an immense canal is being constructed which will convey water enough ... to reclaim 100,000 acres besides providing motive power for an immense amount of machinery," he wrote. With the 35,000 acres already under cultivation, he declared that "When the canal is ready for use, it is expected in the spring of 1885, this valley will be as valuable and productive as any area of equal extent in America." He encouraged settlement: "The possibilities for the immigrant in this and adjacent valley of the Gila are wonderful. . . . Land can be had reasonably cheap [sic]; that which has not been improved can be had for $5.00 to $10.00 per acre; improved land for $15.00 to $30.00 per acre, according to the character of the soil and location. This price includes a water right sufficient for crop raising. . . ."[15]

Tritle covered topics like water power, water supply, reclaimed lands, agricultural products, livestock, land grants, and artesian wells in his report. Among other things he observed, "The most profitable cereals are wheat, barley, and oats. The yield this year (1884) is estimated at 340 million pounds. The wheat produced here is of extra fine quality and makes superior flour. The market for these products comprises a radius of four hundred miles of surrounding country. . . . The means to water crops being in the hands of the farmer and with no frosts to interfere, the yield is very certain. There has not been a failure of crops in this valley since its settlement thirteen years ago. It is a notorious fact that in all the countries where lands that are supplied by water for irrigation rate at more than double the value of those lands that depend on rainfall, and this is owing to the larger crops produced and the greater certainty of crops on irrigated lands. In some countries—Spain for instance—this disparity is even greater, the value of irrigated lands being more than three times that of other agricultural lands." The Governor added more detail: "The land is deep alluvial soil of surpassing fertility. The surface remarkably even, being free from elevations and depressions, with an even grade of about ten feet to the mile from the foothills to the river, rendering it perfectly adapted for irrigation. . . . The yield per acre of wheat and barley is from twenty-five to thirty-five bushels and after this is harvested corn can be planted on the same ground, and a fine crop raised the same season. . . . Apples, peaches, pears, plums, figs, quinces, apricots, and nearly every other variety of fruit yield largely." In anticipation of introducing citrus into Salt River Valley agriculture, Tritle sometimes overstated the virtues of the region: "Lemons, oranges, and olives can be raised with profit, and finer grapes cannot be produced anywhere. Sugar cane and cotton have also been grown successfully."[16] Tritle often repeated his effusive praise to potential investors throughout the eastern seaboard and effectively sold the Salt River Valley as a Garden of Eden and potential agricultural empire.

But the environment needed to be harnessed to the benefit of man, and often Salt River Valley residents experienced the humbling power of the natural environment. In December 1884 the river's natural flow interfered with dam construction. On December 13 a steady, hard rain lasted for twelve hours, and by evening the river had risen two feet. The mountains were white with snow, and the current was so rapid that it carried away two days of work. On Christmas Day 1884, melting snows in the Salt River watershed left the river "booming." More of the dam was swept away. Yet by January 22, 1885, construction of the dam seemed "under control," and the workers labored away "day and night." The carpenters finished their work on January 24, and Norton expected the dam to be completed on January 31—if "there was no big rise in the river."[17] On January 28, the crew had been cut to thirty men and ten mule teams, and Norton and one of his close friends on the crew, James "Jim" Cashion, brought in the last load of lumber to complete the bridgeworks. Over the next month the crews completed final touches on the dam and bridges and repaired leaks. By April 30, 1885, the work on the canal and its attendant

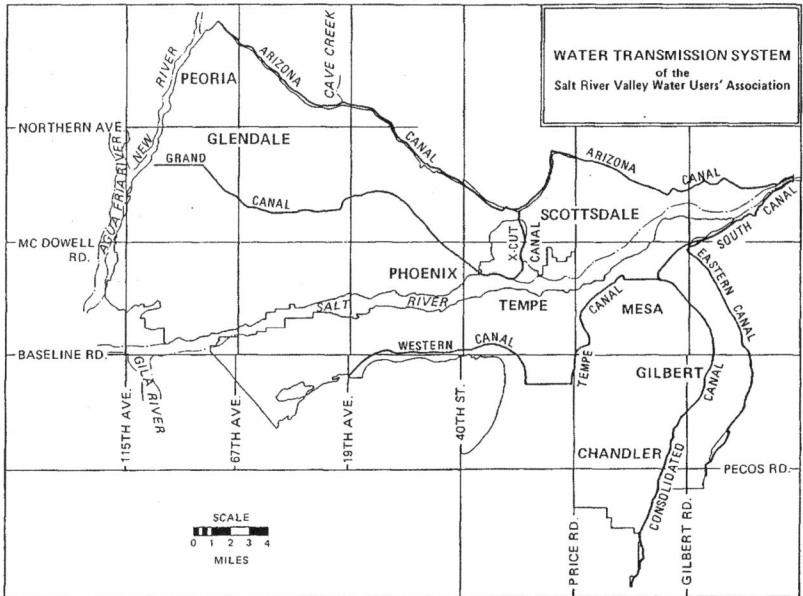

Map of the Arizona Canal, from the Historic American Engineering Record (HAER AZ.19), National Park Service, US Department of the Interior. Shelly C. Dudley, historian.

infrastructure had been completed, although the company's financial troubles continued. Murphy was scrambling to pay workers and creditors alike.[18]

"John R. Norton, foreman for W. J. Murphy," according to the *Weekly Phoenix Herald*, headed the final inspection of the completed canal during the last week of May 1885. Along with the Arizona Canal Company president of the board, Clark Churchill, the board director, Colonel F. C. Hatch, and engineer Charles Marriner, Norton set out in an ambulance leading a saddle horse behind it. The entourage first visited the W. J. Murphy campsite about eight miles northeast of town near Squaw Peak (renamed Piestewa Peak in 2003). From Murphy's camp the group headed east and inspected the headgates and the dam that conveyed the Salt River's waters into the canal. In order to gain better perspective, Churchill mounted the horse and rode on top of the canal bank. The tour ended with a return to the dam and headgates and Norton, Churchill, Marriner, and Hatch camped the night near the home of the gatekeeper, Mr. Stelzreide, and amused themselves by fishing and swimming in the river.

The next day the party returned to Murphy's camp, Churchill riding the horse along the canal bank the entire distance. On the third day of inspections Colonel Hatch returned to Phoenix, while Norton and the rest of the group continued their inspection to the west, covering nineteen miles from Arizona Falls. Cuts in the rock in the eighth and ninth miles required blasting, and in miles twenty-six and twenty-seven were deep cuts in cemented material, which was more difficult than solid rock. The *Herald* reporter had kind words to say about Murphy and, by inference, Norton. "A master stroke was the selection of W. J. Murphy by General Churchill as the party to execute the work of construction. Our people have good reason to know this." He added, "The contract for the entire work having been let by him. . . . At times as many as five hundred men and three hundred mules have been employed on the work, but in all this work there was no confusion. . . . The most perfect order and discipline," he averred, "prevailed all along the line and Mr. Murphy has not only performed the work but has also performed it faithfully, and paid all his workmen and others for their labor, materials, and supplies consumed."[19] Accounts were paid up—at least for the moment.

The *Herald* reporter spared no verbiage in describing the physical appurtenances of the new canal, dam, and related infrastructure. His description provided future historians with a clear snapshot of the remarkable achievement that shaped the valley's economic, social, and cultural future. The upper canal was cut in solid rock, "excavated through solid granite." "The dam," he wrote admiringly, was "constructed of fascines supported by cribs filled with heavy rock and extends across the Salt River from a point . . . some three-quarters of a mile below the Verde River so that the projecting natural rock of the mountain practically forms a section of the dam itself." He observed that the waste weir (a structure to divert excess water during times of high water) was built of heavy timber about forty feet wide and sat at the height of the dam. Just to the west of the waste weir sat the headgate. "This structure," he wrote, "is a model of mechanical skill, strength, and symmetry . . . built of exceedingly heavy timbers and solid masonry laid in Portland cement. It is forty feet wide and so contrived to let six feet in depth of water in to the canal. Its bulkhead rises over the gates to a height of twenty-three feet."

On June 1, 1885, the *Arizona Gazette* trumpeted that the canal "stood the test and . . . was finally accepted by the company. Water flows gracefully and evenly through the entire length of forty-one miles." The reporter wrote effusively of the sheer size of the project, noting that "this canal will carry as much

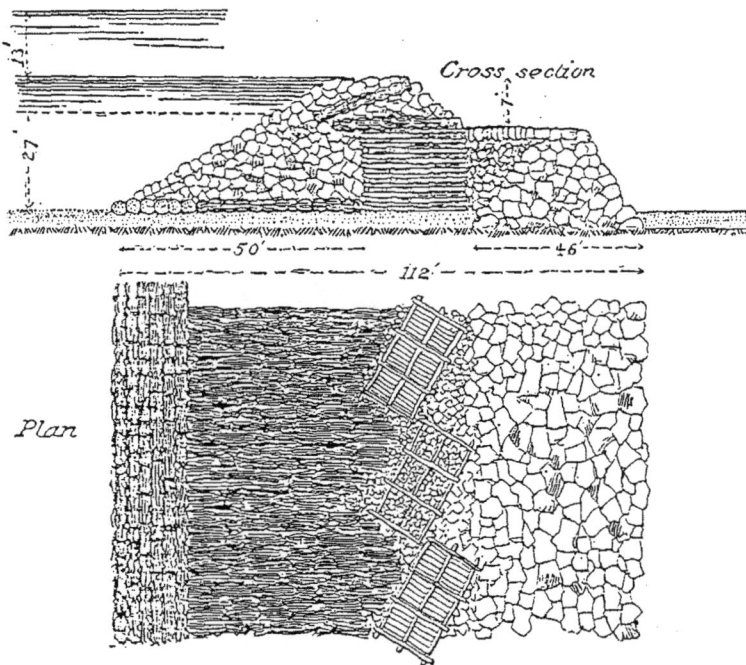

Cross section of first weir, Arizona Canal (HAER AZ.19).

water as the Erie Canal in the state of New York. . . ." The Arizona Canal was "a grand improvement and although of a public nature it has been constructed entirely by private means and great business capacity of a few men. . . . It will furnish water to reclaim a very large tract of land . . . to supply thousands of farm, vineyard, orchard, and stock-growers homes, upon lands which have been an unproductive desert of no value for any purpose. It will be of incalculable benefit to this valley and the whole territory. . . . No improvement in any part of the country originating in private enterprise and involving so much expenditure has, to our knowledge, ever been brought to completion."[20]

The completion of the Arizona Canal in 1885 served as a powerful example to local citizens; as more land was brought under cultivation, most of the canal companies in the valley became corporate associations issuing capital stock. The informal operating model and maintenance procedures, and the impressionistic management of shortages and distribution that marked the

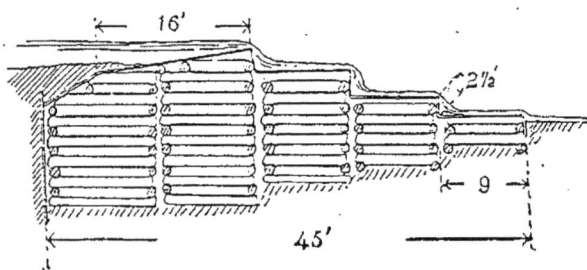

Cross section of second weir (HAER AZ.19).

early ditch days, were no longer sufficient. Too many people were dependent upon water delivered through the canals, and their numbers were increasing. The emerging corporate canal companies leased and assigned water rights within the service limits of an irrigation ditch with little regard for the primacy of prior appropriation, thus setting the stage for interminable lawsuits. Water certificates, in fact, became common currency in the Salt River Valley in the 1880s, as increasing numbers of them found their way into the hands of money lenders collecting defaulted loans from landowners. Water rights were becoming divorced from the land and falling under corporate ownership.[21]

The Desert Land Act provided for would-be settlers to claim 640 acres of public land if they could irrigate the total acreage within three years. Between 1883 and 1893 a total of 277 people filed entries under the act for land along the canal, and many were Murphy-Fulwiler family members who had been signed up by Murphy and Churchill when the canal project first began. Murphy actively promoted the Arizona Canal Company in his home state of Illinois and recruited allies and confederates from the US Congress, Tucson District Land Office, and other strategically placed bureaucrats in the federal government. Instead of actually selling those claims made on behalf of his friends and relatives, Murphy merely informed potential buyers of land and water that was available through the Arizona Canal Company Project. If someone was interested in filing a claim under the Desert Land Act, Murphy likely collected a "finder's fee" or had the "buyer" sign a mortgage—unrecorded, of course—in Murphy's favor. Murphy then sent a "relinquishment" of the rights in question to the General Land Office and arranged for the buyer to be the next entrant for the relinquished parcel. If, in fact, the buyer intended to occupy and farm the land, he either purchased a water right or an Arizona Canal Company bond.[22] After May 1885, when the canal was completed, and "at any time before one thousand water rights shall be sold," each $1,000 bond could be converted into two water rights (enough to irrigate 640 acres). Money derived through the Merchants' Loan and Trust Company of Chicago was to be used to pay off the canal company's bonds.[23]

Meanwhile Norton saved his money from the railroad grading, canal company project, and other ventures, and on November 12, 1886, the thirty-two-year-old Kentucky native purchased an eighty-acre parcel of land south of Indian School Road between what is now Tenth and Twelfth Streets in Phoenix. This action solidified his stake in his new southwestern home; he was a permanent resident, although he continued to range the valley far and wide. He lived on the parcel for thirteen years, improving the property, harvesting grains, and grazing livestock. He raised mules and fattened two thousand head of cattle per year on his feedlot, and indulged his penchant for ranching and horses. In short order he acquired some of the fastest quarter horses in Maricopa County and was on the Jockey Club program charged with making arrangements for the winter races in 1891.

As Governor Tritle anticipated, citrus was quickly becoming a major cash crop for the valley, despite the predictions of skeptics that the area would prove too cold for citrus. W. J. Murphy planted twenty acres of citrus trees at Ingleside, on the south side of the river, and fellow pioneer Charles Trumbull Hayden followed suit. The trees bore fruit. Five years later Murphy laid out the Orangewood section, and using two parcels on opposite sides of North Central Avenue he planted thirteen hundred acres in citrus. In the 1890s John Norton experimented with citrus on his property, as well. All were part of the early history of what became an extensive and profitable enterprise in the Salt River Valley.

Throughout his life John Norton maintained his close personal and professional ties to W. J. Murphy. In May 1887 Murphy, in order to generate capital and obtain land for himself, formed the Arizona Improvement Company (AIC). In effect, Murphy sought to gain control of the Salt River, Maricopa, and Grand Canal Companies north of the Salt River. Norton became the superintendent of AIC and the water master for Murphy's newly configured Arizona Canal Company consolidation.[24] Under the aegis of the AIC, Norton oversaw the building and construction of the Crosscut Canal in 1888, which successfully brought all canals on the north side of the river under one management and operational system. From that time until the turn of the century, he assisted Murphy in developing not only strategies and plans for future land acquisitions and internal improvements, but also represented him at various business and governmental meetings and official functions. Performing administrative duties for AIC and the Arizona Canal Company, while at the same time maintaining his own farm and livestock operation, consumed Norton's time for sixteen hours per day or more. Ultimately, he resigned his water master post on January 1, 1899, turning it over to his trusted friend, Dan McDermott, who had worked for many years as a *zanjero* for the Arizona Canal Company. As the *Arizona Republican* described this succession, "[McDermott] is a capable man and will make a good superintendent."[25]

As superintendent of AIC, Norton oversaw the grading of the eighty-four-mile Maricopa and Phoenix Railroad right-of-way, using the men, teams, and scrapers from the recently constructed Arizona Canal. Even though he had purchased property in north Phoenix, for the interim Norton's home became his tent, as it was during his two years laboring on the canal. The *Phoenix Gazette* described the grading project as it reached a point four miles east of Tempe: "Their tents are pitched about a mile and a quarter from the river. The outfit consists of a commissary tent, sleeping tents, and two others occupied as a dining room and kitchen. A great feed bin for the mules flanks one side of the encampment, and wagons and the usual incidentals of a mechanical camp are scattered about." Under Norton's experienced supervision, the crew completed the roadbed grading in late June 1887, in time for the first passenger train to carry fifty people over the completed line from Tempe to Maricopa.[26]

The implications of the completed railroad spur were significant and further transformed the valley's agricultural economy, triggering the crop diversification that marked the 1890s and the first decade of the twentieth century. As Maricopa County's population grew from 235 souls in 1870 to 11,000 by 1890, the Maricopa and Phoenix spur line opened the valley to unprecedented opportunities. Farmers knew prices of wheat and barley fell in the late 1870s, after a brief boom in the earlier part of the decade. Wisely, they diversified their crops in the 1880s, as they discovered that citrus and some other fruit trees flourished in the valley's mild climate. They planted peach, fig, apricot, and a variety of citrus trees next to the traditional fields of alfalfa, clover, and grains. On the north side of the river, along the fifty thousand acres of land irrigated by the Arizona, Maricopa, Grand, and Salt River Valley canals, farmers planted alfalfa, citrus orchards, and grains. The completion of the railroad spur in 1887 allowed the Salt River Valley to ship its burgeoning produce via the mainline Southern Pacific railroad. As local boosters spread the word, more settlers came to build homes and farm the rich, alluvial soil of central Arizona.[27]

Among the investors in Murphy's Arizona Improvement Company were such wealthy outsiders as Francis G. Newlands and Frederick Sharon of San Francisco, as well as Andrew Crawford of Chicago. (Newlands, who later became a member of the US House of Representatives from Nevada, would play a major role in bringing the federal government into the reclamation of arid lands fifteen years hence.)[28] John Norton spearheaded the newly constituted company's first set of development projects, including the planting, harvesting, and marketing of citrus, and the development of Grand Avenue, then on the northwest corner of Phoenix, in order to promote land sales both to the west of town and in the northern agricultural reaches of the valley. As AIC superintendent, Norton oversaw important aspects of other transportation projects, like the Santa Fe, Phoenix, and Prescott Railway right-of-way project of the late 1880s and early1890s (the spur to Ash Fork), and the effort to build a Phoenix street railway.[29]

In spite of Murphy's and his partners' frenetic activity with outside investors in their never-ending efforts to capitalize and stabilize the Arizona Canal Company, the unremitting issues of land speculation, flood damage, lawsuits over water rights, and a host of related problems cast a pall over the system into the 1890s. Not surprisingly, eastern capital and credit tightened, and the national depression of the 1890s further discouraged potential investors.[30] It surprised few local observers that, in spite of a Herculean effort, the Arizona Canal Company went into receivership in the late 1890s. No longer able to pay interest on its bonds, by 1896 the company was claimed by eastern bondholders who reorganized it as the Arizona Water Company. In that year, the northside system had 106 miles of main canals, 200 lateral ditches, and the potential for irrigating over 150,000 acres of land.[31]

At this point, however, John Norton had already carved out his individual identity and established his sterling reputation. Over time he had separated himself from the fate of the canal company, and by the late 1880s his increased role in public policy issues—including encouraging the federal government's developing interest in irrigation, public lands, and social reform—were widely known. His inclusion as one of the three men who conducted the Breakenridge Survey of 1889 surprised no one in the Salt River Valley.

The Breakenridge Survey was one small piece of the federal government's participation in the great westward expansion of the late nineteenth century. It was prompted, in part, by the findings of Major John Wesley Powell, the heroic explorer who was the first to descend the last fifteen hundred miles of the Colorado River in 1869. Even though Powell had lost an arm in the Civil War, he led an extremely dangerous three-month expedition which was, according to the Pulitzer-Prize winning author William H. Goetzmann, "the climactic event of the late-nineteenth century exploration [that] would mark the opening of the last completely unknown territory in the continental United States."[32] According to Goetzmann, the major was "a casually educated and self-made scientist with a driving ambition. Powell was perhaps the outstanding representative of a breed of political men who came to the fore in the late nineteenth century." He proclaimed Powell "the greatest explorer-hero since the days of Fremont. . . ."[33]

Major Powell, a scientist and geologist who carefully documented his explorations, was named director of the United States Geological Survey (USGS) in 1881. Three years earlier, in 1878, he published his *Report on the Lands of the Arid Regions of the United States*. His argument for a more rational approach to problems posed by the arid West undoubtedly stimulated the imaginations of countless Americans. In the report Powell argued that a great number of farms could be created out of the public lands west of the 100th meridian if water were made available. The report, coming on the heels of Powell's celebrated explorations and writings on unexplored reaches of the arid West, cannot be overestimated in terms of its impact on Norton and his fellow citizens in the Salt River Valley. With federal funding of the Breakenridge Survey, Norton and his fellow Salt River Valley farmers and landowners whiffed the "scent of government appropriations" to come.

Powell and those like him labored primarily in the cause of reform; they were intellectuals and scientists, which sometimes made their concepts and writings difficult to understand. Their knowledge was gained by hard experience as field explorers. If Powell appeared overconfident as he unrolled his maps before Congressional committees and federal bureaucrats or published his massive, erudite tomes in obscure series, he nevertheless had the good of the nation on his mind. Further, as scientists and public men, Powell and his cadre sought to put the explorer and his practical experience to the highest purpose by promoting the fair, efficient, and socially useful development of the American West. Twenty years before Theodore Roosevelt and the Progressives ushered in an era of conservation and social reform, Powell and his like-minded scientists had already developed the framework along which such reform could be accomplished. They had already formed the bureaus and commissions, trained the staffs of experts, and streamlined the dissemination of information whereby the reform ideas could reach the mainstream of American society. Powell and his associates were figures grappling heroically with the vast forces of nineteenth-century political and economic darkness that surrounded them.

The significance of Powell's work went far beyond adventure into the unknown. He and his men made their work meaningful by focusing on issues of social and scientific significance. He remained a scientist with a clear view of the single central problem that lay behind all his study of the West, namely the problem presented by the environment to people who wished to settle and make use of western lands. Powell's mission was to describe accurately the new environment for an onrushing civilization. It was, in essence, a pedagogical mission, and as civilization invested in the West, Powell's mission ultimately became a reforming one. He was a teacher and social reformer.

A central theme ran through all of Powell's writings: that institutions and techniques devised in what he called the humid eastern sections of the United States could not be successfully transplanted to the new and challenging western environment. As early as 1873, in his annual report on the surveys of the Plateau Province, Powell emphasized the extreme aridity of the region and the necessity for some sort of rational land classification system. Powell's *Report on the Arid Regions of the United States* was his most famous work and arguably the first modern treatise on political reform as it related to the American West. It was part of his campaign for the consolidation of the various government surveys and land office projects into one responsible civilian authority. In it he clearly stated that much of the region was unsuitable for settlement and farming along patterns devised in the comparatively humid East.

Irrigation and cattle grazing, topics that tied together Norton and his progeny, were economic activities that drew the most attention in the famed report. Powell warned that two-fifths of the United States was arid and stated flatly that a scientific and environmental approach to the West and its resources was imperative. Powell offered innovative solutions: First, the country had to be mapped and the lands classified as mineral, coal, pasturage, timber, and irrigable. He then proposed two new land laws that would organize irrigation districts and pasturage districts. As Powell drafted them, they would have provided sweeping changes to existing laws. No longer, for example, would the traditional Land Office grid pattern of 160-acre farms be laid out across the West. Instead, he argued, irrigation districts, similar to the Mormon colonies he had observed in Utah, should replace the existing system. The unit for each irrigated farm would be 80 acres and not 160, and all water rights would inhere to the land. Powell went further. Groups of farmers, he urged, should come together to form irrigation cooperatives, thereby sharing in the enormous expense of the dams, flumes, and ditches. His vision: that these districts would be governed democratically and, if possible, locally, in the public interest. Grazing, he proposed, should be organized in units of 2,500 acres, twenty acres of which might be irrigated farmland used for growing winter hay and various other necessities. In this instance also, all water rights would inhere to the land. Cattlemen were also encouraged to cooperate in grazing districts of adjoining and, importantly, unfenced land.

These proposals, the capstone of Powell's towering career as an explorer and scientist, were intended to avert a number of problems that he had observed in his extensive travels in the West. Important to Norton and the Salt River Valley was the issue of monopoly control of water rights—monopolies that resulted from traditional Anglo-Saxon settlement patterns and law. Typically, individuals settled along the streams and dammed up the flow for their own use, as reflected in Phoenix and other arid areas of the West. Then, Powell observed, because of the expense involved, water companies (like Murphy's ventures) were formed. Using loopholes in the Timber Culture Act of 1873 and the Desert Land Act of 1877, these companies were able, in many cases, to gain a monopoly on much of the available water in the West. They could use it for mining, sell it to the highest bidder, waste it, or use it to force settlers out of a certain region. With the formation of irrigation districts and water rights inhering to the land, this undesirable outcome, in Powell's opinion, would not happen. In the end, Powell's arid land report was one of the most important books to come out of the West. It focused on fundamental challenges and problems, and significantly, it put forward creative answers. Powell's detractors—spokesmen for land rings, water monopolists, cattle combines, speculators, railroad promoters, mining tycoons, and local political bosses—derided his prescient conclusions. Still, Powell's impact, especially in territorial Arizona, became evident as residents moved to develop coherent systems of land use and water resource conservation, development, and distribution.[34]

Richard Hinton, the widely read western journalist, carried Powell's arguments further. In uncompromising terms, Hinton urged massive federal spending for devising comprehensive irrigation works in the region. And when Powell was appointed to direct the USGS in 1881, he urged Congress to deal directly with irrigation policy. He continued to pressure lawmakers, and by 1888 the Senate Select Subcommittee on Irrigation, chaired by Senator William Stewart of Nevada, had held fifty-three sessions throughout the West to hear testimony.

The committee arrived in Phoenix in the fall of 1889, and local officials handed the senators the notes generated from the Breakenridge Survey. Norton, McClintock, and Breakenridge discussed their findings with the visiting lawmakers.[35]

But opposing arguments—that federal intervention could disrupt private efforts to build profitable irrigation works—gave Congress pause. Support for massive federal intervention, as advocated by Powell and other government reclamation proponents, cooled temporarily, and instead Congress focused its efforts on assisting the states to develop projects in conjunction with private enterprise. Arizona, of course, had not attained statehood, so consideration of the various proposals floated during the early 1890s were neither relevant nor applicable to the Salt River Valley or the Territory of Arizona. But there was more than a laissez-faire constitutional argument underscoring the concept that this was a job for the states, and not the federal government. The colonial tension between the underdeveloped West and the developed East often caused westerners to resist what they perceived to be eastern control of their right to self-determination.[36]

Indeed, Norton and others heard the repeated argument that the federal government was too far away and the arid regions too dissimilar, while easterners argued that it was not fair that their taxes could underwrite internal improvements and infrastructure, thus developing the lands and natural resources of another region. In the end, the philosophy of state control over potential government irrigation projects, favored by westerners, and the simultaneous resistance on the part of easterners to use federal taxes to pay for western development together found a legislative voice in the Carey Act of 1894. This law enabled the president to authorize an allocation of one million acres within each state's public lands for irrigation, reclamation, settlement, and cultivation; surplus funds were to be used for reclaiming other land within the state. The Carey Act—a compromise—was largely ineffective, however, because it was partly dependent on private financing, and little public domain was reclaimed under this legislation.[37]

John R. Norton, meanwhile, continued to establish himself in the community, and sometime before 1891 he proposed to Etta Willey Wright, who was fifteen years his junior. Etta Wright was born on October 14, 1869, in a covered wagon. J. C. Wright, a Confederate veteran and an impoverished farmer after the war, left Arkansas seeking a better life in California. The Wright caravan crossed the Colorado River at Yuma and camped on the California side, where Etta was born. Later, she commented that she regretted not being born on the Arizona side of the river, since the family stayed in California for only a few years, then returned to Arizona to settle and ranch in Skull Valley, southwest of Prescott.[38]

On January 14, 1891, Etta married then thirty-seven-year-old John R. Norton, described by the local *Arizona Republican* as "well and favorably known here . . . and a prosperous rancher under the Grand Canal." It was the social event of the month. The ceremony took place three miles west of Phoenix at the home of Etta's parents, Mr. and Mrs. J. C. Wright, well-known residents of the valley. Eli G. Roberts, elder of the Methodist Episcopal Church in south Phoenix, presided over the matrimony, and friends and relatives filled the Wright home to capacity. "The only regret," the newspaper reported, "was that limited accommodations prevented the number of invited guests being larger. . . . The bride in white albatross with train, white veil, and orange blossoms, and the groom in black made a handsome couple." The remainder of the evening, the reporter wrote, "was interspersed with music and passed pleasantly and quickly away." The newlyweds happily accepted gifts of two bisque lamps, a handsomely framed engraving and easel, an oil painting, and a china dinner set.

Married and planning to begin family life, Norton busied himself with AIC and farming and ranching pursuits in the Salt River Valley. Yet environmental and political forces altered the local, regional, and national landscape and impacted not only Norton's personal and professional life, but also the lives of all who called the Salt River Valley home. Shortly after his marriage to Etta in mid-February of 1891, Norton and his wife and neighbors witnessed the destructive power of the unregulated Salt River. The flood of 1891 dwarfed the floods of 1884 and reframed the substance, nature, and tone of the overall water resource development debate in the valley. Heavy and continued rains from late January through February on the Salt and Verde watersheds caused the "biggest flood of the Salt River that had ever been known," and erased years of human effort and toil in central Arizona. Although the flood crest did not reach the valley until February 20, the railroad bridge over the Salt washed out two days earlier, leaving Phoenix without a railroad connection to the outside world for three months. According to one victim, "white people and Indians" scrambled to high ground around Mesa and the refugees witnessed the destruction of bridges and weir dams that had previously laced the agricultural communities together. Letters to worried relatives and friends took weeks to reach their destinations. In March, when Norton and his neighbors looked out upon the regional ruin, they saw unprecedented damage and destruction.[39]

The community attempted to regroup. In 1893, when Wells Hendershott, a lawyer and promoter from New York, filed claim to the reservoir site upon which the Breakenridge Survey had recommended construction of a storage dam, Murphy, Norton, and valley leaders were, at first, delighted. Hendershott, who had formed the Hudson Reservoir and Canal Company, told local farmers and businessmen that he planned to turn the Tonto Basin reservoir site into a lake, the object being to store floodwater. Water from the reservoir would be used for the irrigation of a large expanse of dry land east of Mesa. The company was financed largely by Man and Man, a New York firm of lawyers and engineers, and a journalist, writer, and businessman named Sims Ely, who later wrote a popular book on the Lost Dutchman Mine. The Hudson Reservoir and Canal Company opened offices in Phoenix in 1894 and developed engineering plans for a masonry dam of 225 feet to create a thirty-square-mile reservoir.[40] Part of the plan was to construct a diversion dam at Granite Reef, a site which could divert water into the north side and south side canal systems. Throughout the 1890s the company studied, measured, and calculated stream flows; figured depth of bedrock, and surveyed potential lines for a canal route to the lands near Mesa that Hendershott favored. Demonstrating the company's seriousness of purpose, Hudson applied for a right-of-way through the Gila Indian Reservation in exchange for supplying water to those Indians living along the proposed easement.[41]

But the Hudson Reservoir and Canal Company could not close the deal. Despite reasonable plans and solid engineering information, Hendershott and his confreres could not convince outside investors that the $3 million project was feasible. Like Murphy and the Arizona Canal Company, the Hudson Reservoir and Canal Company ran headlong into tight money, skepticism, and narrowing credit markets. In fact, the two projects landed on common ground when the Hudson River and Canal Company attempted to purchase from the

Arizona Improvement Company one of Norton's most important charges—north side water rights and land claims—as collateral to borrow upon, but the possibility of liens against the financially wobbly AIC prevented this transaction. Henry Man, one of the chief investors, insisted that all of the acreage in the valley had to be in private hands and the rate for water rental doubled in order for the Tonto Basin Project to be profitable.[42]

In what became a cruel irony the devastating flood of 1891 was followed by a decade-long drought. One Norton contemporary suggested that "the big drought started in the 1890s and extended into 1903, practically ten years." Another chronicler interpreted that ten-year period as "the blackest period in the history of the Salt River Valley." And the nationwide depression that began in the nineties and lasted well into the new century compounded the human suffering. Norton struggled to help Murphy maintain his businesses, and like so many others, he suffered severe financial setbacks and tightened his fiscal belt. By 1897, one of the worst years of the drought, water scarcely trickled in the Salt River, crops failed, water shortages for livestock and domestic uses grew acute, and the local economy ground to a halt. The following year Norton saw the desert begin to reclaim the desert empire: livestock died, settlers abandoned once-prosperous farms, and those who stayed hoped and prayed for a change in the weather. Desperate residents from many districts throughout the valley called meetings to take action against the environmental extremities of flood and drought.[43]

The nineties also brought personal tragedy to the Nortons. On January 26, 1896, Etta Norton gave birth to the couple's first child, a girl, but two months later the infant died of pneumonia.

The residents of the Salt River Valley were not alone in facing enormous economic and environmental challenges as the century turned. Managing water resources was a major challenge throughout the West, and the National Irrigation Association, headed by reclamation pioneers William Ellsworth Smythe and George H. Maxwell, formed in response to this broad-based cause. Their annual National Irrigation Congresses, which began in the early 1890s, helped shape the debates in John R. Norton's changing world. Norton, like so many others, became embroiled in several trying water rights suits. Water scarcity, naturally, led to overdrafting of scarce supplies, and Norton spent sums of money on what he believed was unnecessary litigation. He began reading any material available on irrigation and reclamation, including Smythe's magazine *Irrigation Age* and his book *The Conquest of America*, along with Maxwell's published tracts, which oftentimes resembled diatribes. The national press covered National Irrigation Congresses where Smythe, Maxell, and their supporters held forth, and Norton read with interest the proceedings and resolutions developed at these heavily attended affairs. These late-nine-teenth-century pioneers, dedicated to the "Conquest of Arid America," gave Norton a clear understanding of the problems not only of the Salt River Valley but also of the other western states and territories.[44]

By the mid-1890s, the vicissitudes of flood and drought, coupled with the apparent inability of the private sector to develop the necessary infrastructure to collect, preserve, and distribute water, led many valley farmers to conclude that only the federal government possessed the ability—specifically the money and technical expertise—to put water to use for the public good in the arid Southwest. They began looking to the nation's taxpayers to alleviate their problems. In anticipation of the Fifth National Irrigation Congress—held in Phoenix in July 1896—William "Bucky" O'Neill, a longtime Norton friend, territorial legislator, and Arizona delegate to the Irrigation Congress, submitted a national irrigation bill for review by Congress participants. Adapted from California's Wright Act of 1887, which permitted farming regions to form and bond irrigation districts, O'Neill's plan, in brief, called for the federal government to advance funds to state-recognized irrigation districts in order to reclaim desert lands for agricultural production. The loans, he allowed, were not to exceed $1 million in any given year, and the cost-per-acre could not exceed $25. Loan charges were to be a lien on both the land and irrigation works, and every government-lent dollar was to be repaid in full. O'Neill also prepared a well-reasoned response to the predictable laissez-fair arguments of those who opposed the federal government entering the reclamation business: "After the expenditure of millions in improving eastern harbors and rivers, building post office and other edifices, after guaranteeing the bonds of railroad companies to the extent of millions, with the Republican and Democratic parties advocating the expenditure of $100 million to build the Nicaragua canal, after the payment of over $12 million in bounties to sugar growers of the South, after the establishment of protective tariffs for over fifty years for the benefit of the East, there is no reason why such a bill herein proposed [the national irrigation bill] should not be passed."[45]

Norton attended the Congress and watched O'Neill grow frustrated as delegates scoffed at his proposal. He then observed O'Neill take the floor to wrathfully denounce the delegates for speaking in generalities and evading particulars. O'Neill was booed, and after the session ended many members chastised him for his behavior. Frederick Newell, representing the US Geological Survey at the Congress, said that he doubted the annual meeting would result in securing government reclamation of arid lands, but if the annual meetings persisted, he allowed that the matter might gain traction with federal legislators. Newell later recalled that the congress "favored construction of storage reservoirs by the federal government, where necessary, to furnish water for reclamation of public lands," although the public reaction immediately following the irrigation congress seemed to suggest that the public felt otherwise.[46]

It was 1896, however, and in the Salt River Valley the drought showed no signs of relenting. Immediately after the Congress in Phoenix—and perhaps reflecting O'Neill's frustration—Salt River Valley farmers called a mass meeting to discuss their problems. Significantly, a citizens' water storage committee was appointed to investigate three possibilities for storage in the Salt River Valley: private ownership, corporate ownership, and government construction. Norton served on the committee, discussed the various proposals put forward to the farmers and landowners, and participated in research and conclusions.

In essence, the citizens' water storage committee considered several financing schemes, including the creation of irrigation districts modeled after California's Wright Act; tax levies on lands; exemption from taxation of water storage works; and the creation of new canals to foster further agricultural development. Norton observed that long-time landowners, especially on the south side of the Salt River in Tempe, remained reluctant to participate in a privately-held water storage program without special assurances and provisions. He learned that these farmers wanted to ensure their interests and protections in any canal or reservoir company proposal that related to water distribution. Indeed, the committee report revealed the fractious nature of the water storage debate as it pertained to water rights and privately-held lands in central Arizona. But if nothing else, Norton and his fellow committee members had at least provided a list of reasonable alternatives to valley residents.

Almost imperceptibly, however, sustained and growing support for a national irrigation program had begun to take root. Frederick Newell, the USGS hydrological engineer who attended the Fifth National Irrigation Congress, persisted in his efforts to convince the US House and Senate that some type of national bill should be passed. Famed scientist and army engineer Hiram Chittenden declared, "A comprehensive reservoir system in the arid regions of the United States is absolutely essential." He agreed with Powell and his disciples within USGS—and with journalist Hinton—that the federal government was the best and only place to facilitate this development. Put another way, the "Chittenden Report" (1897), the result of a congressional directive to explore reservoir sites and report upon the feasibility of constructing hydraulic works for water storage, recommended government construction, ownership, and operation of storage works in the West. While most historians agree that the Chittenden Report marked the official beginning of the movement for national irrigation by declaring that the federal government was the only entity capable of implementing the proposed program, the report also concluded, erroneously, that federal reclamation for water storage might be packaged together with river and harbor concerns. Captain Hiram Chittenden served with the Army Corps of Engineers, and the report's sponsor, Senator Francis Warren of Wyoming, placed reclamation legislation within rivers and harbors bills. Therefore the Corps, and not USGS, at first served as the bureaucratic home for national irrigation—a notion that displeased Newell and the US Geological Survey. The movement to lobby Congress to enact national irrigation legislation took on another intriguing dimension as the USGS worked vigorously to regain control of the movement among the stakeholders in the federal government. In fact, Newell played a central role in efforts not only to pass some form of federal legislation, but also to affirm that the Geological Survey—beginning with Major John Wesley Powell's leadership—was at the heart of this progressive reform. It would have a dramatic impact on water resource development and agricultural expansion in the American West.[47]

Between his arrival in the Salt River Valley in 1883 and his attendance at the Fifth Annual National Irrigation Congress in 1896, John R. Norton had lived through a dozen years of remarkable growth, change, and agricultural development. He had labored on projects that fueled this regional economic and environmental transformation. From living an abstemious life with work crews in northern Arizona, tending to mules and livestock and sleeping in the open or in tents, to advancing his professional career as a bridge between the work crew and the corporate board, Norton personified the history of the colonization and settlement of the nineteenth-century arid Southwest. His activities in community-building—securing ownership of and developing prime agricultural land north of fast-growing frontier Phoenix—testified to his resilience and resolve. Norton's life became interwoven with the history of social reform in the United States during the period known as the Progressive Era, roughly 1890 to 1920. He witnessed the false starts and early successes of cooperative water resource development as it evolved into corporate water development during this period. As the nation stood at a crossroads in water and land policy in the 1890s, and as Norton and his fellow Salt River Valley residents groped for answers in a depressed economy and an environment that at once sustained and threatened, he took individual and public responsibilities in his community. Civic involvement, political engagement, and local leadership in the issues of the day framed Norton's character and reputation. Reconciling private initiative with the conservation credo of the greatest good for the largest number of people over the longest period of time—a progressive maxim—informed Norton's world in the last decade of the nineteenth century. The political and social tension of an emerging and modern irrigated agricultural civilization, which Norton and his crew helped open to central Arizona when they completed the Arizona Canal in 1884 and the spur line from Tempe to Maricopa in 1887, fueled a population influx and agricultural revolution by the mid-1890s. The twenty-six-year-old young man who was a subcontractor for the Atlantic & Pacific Railroad in 1881 seemed a lifetime away, and the married man, who was now a prime stakeholder in the Salt River Valley community, girded for the public policy debate that shaped the region's future.

·3·

WESTWARD TILT

◇◇◇

"Arizona is one of the regions from which I expect the most development through wise action of the national congress in passing the irrigation act. . . . I look forward to the effects of irrigation partly as applied by and through the government, still more as applied by individuals and especially by associations of individuals, profiting by examples of the government and possibly by help from it. I look forward to the effects of irrigation as being of greater consequence to all this region of the country in the next fifty years than any other movement whatsoever."

PRESIDENT THEODORE ROOSEVELT, IN A THIRTY-MINUTE ADDRESS AT THE GRAND CANYON, ARIZONA TERRITORY, MAY 3, 1903.

"Not land, but water, is the ultimate value."

BENJAMIN FOWLER, FIRST PRESIDENT OF THE SALT RIVER VALLEY WATER USERS ASSOCIATION AND PRESIDENT OF THE NATIONAL IRRIGATION CONGRESS, IN HIS ADDRESS TO THE NINETEENTH ANNUAL NATIONAL IRRIGATION CONGRESS, CHICAGO, ILLINOIS, DECEMBER 5, 1911.

O**n November 17, 1896, the *Arizona Republican*, an increasingly effective organ of the territory's minority party, grudgingly proclaimed that John R. Norton, Democratic Party candidate for one of the three seats on the Maricopa County Board of Supervisors, had been elected, defeating with ease his Republican and Populist opponents.[1] His victory represented the culmination of a decade-plus of hard work, public engagement, and community service on boards, electoral commissions, and cooperative organizations. When primary or general elections took place, for example, Norton, an avid student of government and its role in the development of water and public lands, could be found serving as an "Inspector" at the Precinct 3 polls located at the Arizona Canal Dam or serving on a committee that evaluated the need for the issuance of municipal bonds or some other public benefit.[2] Governor Louis C. Hughes, on September 14, 1895, tapped Norton, a good friend and political supporter, as one of three members of the Territorial Irrigation Commission.[3] His knowledge of almost every corner of the territory, entrepreneurial nature, and leadership skills boded well not only in this influential post, but also for his overall future aspirations. For his work as supervisor he received the annual sum of $410.40, which supplemented income generated from his various activities as superintendent for the Arizona Improvement Company, water master for the Arizona Canal Company, and his agricultural and ranching activities.[4] As the nineteenth century blended into the twentieth Norton's private sector and public roles placed him at the center of the various debates that comprised water and arid lands policy reform. The national economic downturn that began in earnest in 1893 and crippled local development posed unique challenges for Supervisor Norton. Yet his two two-year terms as supervisor demonstrated to many that he served as a critical link between the county's farmers and working men—from Buckeye to Tempe to Wickenburg—and the business and professional sectors that sought to invest in and shape the region's economic growth and development.

Norton clearly enjoyed the thrust and parry of the political process. On July 24, 1894, two years prior to his election as county supervisor, for example, he was among the thirty-four candidates selected for the Democratic state of electors, which were to be reduced to seventeen by the "voter's own scratching." He lost this round but kept his sense of humor, as the *Arizona Republican* noted:

> One of the best known citizens of Phoenix is John Norton, and his boast heretofore has been that his birth place was in the free and truly Democratic state of Kentucky, where the grass is greener than elsewhere, where the birds sing sweeter than anywhere else on God's footstool. Since the election John has been pensive. He hasn't been saying much, but the other day explained himself in a single word. In a company of friends he was casually asked by one who had not known him long, "What is the state you hail from Mr. Norton?" The answer was ready, though savage: "Arkansaw."

The Razorback State was a frequent target for anti-Democratic party jokes in the *Arizona Republican*; Norton's comment reflected the paper's editorial slant that southern Democrats, like Norton, lacked a degree of social and intellectual acuity. Win or lose, Norton remained an active participant at Arizona Democratic conventions, rubbing elbows with the likes of Carl Hayden, George W. P. Hunt, Marcus Aurelius Smith, and Henry Fountain Ashurst. When he announced in 1896 that he was a candidate for Maricopa County Supervisor, the opposition *Republican* sniffed, "He will doubtlessly get the nomination."

Restive, energetic, and increasingly entrepreneurial, Supervisor Norton seemed omnipresent in every corner of Maricopa County. On May 19, 1899, he purchased Guy Bennett's 360-acre farm at Liberty, twenty-seven miles west of Phoenix. He retained his Phoenix property with its livestock, grains, and pastureland, but decided to embark on a new venture. Described by contemporaries as one of the best alfalfa ranches in Arizona, the "Old Bennett Place" was situated on what is now Southern Avenue between the Buckeye and Extension Canals. His reputation in this enterprise was illustrated best by the March 28, 1900 announcement that "At a meeting of the stockholders of the Phoenix Hay and Grain Company yesterday, J. R. Norton was reelected president and I. M. Christy was reelected secretary." Norton was an active member of the Arizona Agricultural Association and a featured speaker at their annual meeting in December 1900. Soon, local newspapers began praising Norton's agricultural and stock raising acumen. On several occasions the *Arizona Republican* ran stories about his successes. "John Norton came up from Buckeye yesterday with one hundred and fifty head of fine beef cattle which are a credit to that section of the territory," or "John R. Norton shipped a car of beef to the San Carlos Indian Agency." An article published on June 6, 1900, noted that "Supervisor John R. Norton returned from northern Arizona last night where he has been purchasing range cattle. He brought approximately two hundred head and will ship them to the Salt River Valley to fatten." Indeed, Norton, who also raised mules, fattened 1,500 to 2,000 head of cattle on his feedlot each year during the first two decades of the twentieth century. He was forging an identity as a successful and prosperous rancher and businessman while still maintaining a common touch.

He scoured the northern Arizona circuit, purchasing from ranchers in the Tonto Basin as well as in Mayer and Del Rio Springs in Yavapai County. He often rode his horse to Flagstaff to purchase steers from the Babbitt Brothers, who owned large ranches in northern Arizona. When Norton visited the Babbitts, they played cards and drank whiskey for two days. When they finally wearied of the drinking and gambling, they headed to the cattle pens, traded on steers, and completed that part of the process in twenty minutes. Two days of cards and betting and twenty minutes of business typified his working relationship with the Babbitts. Clearly, Norton's reputation as a farmer, rancher, and stockman had grown noticeably during this period. He was an integral part of the agricultural and ranching establishment.[5]

Norton played a central role in organizing the territory's stockmen and was active in the recently formed Arizona Livestock Association. His increasing affinity for equines, especially, reaffirmed his commitment to the association. He trumpeted, "We have the best-bred stallions in the United States in this valley. I mention this to show how we have advanced in the horse line, and I believe we should encourage the same advancement in stock." He supported severe punishment of cattle thieves and championed the cause of Territorial Veterinarian John C. Norton (no relation). At one livestock association meeting on January 10, 1899, for example, County Supervisor Norton stood up and commented, "The past two years has shown that a territorial veterinarian cannot be dispensed with, and since Dr. Norton has proved so satisfactory, I believe that he should receive a salary commensurate with work which he is doing to protect the stockmen of Arizona." Regarding cattle feeding, he suggested research was needed to test his theory that the proper way to handle the feeding of alfalfa in pens was to first let the alfalfa half cure. He suggested experiments in this line so that ranchers might learn whether they were losing or gaining by allowing their stock to graze.[6]

Supervisor Norton gained a territory-wide reputation for his expert knowledge and appreciation of horses. When two prize equines were stolen from his Phoenix ranch on December 15, 1898, the incident took on international significance and was reported in the various valley newspapers for weeks. The loss of the horses, one of which was owned by family friend and fellow valley ranchman C. T. Hirst, was discovered early Saturday morning, December 16, and Norton notified the Maricopa County sheriff's office at once. The deputies, Charlie Slankard and H. M. Davenport, had no clue about the direction of the escape route but they began the initial investigation with a ride to Buckeye. Unfortunately they could not pick up a trail. The thieves absconded with Norton's special forty-dollar saddle, which further aggravated the loss. "As a Kentuckian and owner of many good horses in his time, Mr. Norton said he had never had so good a one," the *Arizona Republican* reported. Many who followed the story surmised that the horses were taken to Mexico because deep snows had blanketed the northern half of the territory. Mexico became a favorite refuge for all classes of malefactors along the border, and Norton worried that the extradition treaty—the *Ley de Extradicion Internacional* of May 19, 1897—would expire in two weeks. According to contemporary accounts, there was little prospect of renewing the treaty. Thus all criminals north of the line, however aggravated their crimes, could be safe in Mexico. Fortunately for Norton and Hirst, a rancher in Benson purchased the two horses, bridles, and saddles for a bargain $75, and the buyer, when he learned of the transgression, contacted the Maricopa County sheriff's office. The stolen property was returned to the two victims, and though the thieves were never captured or held accountable, Norton expressed satisfaction with the outcome.[7]

During the last year of the century and into the first year of the new one, Norton became an increasingly active participant in western Maricopa County water and land development. The Buckeye Valley, like the Salt River Valley upstream, held promise, and Norton saw the area as a latter-day agricultural frontier. The Gila River, as it flowed to its confluence with the Colorado River in the southwestern portion of the territory, sustained settlers there. As an elected official Norton maintained a public profile. He possessed years of relevant experience in water and land development and understood both corporate financing and the role that government played (still limited at that time) in shaping the local economy. In short, he possessed the requisite skill sets for developing a new venture in the Buckeye Valley.

Norton learned of the Buckeye Valley and its potential in the mid-1880s. Just as he completed work on the Arizona Canal in the spring of 1885, three Salt River Valley farmers, Malin M. Jackson, Joshua L. Spain, and Henry Mitchell, loaded a wagon full of food, clothing, and other supplies, and, from their base in Phoenix, headed west along the Yuma freight road in search of land that was suitable for irrigated agriculture. Eighteen miles into their journey they reached the Agua Fria River, which intersected the southwesterly flowing Gila, and at this confluence they discovered a vast expanse of land—nearly twenty thousand acres—that appeared suitable for irrigation. They returned to Phoenix, mused over the findings of this reconnaissance, and planned another visit to the austere desert land they had just informally surveyed. In essence, they hoped to create another Salt River Valley success in the western desert reaches of Maricopa County, and on their second trip to the area, which took place in April 1885, they investigated land on the south side of the Gila. After serious study, analysis, and contemplation, Jackson, Spain, and Mitchell returned to the north side of the river. Using only a handmade wooden triangle with a plumb bob to measure slopes and grades, the men located their project dam site and canal heading just west of where the Agua Fria flowed into the Gila. On June 30, 1885, after more tedious studies of the feasibility of irrigating the adjacent lands, the intrepid trio returned to Phoenix and filed claim at the county recorder's office.[8]

The notice of location affirmed that "the head of said ditch commenc[es] in an old slough immediately under a high bank on the north side of the Gila River and then [runs] westward about one-fourth of a mile and then turns northward, leaving the river bank and running northwesterly passing Taboo Point on the Gila River." The three stated their intent clearly and elaborated in succinct terms; the proposed "canal or ditch was to be for agricultural, milling, or mechanical purposes," and twelve thousand miner's inches of water, a flow of about 240 cubic feet per second, were located and claimed for said purpose. Further, the locators claimed a forty-foot-wide right-of-way to the Hassayampa Creek over the public domain on which to build their proposed canal. They named it the "Buckeye Canal" in honor of Jackson, who hailed from Ohio, the "Buckeye State."[9]

Organizational activities continued apace. In September 1885, under the leadership of M. E. Clanton, the growing group of Buckeye Valley pioneers organized the Buckeye Canal Company, and filed "articles of incorporation" with Arizona Territorial Secretary John Gosper on September 25, 1885. Modeled after the Arizona Canal Company, the newly formed venture paid the three original locators $300 for their interest, and thus the Buckeye Canal Company asserted its place as a privately held entity in the federal territory. With ownership thus affirmed, the company awarded a contract for construction of the first five miles to T. N. Clanton.[10]

The pressing local problems of flooding and drought coupled with the federal government's interest made water resource development a burning issue at the turn of the century, and Norton observed the developments twenty miles west of the Salt River Valley with great interest. With his longtime acquaintance from northern Arizona, William "Bucky" O'Neill, and a group of investors, he formed the Buckeye Irrigation Company—one of the earliest of several incarnations of the organization. On October 28, 1888, O'Neill and his partners entered into a contract to purchase all the right and title to the fifty thousand miner's inches of water, save six thousand inches the company had already appropriated.[11] These previously appropriated waters, moreover, were to maintain prior rights in the canal—six thousand inches (equivalent to a flow of about 120 cubic feet per second) were to be maintained at all times. With dispatch the O'Neill group commenced completing the work of extending the canal from the Hassayampa to the lower end of present-day Arlington Valley. The work took in excess of two years. O'Neill retained the services of Major Edward H. Wilton, who surveyed the possible extent of irrigable acreage under the canal—that is, within its watershed. Wilton asserted that the 50,000 miner's inches (about 1,000 cfs) could irrigate 100,000 acres of land if the canal was "fifty feet on the bottom; six feet deep and a grade of 1.58 feet per mile (about 22 inches) with a bank slope of 1 to 1." The major recommended a new dam a quarter of a mile upstream and claimed "The location of the head of the canal is the best in the valleys of the Salt and Gila Rivers as it catches all the water from the Salt and Verde and Gila Rivers, and also of the Agua Fria and Cave Creek: in fact it catches the drainage of northern and northeastern Arizona, northwestern and western New Mexico and northeastern Sonora (Mexico). [It is] a vast territory larger than all the New England states." Wilton added, "I claim that the land under the Buckeye Canal is far ahead of the Salt River Valley. . . . I look at this place as a better section than around Phoenix, as a fruit country, the soil is better, the altitude is good or better, no winters, and scarcely any frost during the last two winters. This country would have been settled up long ago but for the Salt River washing out canals in 1883 and 1884 on account of their bad location and bad construction."[12]

An unforeseen wrinkle, however, caused the Buckeye Irrigation Company some trepidation and a delay in financing the projected improvements. Irrigation bonds were a new and unusual concept in the late 1880s, and the original deal between the company and Union Trust in New York fell through. Thus in early 1889 William Barnes, president of the Buckeye Irrigation Company (O'Neill held the title of secretary), and Charles N. Fowler, president of the Equitable Mortgage Company of Kansas City, entered discussions about the possibility of financing the proposed enlargement of the canal and related improvements, including a more substantial dam. Negotiations ensued, and ultimately Equitable Mortgage proposed steep terms.

On January 23, 1889, the Equitable Board of Directors informed Barnes, "We have certainly considered the Buckeye ditch matter and while we appreciate the intrinsic merit of the Bonds we feel as though we should also recognize the fact that Bonds of their character are quite unusual and that it may be difficult to find purchasers for them." Equitable went on to say, "We, of course, must look upon them from the standpoint of sellers and compare them with other securities of equal or not very different merit, but about which the public is quite well informed." As a result of these reservations the board offered to advance $30,000 upon its note at 6 percent interest "until sufficient had been advanced to complete the work up to $240,000 when at such time the Irrigation Company should turn over to Equitable the whole issue of $300,000 in bonds and $120,000 in stock as collateral for the loan." Furthermore, Equitable reserved the right to investigate the company's progress and to make sure the money advanced was going into construction and necessary repairs.[13]

Barnes responded almost plaintively, expressing hope for better terms, but he accepted the offer nonetheless because of the need for speedy commencement of the work, which had been delayed due to the financing pitfalls already encountered. Shortly after this agreement in principle, Barnes, who maintained offices in New York City, and Equitable representative H. P. Churchill met in Arizona to survey the landscape and craft a report to their respective boards of directors. Churchill, who drafted the narrative, exuded enthusiasm for the project. "The extent of the area drained—and the known annual rainfall—assures beyond doubt of a permanent supply of water far in excess of the appropriation," he asserted. "The strategic location of the Buckeye Dam at a point just below the junction of the Salt and Gila rivers and the Agua Fria enables the Buckeye Company to command the water of the entire watershed," he continued, and "as the Salt and Gila Rivers are the largest rivers in Arizona and as the Buckeye is the only canal that commands their united volume, it is in my opinion and must remain the most valuable canal property in Arizona." With that compelling assessment of the economic potential, Equitable Mortgage appeared poised to issue the loan as a sound investment.[14]

However, significant congressional intervention concerning land laws and evolving federal irrigation policy delayed the loan package. The events in Washington, DC, resonated in the Buckeye Valley and caused A. W. Chamberlin, the general solicitor for Equitable Mortgage, to take a second look at loaning the money to the private group. In fact agitation throughout the arid West—what some historians have labeled "a perfect storm of indignation and protest"—induced Congress to take up the subject of irrigation legislation during the last year of Grover Cleveland's first administration. Westerners were so vocal that Congress responded by taking up irrigation as a national work. It would have the government construct the reservoirs, ditches, and canals at its own expense, and at the same time dispose of the lands below these sites to settlers. The resulting Act of October 2, 1888, also known as the Segregated Reservoir Sites Act of 1888, assigned the work of selecting the sites to the Geological Survey under the direction of the Secretary of the Interior.[15] In essence, the 1888 Act reserved all the land that might be designated as conducive for irrigation development, and had the practical effect of withholding all of the public lands in the West from public settlement. Homesteaders who had been on their land less than five years at the time the act became law would not be able to gain full title to their property. Certainly this had a chilling effect on private lenders like Equitable.

Congress quickly realized that it had gone too far, however, and it took corrective action during the administration of President Benjamin Harrison. A series of acts, beginning with the Act of August 30, 1890 (which effectively repealed the Act of October 2, 1888), and Section 17 of the Act of March 3, 1891, promulgated a new, more rational policy that better fit the settlement history in Buckeye and other arid portions of the Far West. The 1891 law provided for the reservation of reservoir sites by the government, which would "contain only so much land as is actually necessary for the construction and maintenance of reservoirs, excluding so far as practicable land occupied by actual settlers at the date of location of said reservoirs." Indeed settlers poured into the western United States in the 1890s, and these more business-sector-oriented federal statutes played a significant role in the Buckeye Irrigation Company's canal construction and the development of private lands under that project.[16]

As federal lawmakers struggled to forge a workable formula for supplying water to the West, Buckeye leaders and the Equitable Mortgage Company continued their efforts to finance the Buckeye project, though federal activity tended to dampen the previously heady atmosphere. Nevertheless the parties intended to "supply sufficient money to put in a dam and improve and enlarge the already going canal and that by the time the law [Act of 1888] would be clarified or repealed by Congress and that they could go ahead and finish the work."[17] In other words, the parties intended to move forward no matter which way Congress voted. More financial negotiations, two surveys, and lobbying of territorial officials—including gaining the support of Governor Lewis Wolfley—resulted in a more carefully configured plan for expansion that would bring over 123,000 acres of land under cultivation.

But Buckeye farmers and landowners were soon grappling with nearly overwhelming challenges: floods in 1890, when the Walnut Dam collapsed on the Hassayampa River that flowed into the Gila above Buckeye; the more devastating floods of 1891, and then a decade of drought. Repeated breaches in the canal, frustration with rising water costs, and the resulting economic and environmental instability forced local farmers to seek alternative methods of controlling and delivering water to agricultural lands. The problems were similar to those that plagued the Arizona Canal Company, and because of them discussions between the Buckeye Irrigation Company and Equitable Mortgage ground to a halt. From 1892 to the end of the decade, financial challenges to the irrigation company and the widely cited economic downturn that marked the 1890s confounded Buckeye Valley residents.

A few notable developments took place in the Buckeye Valley during this challenging period. On July 1, 1898, Norton's longtime friend William "Bucky" O'Neill had been killed in the Spanish-American War at San Juan Hill. The bulk of his property, including his ownership share in the Buckeye Canal, went to his widow, Pauline. On October 5, 1899, O'Neill's interests in the Buckeye Canal system were sold to W. Moultrie of Fresno, California. This was one of many equity transfers in the company. J. H. Braley, who negotiated the deal between Pauline O'Neill and Moultrie for $6,400 on June 5, 1900, sold the Buckeye Canal and three thousand acres of land under it to J. Curtis Wasson and J. Ernest Walker for $30,000. This transaction—almost an afterthought in the passing of Bucky O'Neill—provided yet another example of the canal company's various incarnations during this period. It demonstrates the instability in the Buckeye Valley still prevalent when Norton purchased his 360 acres in May of 1899.[18] The new company, rebranded as the Buckeye Canal and Land Company, maintained offices in Phoenix, rather than Buckeye.[19]

Called "one of the darkest decades in this history of the United States,"[20] the 1890s were marked by a depression that was magnified by a number of factors: a deluge of European immigrants that exceeded four million, the growth of America's urban centers, the expansion of tenant farming, and the proliferation of large institutions, including manufacturing corporations. These sweeping nationwide developments deepened fears among the country's social and political reformers. The corporate consolidation movement that began during the nineties reinforced fears of monopoly and the subversion of democracy; the perceived villains were urban bosses, industrial tycoons, and the new immigrants. From 1893-1898 an average of 20 percent of the urban workforce lacked jobs; by 1894 the unemployment rate had peaked at a staggering 30 percent. The Carnegie Homestead plant strike near Pittsburgh in 1892 and the

Pullman strike in Chicago in 1894, for examples, raised fears that the country was in deep social and economic trouble.[21] Worse, the drought in Arizona and many parts of the West resulted in complete crop failures in 1894 and 1895 and reinforced growing demands for federal aid.[22]

Supervisor Norton, who knew every farmer and water rights holder in his second home, assumed his role as an active participant in water affairs in the west valley. His developing influence there would come to mirror his continuing influence in the Salt River Valley. Reading newspaper accounts of strikes, social unrest, and misguided corporatism, Norton discussed with constituents and friends whether Arizona Territory could sustain its mines, ranches, farms, and factories. The ongoing depression suggested to him and others that the nation, let alone central Arizona, could not survive without diffusing urban tensions and returning to family farms. As he well knew, private funding for arid land reclamation had dried up, and the value of his constituents' properties had dropped precipitously. The future of the region that looked so bright when he arrived in the 1880s now appeared ominous and forbidding. More than once Norton told local groups that in years past, the federal government had promoted the development of the West in many ways, including mapping of lands, construction of military installations, and subsidizing railroads. But in recent years, and since his election as Maricopa County supervisor, he had come to believe that the arid Southwest had been shortchanged—especially in the two decades after the Civil War. The two major sources of pork in the federal budget, he avowed, were in Civil War veterans' pensions and in river and harbor appropriations, both of which benefitted the eastern half of the nation and not the western portion of the country.

Over the nearly two decades Norton had lived in Arizona he had grown politically literate and increasingly savvy and outspoken on local, regional, and national developments. He joined the chorus of voices in the movement toward federal reclamation of arid lands, having witnessed and experienced the agricultural and economic contractions of the late 1890s in the Salt River and Buckeye Valleys. Often he equated the national irrigation campaign with the campaign for free silver, which erupted in large part because of the West's increasing political power in Congress. For example, in 1889 and 1890 the western states doubled in number with North Dakota, South Dakota, Montana, Washington, Idaho, and Wyoming entering the union. They joined California, Colorado, Nevada, Texas, Kansas, and Nebraska. The trans-Mississippi west held 10 percent of the population—only a handful of seats in the House of

Representatives, but more than a quarter of the seats in the Senate—a substantial power. Norton realized that significant differences separated the Great Plains, the Pacific Coast, the Rocky Mountains, and the Southwest. These subregions of the trans-Mississippi West, he told Maricopa County residents, vied not only for settlers but also for federal aid.

At the fateful Fifth National Irrigation Congress in Phoenix in 1896, Norton met Frederick Newell, chief hydrographer for the USGS. Newell, like several fellow federal employees, stood at the center of efforts to pass national irrigation legislation, and his not-so-subtle agenda was to make sure that the Geological Survey—not the Army Corps of Engineers—would be responsible for its implementation. Norton, who used the mass meeting to campaign for county supervisor—the elections were two months hence—appreciated Newell's technical expertise and scientific knowledge. He listened carefully as Newell sought to persuade local officials, farmers, and water rights holders of the wisdom of federal reclamation. Newell's influence on Salt River Valley leaders proved hypnotic, and his visits to the valley had an enduring impact on the history of federal reclamation in Arizona Territory.

Newell's background fascinated Norton and others at the Congress. He was born in the small lumber and mining town of Bradford, Pennsylvania, in March 1862. His mother died while he was still a baby, and he was cared for by a series of aunts. His father owned a large number of oil fields in western Canada and Pennsylvania, however, and although Newell was motherless, he was a privileged, beloved, and bright child. Before he finished high school he had the opportunity to learn surveying, mechanics, bookkeeping, and printing, and was better prepared than most youngsters to advance into higher education. His father, however, was anxious that he enter the oil business and forced him to leave high school before graduation. The elder Newell had contracted malaria, so young Frederick, as a high school junior, had the overwhelming responsibility of managing an oil conglomerate. Despite these distractions, Newell was accepted to the Massachusetts Institute of Technology (MIT) in 1880 and pursued his undergraduate and graduate degrees there. At MIT the study of geology captured his imagination, and in need of material for a doctoral dissertation in 1887, he joined the US Geological Survey. There, Major John Wesley Powell succeeded in diverting Newell's interest from geology and mining to developing water resources in the West.[23] In 1888 Newell became an assistant hydraulic engineer responsible for carrying out Congress's newly mandated irrigation survey of the arid West. In this role he traveled widely through the West and saw firsthand the difficulties the settlers in the region faced when they tried to farm dry land. Ambitious and empathetic, he became an effective missionary to the people and an important contributor to the development of federal water policy. His combination of personal ambition and technical expertise were well suited to promoting a national irrigation program.

Congress appropriated another $50,000 to the Geological Survey to determine the water supply in the United States by gauging stream flows, and to prepare a number of reports on the best uses for water in the nation's arid and semi-arid regions. The second of these reports was on irrigation near Phoenix, Arizona. The chief purpose of this and similar investigations was twofold: to gather scientific and technical data for planning reclamation projects and to publicize potential reservoir sites to Congress. While the report focused on the Gila and Verde as well as the Salt—all of which interested Norton—it provided little new information on the Tonto site to the east, beyond the research already completed by the Hudson Company.

Newell and Arthur Powell Davis, John Wesley Powell's nephew, emphasized the role that the federal government could take in reclamation in and around Phoenix. They came to the same conclusions recently reached by outside investors—that irrigation schemes would rarely be lucrative ventures.

At the same time, Newell saw that the extent of federal largesse required for such reclamations was problematic and, since it smacked of social engineering, likely to be politically divisive. But costs for developing the requisite infrastructure could be repaid in full, Newell argued, if they were uniformly shouldered by all who benefited. As the USGS hydrographic engineer, Newell advocated this approach as early as 1897.

A. P. Davis, the nephew of John Wesley Powell and a topographer and hydraulic engineer with the USGS, also took a keen interest in the Arizona Project. In his report Davis considered the Tonto site "one of the most important possibilities for the future of agriculture in . . . Arizona." Davis added: "It would probably be impossible to find anywhere in the arid region a storage project in which all conditions are as favorable as this one. The capacity of the reservoir, in proportion to the dimensions of the dam is enormous. The lands to be watered are of remarkable fertility. . . . " Significantly, public statements of this nature, emanating from federal officials like Davis and Newell, encouraged Norton and other local officials to include government support on their reclamation agendas.[24]

Norton and his constituents, however, had not arrived at any consensus or clear vision about what sort of government aid Salt River Valley landowners should seek. On one side were those who favored the USGS-favored approach, which was to administer a national irrigation program through continuous annual appropriations from Congress. Another approach, favored by a significant portion of settlers, was for the federal government to cede public lands to the states and territories where they were located—a variation of the Carey Act template. Congress tossed the ideas back and forth, and as the nineteenth century closed, little if any progress on this front had been made. Newell voiced his regret in a private letter reprinted in the *Arizona Gazette* on April 5, 1899: "I have long appreciated the necessity of water storage in Arizona and am doing whatever is possible to forward it. We now have a field party along the Gila River under a clause in the Indian appropriations bill. Their work is narrowly restricted by law, so that we can not take up the question of larger public importance. If this could be done I have no doubt that the results would be beneficial. Until the people of Arizona express themselves clearly and emphatically on the subject it of course is improbable that Congress will take action."[25]

Meanwhile, at a meeting of the trans-Mississippi West Congress in Wichita, Kansas in June 1899, the movement for some type of national irrigation program took on another and more assertive form when the National Irrigation Association was organized. In effect, westerners had decided that the work of the National Irrigation Congress needed to be carried on throughout the year in order to bring positive public sentiment to the cause of federal reclamation.[26] George Maxwell, a peripatetic California attorney, was selected to head the Executive Committee of the new association. Maxwell had already made his presence felt in the valley and would continue to influence Arizona leaders on a variety of reclamation issues.

In spite of these developments pointing toward some kind of federal action, Norton, in his leadership role in the county, realized that the serious drought, federal inaction, and the failure of the Hudson Company to construct the Tonto Dam with private funds required some type of local response— the formation of a new citizens' committee on water storage. The Phoenix and Maricopa County Board of Trade created a Water Storage Committee composed of five community leaders: Sam McGowan, superintendent of the Phoenix Indian School, was named chairman of the commission, and he was joined by John W. Evans, Benjamin Heyman, Benjamin Fowler, and Vernon

Clark. These five were charged with investigating the various water storage possibilities for the valley and to recommend to the board of trade a plan on how best to increase the regional water supply. The Water Storage Committee Report, issued on April 10, 1900, advocated what the citizens' storage committee supported in 1896: constructing a storage dam through the issuance of county bonds.[27] The committee members proposed, unrealistically, to purchase the Tonto Basin reservoir site, build a dam, and purchase all the canal systems in the Salt River Valley. The Water Storage Committee favored bonding Maricopa County, which at the time was assessed at $10 million, and at 4.5 percent annually, it could retire an estimated project indebtedness of over $6 million. John R. Norton disagreed, in part, with this proposal, believing that the Hudson Company had lost its right to the site because they had failed to use it.[28]

In one of the numerous meetings of the Water Storage Committee, Norton, in his capacity as county supervisor, offered his observations: "I do not know how the people of the County will receive a bond proposition. They will want to give it a great deal of consideration, and I myself would like to study it carefully before expressing an opinion.... Still there is one thing all are agreed upon. We've got to have a more liberal and permanent water supply than we now have and there are only two ways to get it. One is diverting the waters of the Colorado and it isn't certain that can be done at all. The other is by means of a storage reservoir." In his statement, Norton anticipated the creation of two major water and power projects that played central roles in Arizona's future: the Salt River Project (SRP) and the Central Arizona Project (CAP).[29]

The Water Storage Committee moved forward with their bonding effort. To finance the brash plan, 500,000 acres needed to be charged $1.25 per acre water rent. This meant that every acre cultivated in the Salt River Valley had to join, as well as a significant number of acres—up to 300,000—not in cultivation. The philosophical difference between the county bonding plan and a privately sponsored reclamation program was easily explained: the people would control the project. As Norton and others reasoned, corporations were unlikely to ponder ways and means of alleviating the conditions of its patrons. Indeed, a hint of populist rhetoric tinged the ongoing debate, and Norton was among its most vocal proponents. He agreed with the Water Storage Committee's assertion that "if any corporation has the control of our water supply we may rest assured that the people will be required to pay about all the traffic will bear." Actually, Norton and his cohorts were not antagonistic to business;

they simply objected to an unfair deal.[30]

The committee's rosy if not overly optimistic report did not consider the difficulty of uniting farmers behind a bonding proposal, nor did it comprehend the enormity of the indebtedness. "A little energy and determination," proponents of the plan asserted, was all that was needed to sell the bonds and begin work within a year. Furthermore, there was a political detail that needed to take place before the Water Storage Committee could execute its plan: Congress had to pass enabling legislation in order for Maricopa County to be able to bond itself. It was located within a federal territory. [31]

Little was accomplished between April 1900, when the Board of Trade's Water Storage Committee issued its report, and the fall of that year. Congress had its summer recess, and valley residents—like Norton who enjoyed deep sea fishing each summer off the coast of Santa Monica—tended to head to California or to northern Arizona's mountain country during the summer. This lack of progress on the heels of the optimistic report was discouraging. Another citizens' committee meeting, held in the fall, resulted in the selection of thirty-six representatives from every canal, every section, and every industry in the valley. Norton, not surprisingly, was elected to serve. But the enthusiasm abated significantly when rains fell and the need for water temporarily grew less urgent. Conflict erupted among groups of landowners. "There was talk, talk, talk, and inaction," one reporter noted, and it was coupled with "grumbling accusations." Some accused others of nefarious motives, and others determined that if their plan were not adopted, no other plan could be.[32]

The impasse, Norton knew, was based on old rivalries between lands under the north-side canal system and those under the south-side system. Also, divergent self-interest played a role. Water users in Tempe, who were served by the south-side system, were extremely protective of their vested water rights. Dwight Heard, who owned all the irrigated lands under the San Francisco Canal (roughly eight thousand acres), and Alexander J. Chandler, with eighteen thousand acres of south-side water rights from his Consolidated Canal and power plant, sought to protect their advantages.[33] Norton realized too that canal companies, always concerned with profits, proposed complicated and costly plans for their sale and use.

Into the mix stepped Norton's longtime friend, rancher and businessman Benjamin Fowler. In 1889, Fowler abandoned his career as a successful book publisher in New York and moved to the Salt River Valley. He hoped for a more leisurely lifestyle. The Massachusetts-born and Yale-educated Fowler immediately became a leader in his adopted home and emerged as a skillful negotiator and persuasive advocate who was able to smooth over rough feelings among contending factions. He was a diplomat of the first order. Soon after his arrival in the valley, he was elected president of the Arizona Agricultural Association, the Phoenix Board of Trade, the Associated Charities of Phoenix, and the Phoenix Chamber of Commerce. All sides trusted Fowler, and in many ways he became a de facto "director of the future." He responded eagerly to the charge and took over as Chairman of the Water Storage Committee.[34]

In the fall of 1900 Fowler traveled to Washington, DC, to press for passage of the enabling legislation so Maricopa County could move forward with the bonding plan.[35] There he met and grew close to Frederick Newell of the USGS; George Maxwell, the reclamation propagandist; and fellow Yale graduate Gifford Pinchot, Chief Forester in the Department of Agriculture's Forest Service. Over the next year Fowler, now the Salt River Valley's de facto lobbyist in Washington, DC, grew convinced that federal reclamation of arid lands could solve the nation's social problems by decentralizing population away from urban centers. Inexorably, Fowler, through frequent lunches and dinners with Newell, Pinchot, and Maxwell, began to lean toward supporting a federal project for the Salt River Valley, though he continued his efforts to secure enabling legislation specifically for Maricopa County.[36]

Unfortunately, Fowler was unable to move the enabling bill through Congress. House committee members cited several reasons for its opposition: there was no way Congress would guarantee the interest on the bonds, and it would not allow government engineers to conduct survey work under a citizens' advisory committee as was outlined in the legislation. Tinkering with the language helped little, and by March 1901 the Water Storage Committee had nearly disbanded, and valley residents, including John R. Norton, began to wonder if a bonding bill was the best idea to secure an adequate and assured water supply. The March 20, 1901 *Arizona Republican* inveighed, "The insidious hope of federal aid has stolen in and adulterated the usual judgment of our people and they have fallen. In the hope of getting something for nothing, they, like our Indians, are willing to sit and wait." The report mentioned that the only federal reclamation bill that could be successful in Congress was the so-called "Newlands Bill," named for Nevada Representative Francis G. Newlands.[37]

While Fowler and others like Norton continued to flail away in their various attempts to resolve local water issues, Maxwell and Newell lobbied

Congress through 1901 for passage of the Newlands Bill. The bill, designed to overcome the limits of private enterprise in the various challenges in construction of irrigation works, also provided for the creation of an arid land reclamation fund, composed of the receipts from the sale and disposal of lands in the sixteen states and territories of the arid West. Significantly, the Secretary of the Interior had sole and complete discretion over the use of the fund and, upon recommendations of the USGS, over the selection of projects that would provide water for public lands. Repayment for construction was to be complete, although no interest would be charged.[38] Norton supported the notion of federal reclamation, though he worried that his privately held lands in the Buckeye and Salt River Valleys might not benefit from the legislation.

Several factors seemed to favor passage of the Newlands bill. Earlier versions of the bill were supported by large landowners in the West, mining interests, railroads, and, importantly, by the administration of President William McKinley. The current legislation would likely have the support of Vice President Theodore Roosevelt, a strong supporter of the conservation and reclamation movements at the national level and a friend of Pinchot and Maxwell. Interior Secretary Ethan Hitchcock, at Newell's urging, sent a letter of support to the House Committee on Public Lands. It read, in part, "In some respects, the case is comparable to that of a city whose harbor has been improved. The land values increased . . . but the work carried out by private enterprise may not be remunerative to the builders. It is evident that if further reclamation is to take place it must be through governmental action."[39]

The Newlands Bill reflected "Progressive" reform sentiment at the federal level, a movement that swept through the country at the turn of the twentieth century. Reformers like Newlands, Newell, Maxell, Roosevelt, and others of their ilk sought to inject rational system and organization into American political, economic, and social life. Overall, it stressed science and technology in the development of arid lands for the nation as a whole, not for specific local interests. While the proposed legislation provided for construction of essential reclamation works in the arid West, it also circumvented longstanding regional policies, traditions, and biases regarding irrigation. The bill called for a more closed system of decision making than past internal improvement legislation. It embodied, as Samuel P. Hays wrote, "the gospel of efficiency."[40]

While President McKinley and Secretary of the Interior Hitchcock supported the natural resource policies articulated best by Newell and Pinchot, they did not actively work to see them put into practice. Newell, Pinchot, and Fowler met almost every night during the early months of 1901 to discuss the Newlands Bill and ruminate over lobbying strategy, but their efforts fell short on March 1, 1901, when the Fifty-Sixth Congress defeated the legislation by a narrow margin.

But fate altered the course of history in the fall of 1901. On September 6, 1901, Leon Czolgosz, a self-proclaimed anarchist, walked up to President McKinley, who was visiting the Temple of Music at the Pan-American Exposition in Buffalo, New York, and shot him twice. The President appeared to be recovering from his wounds but took a turn for the worse six days after the shooting, and lingered until he died on September 14, 1901. Vice President Theodore Roosevelt, much to the chagrin of Republican Party regulars, ascended to the presidency. A friend of Pinchot, Newell, and Maxwell, he committed himself to their vision of natural resource policy in general and federal reclamation in particular. In one of his first messages to Congress, President Roosevelt outlined a plan of action; the language was taken almost verbatim from the three reclamation advocates. He said: "Whatever the nation does for the extension of irrigation should harmonize with the trend to improve the condition of those now living on irrigated lands. . . . Our aim should not be simply to reclaim the largest area of land and provide homes for the largest number of people, but to create for this new industry the best possible social and industrial conditions." Roosevelt was the catalyst for the passage of legislation that provided for a national approach to arid lands irrigation.

The reintroduced Newlands Bill was a priority piece of legislation for the new and popular president. His message signaled to Fowler, Norton, and other private land owners in Maricopa County and throughout the arid West that a change in emphasis which would include private lands in the legislation was possible, if not probable. Roosevelt's message went far in diluting opposition from private landholders who assumed that the Newlands Bill would exclude them from federal reclamation projects.[41] Finally, on June 17, 1902, Congress passed Francis Newland's National Reclamation Act—arguably the most important piece of legislation impacting the history of the American West—and the world of John R. Norton and his fellow residents in Maricopa County, Arizona Territory changed forever.[42]

The legislation was national in scope, gave great authority to federal project engineers, and placed government in the unprecedented role of shaping the character and nature of rural western society. Reflecting the essence of Progressive era reform, the law acknowledged the ideals of small-farmer, Jeffersonian democracy and sought to incorporate local water law into its broad-based framework. For example, the Newlands Bill made water rights "appurtenant to the land irrigated, and beneficial use . . . the basics, the measure and limit of the rights." To be sure, as a shaper and subsidizer of the arid West, the National Reclamation Act took its rightful place alongside the Homestead Act of 1862 and even the historic Land Ordinance of 1785 with its promulgation of "grids, townships, and ranges." On a local level, perhaps Tempe's "south side" spokesman, Carl Hayden, put it best when he assessed the region-wide implications for Arizona residents with the passage of the bill. He prophesied that a reservoir would appear at the Tonto site, and declared, "All of the valley . . . has very little to lose and much to gain by the building of the reservoir. . . . The boom time is coming before very long."[43]

Quickly, Frederick Newell and his USGS brethren took action and proposed to Interior Secretary Hitchcock that he create a Reclamation Service from within the Survey's hydrographic branch. This proposal thwarted the rival efforts of the Department of Agriculture's Irrigation Division and the Army Corps of Engineers, so that control of the new program would rest solely in the hands of USGS engineers. The initiative was successful, and Newell was appointed chief engineer of the Reclamation Service on July 9, 1902, three weeks after the passage of the Newlands Reclamation Act.

As the National Reclamation Act gained traction with Congress and the Roosevelt administration, Fowler abandoned the enabling legislation for Maricopa County. As he told Norton and his Salt River Valley constituents, he and Newell spent a considerable amount of time together in June and July 1902, and Fowler had convinced Newell to come to Arizona. Clearly, the two had discussed the Valley's developing irrigation plans for eligibility under the new law, and Newell thought it important that a proposed project fulfill the financial provisions of the Reclamation Act, that the engineering and hydrologic aspects be sound, and that the water rights be adjudicated. Furthermore, tucked neatly away and in small print, and partly due to Fowler's efforts, was language that "extended coverage past development of new lands not under cultivation to include land in private ownership, already under cultivation."[44]

Norton knew that the private lands in the Salt River Valley fit this technical profile. He realized also that two provisions were problematic. It would be difficult to overcome the long-held traditions of prior appropriation and of individual determination of the beneficial use of water. It would be especially difficult to convince the old settlers who had held vested interest since the days of Jack Swilling—especially "southsiders" under the Tempe and Mesa canals—to share their resources with newer residents of the valley. In order to secure a federal project, he noted, local farmers and businessmen needed to set aside petty differences—something they had been unable to do despite at least three attempts to organize the citizens behind a cooperative plan for water storage. Norton and others looked to Fowler and his well-placed federal ally, Newell, for guidance and leadership in the seemingly unending struggle to achieve adequate water storage, irrigation, and flood control in the Salt River Valley.[45]

As the role of the federal government, through federal reclamation and national irrigation policy, grew in scope and altered local traditions, John R. Norton maintained focus on his varied agricultural enterprises in Phoenix and Buckeye, the latter location suggesting a westward tilt to his multiple endeavors. Water remained central to his daily activities and business success, and his direct involvement in its local administration kept him apprised of the latest legal and policy developments as they pertained to its use and distribution. He purchased the Arizona Water Company for $12,000 in April 1901, keeping one foot in the still vital world of privately-owned water.[46] As a stockholder in the Buckeye Canal Company, he grew weary of that company's inability to repair flood damage to the system. Exasperated, he sued the Buckeye Canal Company for leaving its customers without water for long periods of time. He took further action in 1903 when he and a group of disaffected Buckeye area landowners formed the White Tanks Canal Company, with a capital stock of $50,000. Among other things, the new canal company sought to purchase the Buckeye Canal Company and add a new canal, the White Tanks Canal, to the infrastructure. But as federal reclamation projects superseded nineteenth-century canal company models for irrigating agriculture, the White Tanks Canal Company reassessed its position. Norton stubbornly maintained that "he had some little experience in moving dirt and building canals," arguing that the White Tanks Canal could be completed in fifteen months. Ultimately, however, Norton and others moved back into the fold, for the lawsuit and the formation of the White Tanks Company spurred the Buckeye Canal Company to take appropriate corrective action. By 1907 the landowners, including Norton, signed on to a reorganized and reinvigorated privately run company that operates to this day.[47]

In the meantime, on June 14, 1901, during the height of the debates over the Tonto Basin storage reservoir, the Nortons, John and Etta, hailed the birth of a son, their third child, and named him John Ruddle Norton Jr. The *Arizona Republican* announced the birth with the unusual headline "Beaten by a Neck" and went on to say "Jake Miller was in town yesterday and announced the arrival of a boy at his home three days ago. A similar event happened in the family of Captain John R. Norton, but one day later. Captain Norton admitted he had been beaten by a neck and that the drinks were on him." [48] Norton Jr. entered the world of Arizona territorial growth and expansion in the best of circumstances. His father was at the center of public policy developments, economic growth and expansion, and promised to play a chief role in political and economic leadership.

John Jr. and his elder siblings grew up before the advent of air conditioning, automobiles, and other amenities developed via technological innovation. During his earliest years, Norton Jr. observed his father commute between the farm in the West Valley and Phoenix, where the family always maintained a home and where the elder Norton remained a public figure and prominent businessman. As he matured, John Jr. accompanied his father to the Buckeye Valley farm and learned the arts and labors of farming, stock raising, and water resource conservation and development.

After selling the Phoenix house at Indianola Avenue in 1904, the Norton family moved several times within the city, finally settling in a spacious home at 351 West Portland in 1915. It had a wide porch that provided a degree of respite from sweltering summer evenings. [49]

John Jr.'s earliest memories were of his father attending seemingly interminable public meetings concerning Salt River Valley water storage. The elder Norton was routinely present at town halls where valley farmers and landholders debated the virtues of federal reclamation. Shortly after President Roosevelt signed national irrigation legislation into law, stakeholders needed to put together an organization that could deal with the changes the National Reclamation Act would bring to the Salt River Valley. The task commenced with a twenty-six-member Water Storage Conference Committee, chaired by Fowler. After Judge Joseph H. Kibbey drafted a remarkable plan for harmonizing the differences among the valley's water users, the committee elected an executive committee to prepare articles of incorporation. The committee, in turn, provided Kibbey with its ideas, and he drafted the articles of incorporation for the Salt River Valley Water User's Association (SRVWUA). The executive committee scoured the articles section by section, and discovered numerous disagreements that took weeks to resolve.

Meanwhile the articles were circulated among Phoenix lawyers for review, and then submitted to the full conference committee. Significantly, they subordinated local interests to national ones, making every subscriber subject to the rules and regulations of Congress or any executive department of the federal government. Indeed, the articles of incorporation for the SRVWUA were masterfully drawn, anticipating problems between local custom and federal law where no precedent existed.[50] Not all differences were settled, but the articles were filed with the Maricopa County Recorder.[51]

SRVWUA officially incorporated under Arizona law on February 9, 1903, but Newell had received a copy of the preliminary articles of incorporation when he was in Arizona in January to meet with survey crews. Aboard a train on his return trip to Washington, DC, from Arizona, he read the draft and found the articles more than acceptable—a "model" for others as he later put it—and moved quickly to select the preliminary projects for Secretary Hitchcock's approval. On March 6, 1903, Newell and one of his chief assistants, Charles Walcott, presented five projects to Hitchcock for selection: Milk River, Montana, Gunnison, Colorado, Sweetwater, Wyoming, Truckee, Nevada, and Salt River, Arizona. Walcott wrote Secretary Hitchcock about the Salt River: "The conditions [of the proposed project] are typical of those which must be made elsewhere, and in considering this project and regulations it is necessary to create such precedents as will be desirable for other parts of the United States."[52] The Salt River Valley appeared ahead of every other section of the country in its preparation for federal reclamation.

Interior Secretary Hitchcock approved the articles in April 6, 1903, thus assuring construction of the Tonto Basin Reservoir. Fowler was elected the first president of the SRVWUA, and he continued to wrestle with every question and disagreement within the new organization—a tedious process. Among the most difficult were acquisition of the dam site from the Hudson Reservoir Company, purchase of the canals so that SRVWUA would possess a delivery system, and construction of a road from the valley to the Tonto Basin.[53]

After these signal developments, Fowler and others, including Norton, addressed the equally tedious task of explaining the articles to the farmers and landowners. Farmers using specific canals formed committees to further the goals of the organization. It was a tall order for a pioneer community to change its western predilection of "every-man-for-himself" into a spirit of mutual cooperation and trust. In this context John R. Norton was called again to public service as Chairman of the Joint Committee of Water Users under the Grand,

Maricopa, and Salt River Valley Canals. Frank Parker was secretary; the other committee members were Dwight B. Heard and John Orme.[54] Chairman Norton worked tirelessly on this committee to convince landowners to sign their lands to membership in the Water Users' Association. He led by example, transferring his interest in a claim to a dam site on the Verde River to SRVWUA at cost. Dissenting farmers expressed reluctance to sign on to the new program. They needed assurance that the SRVWUA would eventually own the canals through which they were to receive water. At one especially contentious meeting, Norton struck a conciliatory tone between those who wanted nothing to do with the federal program and others who asserted that it was the only way to create a stable water supply for the Valley. "It's now time to bury the hatchet," Norton averred. "The government can build a reservoir for us all. Every man can and will get his rights. The older ranchers who have stood back are not apologizing for their honest convictions, but they want a reservoir and [they] want it bad."[55]

Norton realized that the national government provided a great opportunity for the landowners to build their long-awaited water storage project. He counseled farmers that the new law promised fiscal subsidies through the use of interest-free federal money. The absence of adequate private capital for construction, he stated repeatedly, had been the primary obstacle to a successful locally sponsored project. Still there was hesitation, because what appeared to be an act of benevolence could carry with it onerous federal oversight. Local control of the project, a burning issue in the 1890s, contradicted President Roosevelt's and the Reclamation Service's current emphasis upon the importance of central authority. In essence, local practice would now be adjudged against the conservation axiom of the greatest good for the greatest number over the longest period of time—a Progressive credo reflected in the reclamation law. The fifteen-year struggle for a storage reservoir in the Salt River Valley became enmeshed in the movement for national social reform, as Frederick Newell and the USGS Reclamation Service sought to make reclamation of arid lands the cornerstone of a new natural resource policy within the federal government. John R. Norton ultimately concluded that government subsidy meant federal control, and territorial residents must live with it.[56]

And so in May and June 1903, Norton began to spearhead discussions aimed at turning the tide among landowners who had resisted this new order. He repeatedly told audiences that at first he had some trepidation about the SRVWUA and its articles of incorporation, but upon reflection, reason won him over to the cause. He advised all landowners to sign up for the Water Users. As chairman of the Joint Committee of Water Users under the Grand, Maricopa, and Salt River Canals, he crafted a resolution on June 7, 1903: "Now, therefore, be it resolved that it is the best sense and judgment of the committees so appointed that the best interests of the water users under the Salt River Valley Canal, the Maricopa Canal, and the Grand Canal, will best be served by the immediate signing up of all land owners and water users under the said canals who desire water storage under government control and supervision, immediately sign their lands under the Articles of Incorporation of the Salt River Valley Water Users' Association in order that the proposition of this valley may be presented by the Secretary of the Interior of the United States within ten days. Every delay from this time endangers our project."[57]

On June 11 Norton announced that nearly ten thousand acres were signed over to the Water Users. Ten days later the amount had grown exponentially to 153,000, and canvassers were still lobbying in Mesa and Tempe, the areas most resistant to change. Dwight Heard, a community leader who owned 7,500 acres south of Phoenix, fought to preserve the advantages of big farmers and canal companies under the old doctrines of prior appropriation and local autonomy. Norton knew he was a formidable force and the de facto leader of the opposition to the new program. In fact Heard took his objections to Washington, where he met with Secretary of the Interior Hitchcock in an effort to amend the SRVWUA Articles of Incorporation. The meeting took place on June 17, 1903, in Hitchcock's office. It was Hitchcock, however, who proved to be persuasive, and immediately after the meeting, Heard wired Norton: "Satisfactory conference with Secretary Hitchcock. Confident reservoir is assured. Return home next Saturday with definite plans to stop friction and ill will." The contents of the Heard-Norton telegram were made public and "before night it had become the sensational feature of the day in respect to the storage campaign." The June 20, 1903 *Arizona Republican* informed its readers: "A congratulatory telegram [sent to John Norton] was received yesterday from Dwight B. Heard, in which he gave assurance of the building of the Tonto Reservoir. He also said his conference with Secretary Hitchcock had been quite satisfactory.... Heard accepted the situation and receded from his position of antagonism and [said] that his followers in the opposition might be expected to join in the popular movement for the building of the reservoir. ... We have always believed that if the matter were fully understood by Mr. Heard that he would not try to turn aside the greatest opportunity that was ever presented to the valley and one which in the nature of things would not be presented again for years." Norton breathed a sigh of relief; Heard had finally arrived at the conclusion that neither he nor any other major landholder could gain anything further in their opposition to the enterprise, and in July Heard subscribed 720 acres personally and 640 acres of the Bartlett-Heard Land and Cattle Company.[58] When the subscription books closed, on July 17, 1903, landholders had committed 200,000 acres.

By 1904 Norton assessed the water storage movement and found that, remarkably, the association encompassed all landowners in the valley. Members of the association received shares in the new water storage system, which would be built over the next five years. The water rights would be "perpetually and inseparably" tied to the land. Farmers like him, and not speculators, Norton realized, would control the organization, and the contract with the government would ensure the construction of the reservoir. Canal companies, for which he worked for much of his adult life, would pass into history.

On March 5, 1904, in his guise as a member of the Phoenix Board of Trade, Norton paid homage to the leadership of SRVWUA President Benjamin Fowler. Norton helped organize a "triumphal entry" and a "general jollification" meeting to celebrate the contract executed between the Water Users and the Secretary of the Interior—the capstone achievement that ensured the construction of the reservoir. The *Arizona Republican* called the event "the celebration of an epoch in the history of the valley." It took place at the courthouse plaza and included music by the Pioneer and Indian School bands, along with the usual round of speeches.[59]

At this point in his career, John Norton had become a well-known community leader, and in fact he harbored further political ambitions. He made two attempts, in 1902 and 1904, to run for the office of Sheriff of Maricopa County. Because the shrievalty was one of the more lucrative county offices—the sheriff received travel money, a percentage of fees collected, and other perquisites—the office was highly prized. On May 2, 1904, the opposition *Arizona Republican* took a political shot at John 'Are' Norton. The humorous yet cynical column read: "Our esteemed correspondent, 'Little James,' seems to be very well acquainted with the mental equipment and characteristics of the Hon. John "Are" Norton, Democratic candidate for the [city] council and prospective Democratic candidate for sheriff. It must be conceded that Mr. Norton has a remarkable penchant for getting in front of the horses in every movement for the public benefit, and then, at the last moment, strug-

Etta Wright Norton, wife of John R. Norton,
mother of John R. Norton Jr. and John R. Norton III's
paternal grandmother.

gling for a place in the rear of the bandwagon." The invective continued on the
following day: "Republicans have no objections at all seeing John Norton made
the Democratic candidate for Sheriff but they object to making him a City
Councilman as a step toward the Sheriff's office. John will have to pay his
own political expenses." And just prior to the primary election in August, the
Republican editorialized: "The *Arizona Republican* is not an organ of any faction
of the Democratic Party but it has a duty to announce the Democratic Party
candidates for the responsibility devolves upon it. It has to announce that John
R. Norton will be in the field. He has been in the field before. He was there
for a time two years ago very early in the campaign. It was said by the friends
of Mr. Norton that he had no official aspirations this year. But Carl Hayden
has sprung the contagion and Mr. Norton caught it, then other Democrats in
whom the infection has found lodgment. The symptoms are becoming more
and more marked and any day may discover an eruption." Ultimately, John

Elliott Walker won the Democratic nomination and the general election, and with the election of 1904, Norton's electoral aspirations in local and county politics ceased.[60]

During the last half of the decade John Norton continued to increase and improve upon his agricultural, ranching, and related endeavors. His Buckeye Valley agricultural enterprise amounted to over five hundred acres by the end of the decade, and by that time he knew that he needed to find more efficient ways to get his alfalfa, stock, and other goods to market. Improved transportation was the key. In late January 1903, Norton opened his ranch for Department of the Interior researchers in order to publicize the area's livestock and lifestyle. His children, Edith, Fred, and three-and-a-half-year-old John Jr. looked on as the events unfolded, and an *Arizona Republican* reporter documented the spectacle: "J. C. Boykin, special agent of the Interior Department, is here to secure moving pictures for the St. Louis Exposition. . . . [He] got a good picture of Deputy Sheriff Oscar Roberts in the act of roping a steer. The biograph picture machine has a capacity of eight hundred pictures a minute. There were two trials. The steers were strong and wiry animals and the exhibition took place at John R. Norton's ranch as his wife and three children looked on. . . . The first steer was tied in forty-three seconds out of the gate, the second in forty-one." Thus Norton's family and ranch provided the setting for one of the earliest attempts to film and document a dimension of ranch life on the southwestern American frontier.[61]

Norton embraced other technologies to promote his region and its virtues. His efforts in fostering transportation links throughout the Salt River Valley were essential in building upon earlier successes. In fact much of the economic development in the Southwest between the turn of the century and 1920 hinged on the development of transportation technologies.[62] Norton worked on and witnessed the expansion of railroads and roads during his years in the Salt River Valley, and he later marveled at the completion of the Panama Canal. During the first decades of the twentieth century railroads remained the major instrument for economic development in the region. Especially relevant to Norton and his fellow agriculturalists was the development of the refrigerated railroad car service pioneered by the Santa Fe, Union Pacific, and Southern Pacific railroads, which carried fruits and vegetables from the West Coast to eastern urban centers. Norton took notice when, in 1906, the Union Pacific and Southern Pacific pooled their resources to organize the Pacific Fruit Express, a joint fleet of refrigerator cars that carried the bulk of

Pacific produce eastward.

The extension of railroads stimulated increasing demands for better roads and highways to connect city and country, and to serve as feeders for the railroads. Norton became an especially avid supporter of the good roads movement. He began to participate in "auto runs" as early as 1905 and in 1910 joined the Maricopa County Automobile Club, modeled after the California State Automobile Association established ten years earlier. He lobbied for improved roads in the Buckeye area, extolled the virtues of Buckeye Valley livestock, and explained that good roads could get these commodities to market. Unquestionably, Norton understood and worked for the expansion of transportation to promote regional economic growth. It was, he said, "a key to economic maturity."[63]

He also continued his activities in various livestock organizations. In 1905 and 1907 he spoke at the annual meeting of cattle growers, held in the Commercial Hotel in Phoenix. He addressed issues like range regulation, horse and cattle thievery, and breeding. In 1911 he was again appointed to the statewide livestock committee along with longtime acquaintances Dwight B. Heard and Charles J. Babbitt of Flagstaff. His deep interest in horses and improved breeding techniques marked his career as he spent increasing amounts of time in Buckeye.

His sense of humor was manifest, and even though the *Arizona Republican* rarely missed commenting on John's political error in harboring southern Democratic proclivities, the editors never hesitated to frame him in a humorous light. On August, 10, 1905, the *Arizona Republican* ran a headline "J. R. Norton Home: Had a Profitable Vacation and Health Greatly Improved." According to the reporter, "J. R. Norton returned home yesterday morning from a five-week stay in Los Angeles. He went away a very sick man in charge of Dr. Stroud, but returns feeling as well as ever. . . . On his way back he stopped over a day in Yuma in the company of Dr. J. C. Norton, the territorial veterinarian, with whom he returned to Phoenix. He was somewhat loth (sic) to come in on the same train with Dr. Norton fearing unkind criticisms from his friends, for he says having gone away with an MD and having returned with a VM, his friends will cast into doubt as to whether he had a kidney or a mule collar and he is not certain himself whether he is a man or a jackass." Another story, entitled "First Snake of the Season," described Norton telling a story that entitled him to "more than passing notice." "In the first place," the *Republican* allowed, "Mr. Norton is a pioneer of the valley and a very close observer of

Tonto Reservoir site. Looking west across a lazy Salt River about 1905. The dam site is in the canyon at right, with the town of Roosevelt spread across the only flat ground above the river. In another three years it would be inundated by the reservoir. Above the lake level, on Government Hill at left, are Reclamation Service buildings and tents for unmarried engineers. The largest building in front is the dining hall. The power canal is winding around the hills toward the dam site. On the hill at center is the cement plant, and across the river, on the right, is dam contractor O'Rourke's camp.

even small things." Evidently, Norton was riding his horse along the bank of the Agua Fria River north of his Buckeye Valley land. He was looking for a proper place to water his cattle. When he passed a clump of bushes, he heard a noise that sounded like a rattlesnake. "It is a good luck charm to kill the first snake of the season," the newspaper added, "and Mr. Norton went back to look for the snake and found a monstrous rattler lying partly out of a hole waiting for sun to shine." Apparently, Norton killed the snake with little difficulty, and the article concluded, "This is the first snake story of the year and Mr. Norton should have first place."[64]

The overarching theme that tied Norton's public, professional, and family life together was the quest for a reliable and sustainable source of water—the lifeblood of civilization in his world—and he, his wife, and children watched intently as the building of the great dam at Tonto Basin unfolded. He often told his children about the legendary three-man survey in 1889 and explained to them how much he enjoyed the adventure. The site of the proposed new dam was sixty miles northeast of Mesa and forty miles northwest of the bustling mining community of Globe. John Jr. recalled his father telling him that these were leg-slashing, hoof-bruising miles across some of the toughest country in Arizona. The Sierra Anchas bracketed the northern reaches of the area, while the Mazatzals curved away to the northwest. To the south loomed the Superstitions. Norton knew there were no nearby settlements where workers could live, and in the first decade of the twentieth century only one road twisted through the mountains to Globe. Everything else, the elder Norton told his son, was rocks, thorns, and heat. Thirty years earlier, Generals George Stoneman and George Crook pursued Yavapais and Apaches across these reaches, but now San Carlos Apaches from the nearby reservation comprised many of the work crews who would build this Anglo monument to progress.[65]

The first order of business was to connect the dam site to the outside world, something the Hudson Company could not or would not do. Government workers blasted a road to Globe, then began the famed Apache Trail that enabled two rail lines from Mesa to access the site. The road gave engineers nightmares and threatened workers' lives. It wound along the Salt River, clung to steep cliffs, and ascended and descended mountains like Fish Creek Hill in 10 percent grades. Road builders, to Norton's amazement, had to use lifelines while they hacked twenty- to seventy-foot-deep cuts.

The next challenge was to minimize costs for supplies. Stonecutters, for example, quarried the masonry blocks of the dam from sandstone cliffs in the area, but project engineers calculated that buying and freighting cement to the dam site would cost an astronomical $9 per barrel. Norton informed Fowler, who in turn informed the Reclamation Service engineers, that in his 1889 survey of the area, he had located deposits of limestone and clay north of the Salt River. These sources enabled government workers to produce cement at $3 per barrel. Though the cement trust protested vigorously, the Reclamation Service continued to manufacture cement at the site. In another instance, private lumber contractors were unable to meet their quotas, so government engineers took over a private sawmill in the Sierra Anchas and raised production from 119,000 board feet per month to 214,000. The onsite brain trust, "Visionary Technocrats," according to one writer, was led by thirty-eight-year-old Louis C. Hill, a professor from the Colorado School of Mines. He and his crews demonstrated that the "gospel of efficiency" could work in the Tonto Basin.[66]

Construction of the dam began in 1905, the wettest year in memory. The rains triggered unprecedented flooding throughout the Colorado River Basin.[67] The rains wreaked havoc in the Salt River and Buckeye Valleys, but J. M.

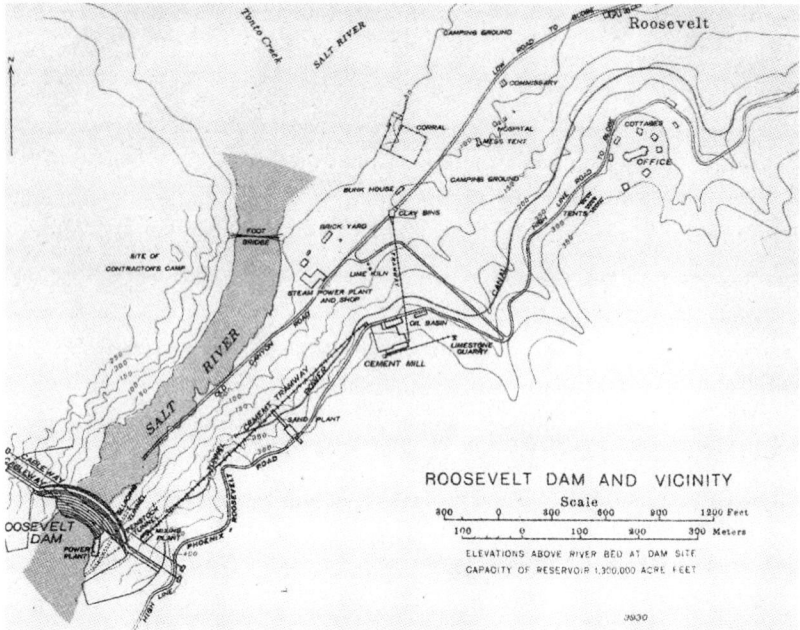

ROOSEVELT DAM AND VICINITY
Scale

This map, prepared in 1910 by the United States Reclamation Service, shows the site about 1907, but with a completed dam as it would appear in 1910. The main street of Roosevelt is at upper right, with the USRS office on Government Hill just below. Dam contractor O'Rourke's camp is across the river by way of a suspension footbridge that washed away in 1908.

O'Rourke and Company constructed a camp for workers, set up a machine shop, quarried stone, and drove piles to erect two temporary dams that would divert the Salt while the permanent structure was built. Then on November 26, 1905, a warm rain melted snow in the mountains, and the river rose thirty feet in just over fifteen hours. Water roared down the canyons at 130,000 cubic feet per second and ripped piles out of the river bottom. It swept away diversion dams and flumes, and more floods through the winter of 1905-1906 prevented construction from resuming until March. O'Rourke and the crews returned and hacked away at earth with picks and shovels because engineers had concluded that too many explosives could weaken the rocks. Then the crews used wooden derricks to hoist stones that weighed up to ten tons, filling the joints with concrete that had to be kept wet for six days to prevent cracking. Slowly the structure took shape, curving upward between the cliffs. By the

LOWER VIEW OF ROOSEVELT DAM

1328

A view of the dam from downstream in late summer 1908 shows the south side rising above seventy feet and work restarted on the lower north side after the sluicing tunnel reopened. The powerhouse is enclosed and water can be seen exiting the tunnel below the building.

time of the next floods, in 1908, the south end of the dam was high enough to force the water over the northern end. This development, perhaps, was the first sign of flood control for the Salt River Valley.

Masons laid the last block on February 5, 1911. The structure rose 284 feet from the bed of the river and was 184 feet thick on the bottom, tapering to 16 feet thick at the top. It arched 1,000 feet from canyon wall to canyon wall in a great concave bow. The dam, which cost over $10 million—triple original estimates—inundated more than 16,000 acres, making it the largest artificial lake in the world at that time. The Reclamation Service hailed it as a "monumental triumph of the skill and genius of [its] scientist creators," with its massive, rough-hewn wall and its three towers.

Five weeks later, the dedication took place. On March 18, 1911, former president Theodore Roosevelt, in a Kissel Kar driven by Tempe resident Wesley A. Hill, a former Rough Rider, led a caravan of twenty-three other cars. They headed up Center Street (today's Central Avenue), where hundreds of people had gathered to see and cheer the former president. "Colonel Roosevelt bowed left and right as he passed through the business section" of town.[68] After visiting the Phoenix Indian School, the caravan traveled east and then north along the Crosscut Canal to the Arizona Canal. They then proceeded along the bank of the Arizona Canal to the Granite Reef Diversion Dam, where they crossed the river and followed the desert trail to the Roosevelt Road (Apache Trail) and the new Roosevelt Dam. John Norton, his wife, and three children rode in the caravan. More than two hundred more cars joined them from the Salt River Valley and Globe, even though a stagecoach had plunged off Fish Creek Hill, killing a woman passenger, the day before.

At precisely 4:16 p.m. an auto shot around the bend of the big road, according to the *Arizona Democrat*, and in the front seat was a figure in a long white Ulster, wearing a black slouch hat. The appearance of Roosevelt's car was the signal for the discharge of eleven guns at the dam, followed by the cheers of at least a thousand onlookers. The reverberations of the guns fluttered the large American flag and the blue Reclamation Service flag floating above the parapets of the dam. Norton took his seat on stage along with other dignitaries present. Prior to Roosevelt, several notables addressed the crowd, including Governor Richard Sloan, chairman of the dedication, who said the dam was "a vindication of the wisdom and foresight and a justification of the effort and labor of all those who were instrumental in bringing about a national irrigation policy." Louis Hill followed, pointing out that even though the reservoir was less than half-full, all of the land served by the Water Users' Association was "safe from drought for two years, even if no flood comes." Without Roosevelt Dam, all floodwaters would be lost. Benjamin Fowler, now president of the National Irrigation Congress, said the dam "will stand as an everlasting testament to serve, conserve, and safeguard the interests of all people alike. To its builders it is a splendid monument. To a great and growing community in an arid region, it is a guarantee for all time of prosperity, happiness, comfort, and peace." Roosevelt followed and spoke extemporaneously. He said that the structure, which was officially christened Roosevelt Dam in his honor, was one of the two "greatest achievements of my administration," the other being the Panama Canal. He predicted that the Salt River Valley would become one of the richest agricultural areas in the world. Then, the former president pushed a button that raised three huge iron gates. Water spouted from the dam and headed for the fields sixty miles away, and according to the *Arizona Republican*, "A mighty roar of water rushed through the canyon and the dedication of the greatest storage dam and reservoir on earth was an accomplished fact."[69]

During the construction of the dam and into its early years of use in the teens and twenties, Norton paid close attention to evolving water law in

Former president Theodore Roosevelt, who had signed the Reclamation Act in 1902, and Territorial Governor Richard Sloan at the dam site, March 18, 1911. Roosevelt had arrived from Phoenix by car to dedicate the government dam named in his honor.

the context of federal reclamation. He attended countless hearings in various courthouses as his two friends, Judge Joseph Kibbey and Edward Kent, handed down decisions that affected his lands and water. Nearly twenty years earlier, Judge Kibbey, then chief justice of the territorial Supreme Court, tried to resolve a number of water rights and canal company disputes that had punctuated the increasingly chaotic community of water users. Kibbey's landmark decision, in *Wormser v. Salt River Valley Land Company* (1892), reaffirmed the doctrine of prior appropriation. Norton, as a longtime water user, welcomed the decision at that time. Eighteen years later, as the dam neared completion, Judge Edward Kent essentially affirmed Kibbey's decision in *Hurley v. Abbott* (1910).[70] The decision impacted Norton in that Kent determined the prior rights of all acreage in the Salt River Valley to Salt River water and even adjudicated when each parcel had been first cultivated. The Kent decree, in its determined complexity, took into consideration elements of the federal Newlands Reclamation Act and enabled Arizona to undergo a seamless transition into statehood in 1912, especially as the transition involved water law administration. The decree stands as one of the great early monuments to Arizona's maturing legal system.

Soon a new and more productive relationship developed between local water users and the federal government. Although the bulk of these users

Although it was mailed in 1921, this postcard probably shows the 1915 spill, when water first overflowed the dam. Roosevelt Lake overflowed again in 1916 and 1920 before entering a long period of drought.

were farmers, on October 1, 1910, City of Phoenix officials executed a water contract with the Department of the Interior's Reclamation Service for the delivery of water to Phoenix residents. Judge Kent, to his enduring credit, had recognized the urban dimension of portions of the Salt River Valley in the Kent decree, and this was reflected in the map designating urban areas that accompanied his legal rendering in the *Hurley v. Abbott* (1910) case. Ultimately the annual renewal of water delivery contracts with the Department of the Interior's Reclamation Service became routine until 1917, when the SRVWUA took control of the project. On March 20, 1919, the City of Phoenix executed its first water contract with SRVWUA. Clearly, public ownership and government stewardship characterized this Progressive Era evolution in water policy.[71]

John R. Norton adapted to these changing times with relative ease. He and Etta reared their children in Phoenix, which increased in population nearly three-fold between 1910 and 1920. It was a decade of dynamic growth and change. In 1910 agrarianism, the small town, decentralization, and competition reigned triumphant in America's constellation of values. The romantic, voluntaristic world of the Arizona Rough Riders and the Spanish-American

War gave way to the efficiency-minded state of the Progressive Era and the systematization that characterized World War I. By the advent of the 1920s, the distinguishing characteristics of a new America could be foreseen—urbanized, centralized, industrialized, and secularized.

Norton experienced the boom times in World War I and planted more cotton than alfalfa in order to support the war effort. His enterprises in Phoenix and Buckeye, and his periodic trips to northern Arizona to restock his ranges with cattle and horses continued apace. But like others in the region and throughout the country, Norton had hoped for a millennium at the end of the war that did not materialize. Instead the postwar period was one of crisis, anguish, and tribulation. John Norton welcomed the end of the fighting in Europe, yet he was beset with tensions and frustrations that came with forced adjustments to a peacetime economy. To a considerable extent Norton's fears were fueled by the economic depression that settled on the region between 1920 and the years that immediately followed. Most residents in Maricopa County—farmers, cattlemen, miners, and those in manufacturing—were affected by the cancellation of war orders. Salt River and Buckeye Valley farmers bore the brunt of shrinking markets and plummeting prices. Prices for beef and hogs dropped to one-half wartime levels, plunging numbers of Arizona cattlemen into bankruptcy. Norton struggled to keep his agricultural, ranching, and related enterprises afloat, but the early twenties proved profoundly difficult.

As Norton worked countless hours to recover losses and maintain his properties, he grew weak physically, and as his grandson, John Ruddle Norton III related, "My grandfather just kind of withered away." Contemporary accounts of his passing, on February 5, 1923, suggested that a heart attack felled the indefatigable Salt River Valley pioneer, but family members have since speculated that it may have been an undetected prostate cancer. With his passing, the *Arizona Republican*, Norton's long-time bête noire, eulogized: "John R. Norton, Arizona pioneer and for years a predominant figure in business and civic circles died at his home at 351 West Portland following an acute attack of heart disease. . . . His death was entirely unexpected. He left a sound agricultural foundation for many and a legacy of water in a harsh land." His death brought to a close a career marked by immense achievements closely connected with the development of Arizona. "While extensive cattle and ranching interests occupied the greater portion of his time," the newspaper noted, "he was the leading spirit in several constructive movements in this and other parts of the state." The obituary also recalled that "When Mr. Norton first came to Arizona he was engaged in the construction of the Atlantic-Pacific Railway, the present Santa Fe system, through the northern part of the state. He came into the Salt River Valley about 1884 and for six years was associated with the old Arizona Improvement Company in building the Arizona Canal and several of the irrigation system auxiliaries now in use. He served as superintendant of the company until he terminated his connection . . . to enter the cattle and ranching business" where he was "eminently successful in both endeavors and owned six hundred acres of fine farm land at the time of his death."

Norton's funeral was the largest Phoenix had seen in months. Cars lined up on both sides of the street for blocks in front of the McLellan parlors and the chapel, which was so crowded that only a portion of the crowd could enter the structure. Most came, according to accounts, "to pay the last tribute to the pioneer resident of Phoenix." Tributes and accolades were numerous: "He was regarded as a leader in the agricultural industry of this section and his name was linked as well with many of the beneficial projects inaugurated in the county in the past quarter century."[72]

Had Norton looked back at the broad contours of his agricultural, ranching, and water handiwork, he would have realized that he had shaped and improved the quality of life in the Salt River Valley. He was a visionary who possessed the rare ability to recognize opportunity where others saw difficulty. A rancher and farmer who had dreams of turning arid desert into thriving farmland, he was one of the trio that selected the site for Roosevelt Dam. Moreover, he was a progressive, early community leader in Phoenix who worked tirelessly to develop water resources and an extensive irrigation system that allowed the city to grow into an oasis in the desert. John R. Norton, who arrived in territorial Arizona in 1881, reached adulthood before the twentieth century: before women's suffrage or collective bargaining, before federal reclamation or Arizona statehood. He grew to maturity on the American southwestern frontier and raised his family in an austere desert environment. Indeed, the man who helped forward so many crucial advances in water resource management possessed roots that reached deep into the heart of the American experience of the nineteenth century. A product of the legendary old West, he helped build a dynamic new West. Little did John Jr. know, in the years ahead, that he did not have a name to make, only a name to keep.[73]

·4·

JOHN R. NORTON JR.
AND THE URBAN OASIS

"The Salt River Valley is the glory of Arizona. . . . Wherever the waters of irrigation have moistened the desert and man has planted the seed of grass, flower, or tree, the most luxuriant vegetation has sprung from the soil. . . . The capital city of Phoenix . . . is the pulsating heart of the new life of Arizona. . . . Phoenix is distinctly modern and almost wholly the offspring of irrigation."

WILLIAM ELLSWORTH SMYTHE, IN *The Conquest of Arid America*, 1900.

John R. Norton Jr., born on June 14, 1901, entered a world that brimmed with optimism and potential. As a small child he often rode on a buckboard with his father to the Norton farm and feedlot at Liberty, Arizona, but that bucolic way of life would retreat quickly as young Norton grew up during the first decade of the twentieth century. A mix of urban and rural lifestyles shaped his early years. His father worked tirelessly in both the private and public sectors, and like others he vigorously championed the Salt River Valley's great pyramid, Roosevelt Dam. Dedicated in 1911, it symbolized a robust and limitless future. His father joined other civic boosters who promoted investment in Phoenix and its environs. The Phoenix Board of Trade, for example, publicized the nation's first reclamation project with a "Tell the World About It" campaign. The Santa Fe and Southern Pacific Railroads published and distributed countless articles that detailed the "marvelous future" awaiting those who moved to the emerging desert metropolis and the verdant Salt River Valley. When the Santa Fe Railroad devoted a special edition of its magazine, the *Earth*, to the promotion of Phoenix in 1908, farmers flocked to the area. This reversed a downward trend of a few years earlier. At the end of one of the region's periodic droughts, in 1905—just as John Jr. turned four years old—his father brought home a report noting that cultivated land in the Salt River Valley had dropped to 96,863 acres, down from 127,512 acres in 1896. Four years later, however, cultivated land in the Salt River Valley had risen again to nearly 127,000 acres, an increase in agricultural productivity that signaled a return to prosperous times. It also reflected the positive impact of the Roosevelt Dam.[1]

During John Jr.'s first decade of life, Phoenix grew in population by more than 100 percent, from 5,544 in 1900 to 11,134 in 1910. Maricopa County, similarly, grew at a rapid pace during this ten-year span; from 20,457 in 1900 to 34,485 by the time young Norton reached nine years old. He matured during a time of technological innovation with brick buildings, some four stories tall, lining Washington Street from Third Street to Third Avenue. The business core during 1900-1910 extended north and south of Washington Street between Second Street and Second Avenue, and commerce continued to grow along Central Avenue between Van Buren and Jefferson streets. Meanwhile warehouses, industrial enterprises, lumberyards, and slaughterhouses sprang up beside the railroad depots along Jackson Street in the southern part of town.

The Norton children, Fred, Edith, and John Jr. grew up in an area known as the Roosevelt Neighborhood, bounded by West McDowell Road and West Fillmore Street on the north and south respectively, and by Central Avenue and Seventh Avenue on the east and west. Most of the buildings and homes that comprised John Jr.'s neighborhood were built between 1897 and 1938, and during this period numerous political figures, entrepreneurs, capitalists, and community leaders chose that area as their place of residence. After living at 49 North Ninth Avenue for five years, Norton bought a home at 338 North Ninth, where the family lived for about five and a half years before purchasing the home of James Angus Cashion, an esteemed family friend. The Craftsman Bungalow, located at 351 West Portland Street, exhibited superior craftsmanship, including carefully laid brickwork, decorative carpentry details, leaded and beveled glass, and elaborate interior woodwork.

The Norton home on Portland Street was originally part of the "Kenilworth Addition," an eighty-acre area bounded by McDowell and Roosevelt on the north and south respectively, and Third Avenue and Seventh Avenue on the east and west. It was known as the Hubbard Tract until it changed hands in February 1910. In February 1911 Kenilworth was annexed to the City of

Phoenix. Kenilworth developed into an exclusive residential area because of three major influences: the extension of the Phoenix Railway streetcar line north along Fifth Avenue through the Addition; a vigorous and questionable advertising campaign which claimed "The air is better in Kenilworth," and the construction of Kenilworth School in 1920. The streetcar made the area accessible, and initial development in the area was concentrated along the Fifth Avenue streetcar extension. The primary developer, Home Builders, planted palm trees along the streets, which were graded, lined with caliche, and boasted cement sidewalks.

Indeed, John Jr. grew up around the families of some of the most prominent Arizonans of the era. His father rubbed elbows with Charles H. Dunlap, Phoenix city commissioner and entrepreneur who founded the People's Ice and Fuel Company and the Phoenix Wood and Coal Company; Herman P. DeMund, a prominent businessman who established the DeMund Brothers Lumber Company—the most extensive lumber business in the Southwest during the first decades of the twentieth century; Elizabeth Oldaker, who founded the Arizona Museum and First Families of Arizona; and W. C. Ellis, one of the founders of Arizona Deaconess Hospital, the first incarnation of Good Samaritan Hospital, where he served as chief of the medical staff. Other prominent neighbors included F. A. Reid, Charles Stauffer, and Lloyd B. Christy, among others.

The importance of this district as a residential area for Arizona's most prominent citizens was not lost on John Jr. Carl Hayden, Sheriff of Maricopa County (1906-1911), the state's first Congressman (1911-1927), and longtime US Senator (1927-1969), lived in the area during much of John Jr.'s childhood. So too did Charles A. Akers, Secretary of the Territory of Arizona, and C. M. Frazier, a distinguished attorney and later Attorney General. Within a five-minute walk from either of the two Norton residences on Portland Street were the homes of Richard E. Sloan, governor of Arizona Territory (1909-1912), and Baron M. Goldwater, prominent merchant and father of future Senator Barry Goldwater. [2]

John Jr.'s boyhood coincided with the widespread use of public transportation in Phoenix. Streetcar suburbs, like the Kenilworth Addition, emerged north of Washington Street as Moses Sherman's trolley system—which secured electricity in 1893, thus triggering tremendous expansion of the trolley network—crisscrossed the city. Sherman, a good friend of Norton Sr.'s, promoted his streetcar system tirelessly and in turn, the streetcars promoted his real estate investments on the north side of Phoenix. Other suburban sub-division investors paid Sherman to route his system through their property, thus enhancing the value of their real estate developments.[3]

Automobiles further spurred this suburban growth. After the first mo-torized vehicles arrived in the city during the summer of 1900, well-heeled citizens quickly privatized their commute from home to work, thus inaugu-rating an ongoing love affair with the automobile. For fun, young John Jr. ven-tured north along Central Avenue to the new and exclusive Los Olivos subdi-vision, developed by businessman and civic leader Dwight Heard, to gaze at the newest automobile models sitting in front of the houses of the wealthiest Phoenicians. As he grew older John Jr. played with the sons and daughters of many of these people, usually during the summer at Iron Springs, the resort outside Prescott where his father purchased a summer cottage in 1904.[4] Sub-urban sprawl received a jumpstart during John Jr.'s childhood; by 1913, five years after Henry Ford introduced the Model T and the masses as well as the classes could afford a car, several auto dealerships existed, and urban residents of the Salt River Valley could reject traditional city planning and commence developing neighborhoods and businesses on the city's periphery. By the time he turned twelve, Phoenix encompassed 3.2 square miles; the city limits ex-tended on the west to Twenty-Third Avenue between Van Buren and Harri-son, on the east to Sixteenth Street between Van Buren and Harrison, on the north to McDowell between Twelfth Street and Seventh Avenue, and on the south to Yavapai between Seventh Avenue and Central.

Suburbs and automobiles increased the demand for roads and other im-provements, and as early as 1903 an automobile club was formed in Phoenix that advocated for decent roads, since dust-filled streets that turned muddy when it rained created problems for drivers. In 1907 the Maricopa County Automobile Association was formed to promote good roads and draft rules for drivers. It lobbied the Territorial Legislature to connect the capital city with important towns throughout Arizona—especially the county seats. Prompted by the increasing menace of reckless drivers, by 1910 it advocated city ordinances to regulate traffic. By the time John Jr. reached the fifth grade, Phoenix boasted 382 licensed autos, and traffic problems became an issue de-bated among civic leaders. In fact, in June 1910 young Norton witnessed the first motorcycle policeman in Phoenix begin enforcing the twelve-mile-per-hour speed limit. Two years later, construction crews completed paving the first nineteen blocks in the city center.[5]

Another important improvement was the construction of the Central Avenue Bridge to facilitate traffic over the Salt River. As early as 1908 Dwight Heard, who maintained substantial landholdings south of the river, noted that, along with the construction of Roosevelt Dam, a bridge over the Salt River would help stimulate development of thousands of acres of land on the south side. Completed in March 1911, the bridge enabled the automobile caravan that John Jr. and his father participated in to pass over the river on its way to the Roosevelt Dam for the dedication ceremony.[6]

During Norton Jr.'s childhood, promoters also used the area's climate to attract health seekers and tourists to the oasis in the desert. This dimension of the booster campaign, which utilized testimonials and advertisements praising the area to great effect, resulted in the arrival of more newcomers. In 1907, because of the area's growing reputation as a "Lungers' Mecca," the Episcopal Church opened St. Luke's Home for tubercular patients; it expanded in 1912, the same year St. Joseph's Hospital, the first hospital in Phoenix, moved into a larger facility.[7]

Primary and secondary schools formed an important dimension of the cultural and social fabric of the Salt River Valley during this time, and John Jr. attended two of the pioneer institutions in public education. He went to Adams School at 800 West Adams, one of a handful of primary schools springing up in the city; then to Phoenix Union High School, the principal secondary school in the city, which had an enrollment of over five thousand students. When John Jr. graduated high school at age eighteen in 1919, Phoenix Union boasted the largest number of secondary school students west of the Mississippi River. Like many of his friends, he often rode a horse to and from school, but at times he walked to school in the morning, then floated home in an irrigation ditch that ran through the city center along Van Buren Street. In company with his boyhood friends, John Jr. would float to Ninth Avenue and Portland on an inner tube.[8]

From the age of fourteen he rose early in the mornings during the summer to work his father's farms, where he learned the essentials of agriculture. As he grew older Norton Jr. broadened his knowledge, working as a cowboy at the 3V Ranch during his breaks from schoolwork. The 3V, celebrated in the annals of Arizona ranching history, extended from Highway 66 north to the Grand Canyon, and from Seligman west to the Hualapai Mountains. At the 3V he developed not only a solid work ethic but also a passion for horses, cattle raising, and ranching. When he was sixteen, the United States entered World War I, and the local agricultural economy was transformed. His father and he responded to changing circumstances as did the other five thousand or so farmers in the Salt River Valley.

The area's population nearly tripled between 1910 and 1920, from 11,134 to 29,053, and much of that population growth was tied directly to the economic opportunities created by wartime demand for products produced in the valley. The disruption of shipping lanes and exigencies of World War I sent the prices of domestic agricultural products soaring. Phoenix and the Salt River Valley saw agricultural activity expand to new levels,[9] and John Jr. saw his father reap profits unlike he had in years past. Almost every crop made money.

At the beginning of the war, diversity characterized Salt River Valley agriculture. Alfalfa covered about half the land, including his father's six hundred plus acres near Liberty, but farmers also cultivated wheat, vegetables, melons, citrus, barley, and other crops. Although much of the forage was exported, large herds of cattle and sheep wintered on valley fields, and the local dairy industry flourished. As the *Arizona Republican* put it on August 19, 1917, "Alfalfa was King, Cotton was Queen, and every Dairy Cow a Princess."[10]

But by 1917 the Queen, as historian Thomas Sheridan put it, "staged a coup." When World War I broke out Great Britain, in order to maintain its supplies, placed an embargo on the export of long-staple cotton to other countries. Extra-long-staple cotton yielded a fiber of far greater tensile strength than other shorter staples, which made it very appealing for producers of industrial fabric, particularly for tires. The world's crop came from Egypt and the Sudan, where British capital had financed and developed this unique strain. Tire companies in the United States were cut off from their sources just as the War Department had ordered thousands of airplanes, which needed cotton both for their tires and for the fabric covering their wings. Defense contractors scrambled for new domestic supplies. The humid southern Cotton Belt had too short a growing season for long-staple cotton, but the Salt River Valley had the perfect climate.

Thus cotton joined cattle and copper as the three Cs driving the Arizona economy. At the beginning of the cotton boom, Salt River Valley farmers were growing 7,300 acres of extra-long-staple cotton. Not only did Goodyear Tire and Rubber Company of Akron, Ohio purchase cotton from Salt River Valley growers, it also bought two large tracts of land to grow its own cotton in 1917. Under the direction of Paul W. Litchfield, the two tracts—one

8,000-acre parcel southwest of Phoenix called Goodyear, and another 16,000 acres west of Phoenix named Litchfield—became the sites of two agricultural company towns. Litchfield hired experts, and the enterprises flourished with cotton gins, mills, and thousands of acres of cotton.[11] Goodyear also contracted with local farmers to purchase more than $3 million worth of their cotton, and John Sr. dedicated a large percentage of his acreage to long-staple cotton in 1918 and 1919. Meanwhile, other tire companies like Dunlop and Firestone joined Goodyear in the Salt River Valley as prices continued to rise. In 1920, cotton production peaked. Between 1916 and 1919, prices had risen from $233 to $406 per bale, with Goodyear willing to pay up to $625, and acreage dedicated to long staple cotton rose exponentially—from 7,300 acres to 190,000 acres. The Salt River Valley, including the Norton spread near Buckeye, became a vast uniform grid as everyone plowed under other crops and sold off dairy cattle. As the *Arizona State Magazine* announced in March 1919, "The milk producer and the land owner have yielded to the siren song of cotton, and much good dairy stock has gone to the block."[12] In the April 17, 1920 issue of the *Country Gentleman*, a magazine dedicated to the business side of farming, Richard Wells wrote, "Men who never saw cotton grow in their lives brought their bankroll, their youngsters, and their household goods to find the foot of the rainbow."[13]

Then the war ended. The military canceled a vast majority of contracts in the fall of 1920, and when pickers were preparing to harvest the bumper crop, 450,000 bales of Egyptian cotton flooded the US market. Arizona cotton growers like Norton Sr., who expected to receive $1.50 per pound, were extremely lucky to sell their crop at $0.28 per pound. It cost $0.65 a pound to grow the crop. Many farmers went broke, and the Norton family teetered on the brink of bankruptcy. "White gold" became a "white elephant" in the matter of one month. As one scribe described the disaster, "the roulette faro crop" that led the Salt River Valley "up the easy grade to the very top . . . kicked it over the edge on the steep side." The cotton market, like so many others, collapsed in the fall of 1920. Norton Jr. suspended his plans to enter the University of Arizona and stayed home to help his father contend with the economic tidal wave that swept through the Salt River Valley in 1920 and early 1921.

His father struggled mightily against the downward economic forces, and in the fall of 1921 insisted that John Jr. attend the University of Arizona. John Jr. began studying agriculture and animal husbandry, and was active in the Sigma Chi fraternity and in all manner of athletics. But when his father died unexpectedly at the family home in February of 1923, the family had many personal and financial issues with which to contend, and John Jr. was forced to drop out of the University of Arizona. The situation was grim. When Norton Jr. added together the assets and debits, he discovered that the family was $60,000 in debt. As he and the family wrestled with their collective grief, Norton Jr. saw that the post-World War I bust was bankrupting more than farmers, as businesses related to cotton also went broke, banks failed, and even Goodyear danced around insolvency. His working experiences at the 3V gave him insight into the fact that ranchers were having to pay exorbitant prices for the small amount of alfalfa that was available. He watched as big farming operators took over smaller ones by assuming their loans and taxes. The ripple effect, he realized, led to the most fundamental effect of all: the farm population of Arizona declined by 20 percent over the next five years.[14]

John Jr. moved back into the family home and took over a business that he was not fully prepared to run. He learned quickly, however, and displayed a work ethic that distinguished him from others. By early 1924 he began to see glimpses of a regional economic recovery as Salt River Valley farmers, having learned from the short but deep post-World War I depression, diversified their crops. Though cotton remained an important component of the overall agricultural economy, crop diversification became the rule, and John Jr. and other farmers approached local banks that had begun encouraging crop diversification by offering generous, low-interest loans for farmers who moved into grains, vegetables, and fruits. John Jr. observed everything around him as he surveyed alfalfa and grain fields, citrus groves, and dairy herds. He noticed especially the citrus industry, which made a strong reappearance in the valley especially north of the Arizona Canal, from Scottsdale to Glendale. Refrigerated railroad cars, moreover, facilitated safe shipment of perishable fruits and vegetables, like lettuce, cantaloupe, and vegetables, enabling valley farmers to once again compete not only locally, but also with other producers in far away regions.[15] Alfalfa, he realized quickly, had returned to its pre-World War I prominence by the middle of the decade. Wheat and barley made similar gains, and cattle ranching and feedlot operations returned to previous levels. As he continued to work and learn on the job, the endless opportunities and multifaceted economy that had characterized the Salt River Valley of his teenaged years began to return.

Meanwhile John Norton Jr.'s personal life changed and this transformation took place on Portland Street. Early Clay Phelps, in 1919, left Benton, Kentucky, just outside Paducah, and moved to Arizona, purchasing a beautiful Mission Revival-style home across the street and two houses over from the Norton residence. Phelps and his wife had four daughters, the eldest, a brilliant and attractive girl named Earlene. Born July 12, 1904, in Benton, Earlene benefitted from her schoolteacher father's interest in education. Her father instilled a love of learning and intellectual rigor that resulted in Earlene's remarkable academic achievements. Incredibly, she finished high school before she turned thirteen years old. She was too young to attend college, so her father enrolled her in Ward Belmont School, a highly regarded finishing school for girls in Tennessee. She graduated from there in 1919 at the age of fifteen.

Earlene matriculated at Stanford University the same year her family moved to Arizona. She graduated with a degree in Romance Languages in 1924, at the age of nineteen, and moved to Arizona to be with her family. Accomplished, worldly beyond her years, and with a scholarly temperament, Earlene found no work in her field in Phoenix. In order to gain traction in the local economy, she enrolled at Lamson Business College in Phoenix, learned shorthand, and was employed as a legal secretary with a respected attorney, Stockton Henderson.

When Earlene arrived in Phoenix in the spring of 1924, the neighbor three doors down took notice. Soon, John R. Norton Jr., four years older and working more hours than most people in order to pull the various family businesses out of their downward slide, became a frequent visitor to the Phelps household. His attraction to the Stanford graduate was immediate. On February 14, 1925, within a year after their first meeting, the two were married in Phoenix.

In 1926, married and ambitious, the twenty-five-year-old Norton Jr. scrutinized further the benefits of crop diversification in the valley's recovery. He observed what he considered an innovative operation known as the Arizona Packing Company, owned by Edward Tovrea. Quickly it became the largest and most technologically advanced packing plant between Fort Worth, Texas, and the California coast. The stockyards and slaughterhouses were located about five miles east of the growing downtown Phoenix business district, and the site became a primary destination for Arizona's sheep, cattle, and hog interests.[16]

Phoenix in the mid-1920s offered many opportunities. New public and

Earlene Phelps Norton, John R. Norton III's mother (1931).

private buildings were erected, especially in the downtown area. Department stores, banks, hotels, and theaters became part of the emerging urban landscape. In the 1920s skyscrapers were viewed as signs of progress, and the seven-story Dwight B. Heard Building, built in 1920, and the Luhrs Building, completed in 1924, were emblematic. In 1928 the Hotel Westward Ho was added to the Phoenix skyline. Others, like the Luhrs Tower and the new State Capitol Building at Seventeenth Avenue and Washington, followed in short order and spoke volumes about this up-and-coming oasis in the desert. The trauma of the early 1920s seemed a distant memory.

By 1927, four transcontinental trains were passing through Phoenix's four-year-old Union Station each day. Completed in late 1923, the downtown station was heavily promoted by business and civic leaders because it boosted the city's development. The new Southern Pacific main line formed a more than two-hundred-mile loop as it left the major El Paso–to–Los Angeles Southern Pacific route at Picacho, forty-six miles northwest of Tucson, passed through Union Station in Phoenix, then rejoined the main route at Welton, thirty-seven miles east of Yuma. Norton Jr.'s agricultural products and raw materials could now be shipped by rail to any point in the country. Tourists could now visit the Salt River Valley much more easily and comfortably, and Phoenicians could likewise visit other states and regions.[17]

Just after transcontinental rail service reached Phoenix, in November 1928, Scenic Airways, a private venture backed by Phoenix and Chicago investors, established Sky Harbor Airport, located a few miles east of the downtown business district. Standard Airlines and a few other carriers joined Scenic Airways in the new venture, and soon it provided the best air service in the area, which was dotted with several private airfields. The exciting world of air travel opened Phoenix to the region and the nation, and on September 2, 1929, nearly ten thousand people celebrated at the dedication of Sky Harbor Airport.[18]

The automobile and road construction also played a big role in connecting cities in the region. Arizona's congressional delegation, especially Congressman-at-large Carl Hayden, worked diligently to secure federal funds for interstate highways through the passage of federal highway acts in 1916 and 1921. Cities, counties, and states as well as private entities worked together raising additional funds to help develop this infrastructure, and local organizations such as the Maricopa County Good Roads Association and the Arizona Good Roads Association lobbied heavily for interstate highway construction. The paved Phoenix–to–Los Angeles Highway that passed over the Colorado River at Blythe, California, was completed in 1928.

By 1925, when Norton Sr.'s old friend Moses Sherman sold his streetcar company to the City of Phoenix, Phoenicians (including Norton Jr., who owned a Buick) had already accepted the automobile as the primary mode of transportation. As more automobiles appeared on Arizona's various thoroughfares, citizens called for better roads. One urban history expert calculated that "the amount of pavement in Phoenix rose from seven miles in 1915 to twenty-five in 1920 to eighty-six in 1929."

By the mid-1920s, Phoenix had become the highway hub of Arizona; cars, trucks, and buses (then known as passenger stages) utilized roads that connected the capital city with all parts of the Southwest and beyond. These transportation options, along with railroad and airline service, encouraged more people to visit Arizona, and the consequent demands for better accommodations triggered a boom in downtown hotel construction. In 1928 two iconic downtown Phoenix hotels opened—the Westward Ho and the San Carlos. In response to these bold initiatives the established Hotel Adams underwent considerable expansion and renovation. And in what was then "outside" Phoenix, two first-class resorts joined the parade of new hostelries for visitors: the Jokake Inn in 1924 and the Arizona Biltmore in 1928. That amenities of this nature—reflecting a new kind of economic vigor—could be sustained in Phoenix was not lost on Norton Jr.[19]

Very early in his business career Norton Jr. visited with the main local economic development organization, the chamber of commerce, and a host of fraternal and cultural organizations. At once he was impressed with the chamber of commerce, which encouraged and organized annual meetings of state, regional, and national groups, and he learned that one of the chamber's chief goals was to market Phoenix as an ideal convention city. He knew that competition to attract winter visitors was keen between Phoenix and other emerging Southwestern resort centers. With more people visiting and perhaps moving to the valley, he could sell more fruits, vegetables, alfalfa, cotton, and cattle. The chamber of commerce and other booster organizations like the Phoenix-Arizona Club, he concluded, were essential to increase the popularity of the city and the valley.

In 1926, Norton Jr. recalled that he attended the first Masque of the Yellow Moon festival and pageant, which involved thousands of participants and became, in time, a Phoenix tradition.[20] Phoenix had in fact become a "city of joiners," where professionals enlisted in fraternal organizations like the Masons, the Elks, the Odd Fellows, and the Knights of Columbus, and civic organizations such as the Rotary, the Optimists, the Kiwanis, and the Lions. Numerous clubs, like the Women's Club, "widened the boundaries of cultural development," and members of the Phoenix Country Club promoted and hosted state, regional, and national golf tournaments. The Phoenix Symphony Association, the Phoenix Little Theater, and the Phoenix Fine Arts Association exposed patrons to concert performances, dramatic productions, and museum displays of all types.

As Norton Jr.'s agricultural enterprises stabilized in 1926-27, he began forging his own path within the broader agricultural community. Phoenix and the Salt River Valley maintained direct contact with the outside world and served as a vital business center for a productive hinterland of farming and ranching, and Norton Jr. also concluded that it was a good place to live. He resolved to build upon what his father had left him, and was part of the class of citizens whose actions benefited all of the Salt River Valley.[21]

John Jr. continued to diversify the crops he grew on his acreage—he began raising more cantaloupes in 1928 and 1929. That Spring Earlene gave birth to a son, John III, at St. Joseph's Hospital, on April 10, 1929. The future looked bright for the young couple and child as farming in the Salt River Valley continued to prove profitable. After his father's passing in 1923, John Jr. had, in five short years, turned around the family's fortunes. With ample water supplies in the Salt River Valley and the Buckeye Valley, his unique lifestyle—a combination of urban and rural—held great promise. Indeed, John Jr., Earlene, and John R. Norton III lived in a remarkable urban oasis built upon an agricultural foundation.

·5·

ESTABLISHING
<><><><><><><><><><><><><><><><><><><><><><><><>
PRIOR RIGHTS
<><><><><><><><><><><><><><><><><><><><><><><><>

"An Arizona only approach to Colorado River development would create endless litigation, and unless the opportunity is taken to settle the issue peaceably and fairly it will take its place as the subject of controversy and conflict between the states for the next twenty-five years. . . . There will be no Civil War over this issue but it will be a fine field for demagogues' oratory."

SECRETARY OF COMMERCE HERBERT HOOVER, ADDRESSING AN AUDIENCE OF 1,500 AT THE COLUMBIA THEATER IN PHOENIX IN SUPPORT OF THE RECENTLY SIGNED COLORADO RIVER COMPACT, DECEMBER 8, 1922.

"My attitude on the Colorado pact seems to worry a good many people. . . . Herbert Hoover will run up against a brick wall—the plot thickens."

ARIZONA GOVERNOR GEORGE W. P. HUNT, DIARY ENTRY FOR NOVEMBER 28, 1922.

When Norton Jr. inherited his father's home, land, and debt in 1923, the development of the Colorado River and its tributaries had become the public policy issue that dominated the politics of the American Southwest. Compared to the effort to bring federal reclamation and Roosevelt Dam to the Salt River Valley, development of the Colorado River, one hundred sixty miles to the west, posed far more complicated challenges for Arizona leaders. As the debate began to swirl in the late teens and early twenties, Norton Jr. realized that the problem of dividing the river's resources among seven states and the Republic of Mexico might influence many of his future decisions.[1]

From its source in the Rocky Mountains, the Colorado River drains some of the most arid yet scenic country in the United States. It flows nearly 1,400 miles in a southwesterly direction through deserts, canyons, and fertile valleys and is fed by nearly 244,000 square miles of drainage area. Specifically, north of the US border with Mexico it drains watersheds in Wyoming, Colorado, Utah, New Mexico, Nevada, Arizona, and California. The Colorado then crosses the international border near Yuma, Arizona, and flows its last one hundred miles through Mexico, emptying into the Gulf of California. In spite of its large tributary system, which includes the Salt and Gila Rivers in Arizona, the Colorado is a rather light flowing stream, ranking sixth in volume among the nation's rivers. Average annual precipitation in the basin amounts to less than fifteen inches, and evaporation reduces runoff by a staggering 90 percent. Based on US Bureau of Reclamation records kept since 1922, the remaining 10 percent totals a meager 15.5 million acre-feet, or one-twelfth the volume of the Columbia River. As a result of the increasing demands placed upon the stream flow, the Colorado, according to the distinguished historian Norris Hundley Jr., has been the most litigated, politicized, regulated, and argued-about river in the world.[2]

Norton Jr. and his father knew that other noteworthy characteristics of this interstate and international stream influenced the nature of agriculture in the region. The Colorado flows on a ridge above sea level, and in years past it periodically tore through its banks and poured into lower-lying valleys. Significant consequences of sporadic flooding were the creation of a large freshwater lake, the Salton Sea, and the depositing of rich alluvial soil that reached hundreds of feet deep. Moreover, the Colorado proved to be one of the heaviest silt carriers in the world, transmitting five times that of the Rio Grande and seventeen times that of the Mississippi. Toward the delta its speed decreased, dropping much of its load of silt, thereby causing the channel to rise above the surrounding countryside. By the twentieth century this occasional flooding had created a rich delta in a climate conducive to year-round agricultural production.[3]

Nineteenth-century desert dreamers envisioned all types of diversion schemes. In the mid-1890s, for example, territorial governor Nathan Oakes Murphy proposed damming the river near the Grand Canyon and irrigating the land between there and Phoenix. Murphy, a Republican appointee in a Democratic territory, suffered no small amount of editorial abuse for his "chimerical schemes." One of Murphy's outspoken detractors, Anson Smith, editor of Kingman's *Mohave County Miner*, soon became a supporter of Murphy's plan and later opined that Colorado River development could provide both electricity for Arizona's mines and water for agriculture, thereby turning Arizona into a "Garden of Eden."[4] But John Wesley Powell generated interest nationwide when he called for a more rational approach to problems posed by aridity in the West. In his *Report on the Lands of the Arid Regions of the United States*, he anticipated that reservoirs would enable Americans to divert water to southern California's fertile lowlands. Though he failed to foresee the precise location of Hoover Dam, his ideas struck a responsive chord. As it was with tributaries such as the Salt and Gila Rivers, extracting the Colorado's potential became a regional obsession.[5]

Earlier, several desert land pioneers acted on the notion that the Colorado River was a source of livelihood. In 1849 Dr. Oliver Wozencraft, racing to the California gold fields, first conceived of a plan to irrigate California's Imperial Valley via gravity flow canals. The unfortunate Wozencraft spent the remainder of his life in failed attempts to fulfill his dream, but he did lay the groundwork for future efforts. In 1879, no less a promoter than John C. Fremont, then acting as Arizona's governor, proposed a grandiose scheme that outraged his constituents: he supported a plan to alter the climate of the Imperial Valley by flooding the Salton Sink with Colorado River water. Arizona Territory residents forcefully opposed Fremont's proposal, prefiguring later animosities over the rights to Colorado River water. It took another nineteenth-century visionary to realize Wozencraft's dream. In 1892, Charles Rockwood began investigating the possibility of irrigating land along the border with Sonora and observed, as Wozencraft and Fremont had earlier, that the Imperial Valley could be transformed into a year-round agricultural empire. Four years later, with financial backing from American and European investors, Rockwood organized the Colorado Development Company, and on June 21, 1901, he succeeded in delivering water to the Imperial Valley.[6] As Norton Jr. and his father knew, putting water to beneficial use in the Imperial Valley enabled California users to assert prior rights to the water under the existing laws of prior appropriation.[7]

Introduction of water into the valley triggered a land rush as thousands of settlers formed mutual water companies that purchased Rockwood's water and distributed it among the users. By 1910 roughly 15,000 people farmed over 160,000 acres. Yet problems beset these agricultural pioneers as the delivery system, based on earlier designs by Wozencraft, tapped the river just north of the international border and fed water into the Alamo Channel, which traversed into Mexico for fifty miles before turning north again onto American soil. International legal problems pertaining to the canal, troubles related to the Mexican Revolution, which threatened the quality and quantity of the valley water supply, and the presence of American speculators in Mexican lands, who often worked at cross purposes to Imperial Valley residents, resulted in a series of crises for Rockwood and the Colorado Development Company.[8] Mexico exacted a high toll in water, money, and other concessions for the use of her territory. Wealthy Americans further complicated the situation. A group of Los Angeles-based business interests, led by *Los Angeles Times* publisher Harry Chandler, were the largest landholders in the Mexican Delta. By 1905, Chandler and his partners owned more than 840,000 acres, most of which they leased to Mexican, Chinese, and Japanese tenants. This unwieldy set of circumstances prompted sentiment for the construction of an "All-American" canal.

The movement for an All-American Canal received further stimulus due to the disastrous flood that began in February 1905. Within weeks the Salton Sink became the Salton Sea, and the flood ruined Rockwood and forced his company into bankruptcy. He literally surrendered control of the company to the Southern Pacific Railroad in exchange for assistance in controlling the river, and by 1907, after two years of continuous flooding, work crews had the river back within its banks. Now Imperial Valley land users had two masters over their water supply: the Southern Pacific Railroad and the Mexican corporation, *Sociedad de Irrigacion y Terrenos de la Baja California*, which controlled the water supply south of the border. Residents concluded that they needed to rid themselves of two receivers and extract their water supply from Mexican control. As a first step, Imperial Valley residents needed to establish public control over the irrigation infrastructure on the American side of the border. In 1911, Phil Swing, an Imperial Valley lawyer, and businessman Mark Rose stepped into the fray and organized the Imperial Irrigation District (IID). This agency gave farmers the ability to elect directors, issue bonds, levy assessments, condemn property, and purchase and operate the valley's irrigation system. Finally, in 1916, the Southern Pacific, anxious to extricate itself from the irrigation business, sold the assets of the Colorado Development Company to IID for $3 million. IID then focused its energies on lobbying Congress for a water delivery system entirely within the United States—an All-American Canal.[9]

The flood of 1905-7 had an additional consequence. It served notice that a pressing need existed for flood control and water storage on the Colorado River, and Arthur Powell Davis of the newly created US Reclamation Service in the Department of the Interior emerged as the most vocal advocate for a federally sponsored mainstream reservoir. He not only championed the idea of flood control but also argued for comprehensive development of the Colorado River system. As early as 1902 Davis had outlined a plan for the "gradual comprehensive development of the Colorado River by a series of storage reservoirs." He had the ear of President Theodore Roosevelt, who shortly after the great flood called upon Congress "to enter into a broad comprehensive scheme of development for all irrigable land" along the river. Davis convinced Interior Secretary Franklin Lane to earmark funds to launch an extensive investigation of the Colorado, and after 1914, when Davis was appointed Director of the Reclamation Service, he oversaw the survey. Imperial Valley residents and Californians in general viewed the study as an affirmation of their ideas for comprehensive development in the realms of irrigation and flood control.[10]

Supporters of comprehensive development of the Colorado received unyielding support from the League of the Southwest, a regional booster organization. Largely a California-sponsored group, the League, founded in late 1917, represented business and local governments dedicated to the commercial and social interests of the southwest quarter of the country. At its initial gathering at the Hotel Del Coronado in San Diego, representatives from the seven Colorado River Basin states, Oklahoma, and Texas joined representatives from Washington, DC and Europe to discuss the possibilities of this burgeoning region.[11] Arizonans expressed gratification that the president of the University of Arizona, Dr. Rufus B. Von KleinSmid, was selected as the League's first president. In 1918 the League met again in San Diego, then in Tucson, where discussions centered on the development of the Colorado River system and how tourism, commercial development, and transportation could be further expanded. By 1920, the organization had focused its mission to virtually one purpose, which was reflected on its letterhead: "The League of the Southwest holds as axiomatic that the development of the resources of the Colorado River Basin fundamentally underlies all future progress and prosperity of the Southwest."[12]

Ongoing Imperial Irrigation District lobby efforts wrought success when, in November 1917, the District convinced Interior Secretary Lane to conduct a feasibility study to determine if it was possible to build a proposed All–American Canal connecting the Imperial Valley with Laguna Dam, a diversion structure that helped supply water to the recently completed Yuma Reclamation Project. The caveat: IID had to pay for two-thirds of the cost. Imperial Valley residents quickly assented to these terms, and a committee of three engineers, the so-called "All–American Canal Board," conducted a study that lasted most of 1918. In December they issued preliminary results. The board recommended the construction of a sixty-mile-long canal that would cost $30 million. IID leaders moved quickly, and in a January 1919 referendum, residents voted overwhelmingly to endorse the project. With the report and the voters' mandate, IID attorney Phil Swing ramped up pressure for congressional support of a federally funded All-American Canal.[13]

Most Arizonans, however, were concerned with flood control rather than an All-American Canal that would benefit the Imperial Valley in California, and Arizona's lone congressman, Carl Hayden, sought to influence the direction and nature of any federal action. Thus he joined California congressmen William Kettner and Charles Randall in submitting flood control bills in addition to the bill calling for the construction of the All-American Canal. Submitted between February and June 1919 during the third session of the Sixty-Fifth Congress and the first session of the Sixty-Sixth Congress, these earliest Colorado River reclamation bills garnered little, if any, traction, but focused national attention on Colorado River issues. HR 6044, one of Kettner's bills submitted on June 17, 1919, called for construction of an All-American Canal without a storage dam further upstream. The House Committee on Irrigation of Arid Lands held extensive hearings on the bill, but Arizona's Hayden, as well as Reclamation Director Arthur Powell Davis, opposed the measure. They had concluded that California was putting the cart before the horse. As Davis pointed out, putting additional land under cultivation "will threaten the water supply of the whole valley." Hayden argued that there should be reservoir construction first and suggested that California pay for an "equitable part of water storage." He admonished the Californians, "You are now coming to Congress asking that an extraordinary thing be done by the passage of this legislation, and Congress must look not only at the Imperial Valley, but the Colorado River valley as a whole, and that can only be fully developed by storage." The Arizona congressman served notice that he opposed California gaining prior rights to Colorado River water with the aid of the federal government.[14]

The unprecedented growth of the Imperial Valley and California's aggressive moves to utilize the river produced significant political reactions in Arizona. Reckless water diversion practices had already irked farmers on the Arizona side of the river. In one instance IID had constructed a small weir dam, Hanlon Heading, which raised water levels high enough to permit diversion, and repeatedly caused flooding on valuable Arizona farmlands abutting the river. In 1916 the practice flooded all of Yuma. Moreover, preliminary engineering studies confirmed the long-held notion that the only adequate storage sites on the lower reaches of the Colorado lay in Arizona. This bolstered theories among many Arizonans that California held grand designs to benefit from Arizona's resources. One reclamation activist advised a special joint session of the Arizona legislature in 1919: "Don't sell your birthright for a mess of pottage. Arizona is in danger, and unless you become alive to the situation, you will wake up one of these days to find Arizona on the dry side of the river."[15]

By 1920 Congress realized that Colorado River development posed problems of enormous legal, engineering, diplomatic, and economic complexity. Reclamation Director Davis emphasized the need for further engineering and scientific studies, adding, "The most feasible point for storage is in Boulder Canyon" in Arizona. But the Reclamation Service required more investigations. Arizona leaders, wary of an informal but powerful alliance developing between IID and the Reclamation Service, were soon joined in their concerns by leaders throughout the Colorado River Basin. The Upper Basin states—Wyoming, Colorado, Utah, and New Mexico—developed a visceral fear that the Lower Basin states, especially California, would develop more rapidly and lay claim, by prescription and prior appropriation, to an inequitable amount of Colorado River water. To add to these fears, the City of Los Angeles joined the Colorado River sweepstakes, making known its desire to secure power and water for its rapidly expanding population. Upper Basin states viewed southern California urban and agricultural water users as threatening their future development, since California projects would put a large share of the river's flow to prior use. The editor of the *Tombstone Epitaph* captured existing sentiment throughout the basin states with a pungent comment in his editorial of June 8, 1922: "California was in the game to hog it all."[16] Clearly Arizona's half-century-long fight with California over the Colorado had commenced in earnest.

As engineering investigations continued, Upper Basin fears concerning

California surfaced with surprising intensity at a portentous League of the Southwest meeting in Denver, Colorado.[17] Four months after a harmonious gathering in Los Angeles at the August 1920 meeting, Upper Basin attendees raised pointed questions about California and its motives in aggressively pursuing federal action on the river. The governor of Colorado, Oliver Shoup, announced: "It is not time for the western states holding the headwaters to lose any of the rights for any reason whatsoever." Colorado state engineer A. J. McCune warned that construction of the reservoir and beneficial use of water stored behind it would give California prior rights over Upper Basin states. McCune stated, "Our main fear is that Los Angeles and the people of the Imperial Valley will get the government committed to a policy that will interfere with our development." These states shared Arizona's trepidations, and called for some type of protection.[18]

In addition to issuing provocative and candid statements about California's water greed at the League of the Southwest meeting in Denver, Governor Shoup introduced his legal adviser, Delpheus "Delph" Carpenter.[19] Carpenter offered a recommendation to assuage Upper Basin fears; his proposal led ultimately to the creation of the Colorado River Commission and the Colorado River Compact negotiations, one of the most significant developments not only in the legal and political history of the river but also the history of the American West.[20] At the time of the meeting in Denver, Carpenter was serving as a member of the defense counsel in the eight-year-old *Wyoming v. Colorado* (1922) Supreme Court case that centered on the rights to the waters of the Laramie River, which rose in Colorado and flowed northward into Wyoming. Hoping to spare the seven states that shared the Colorado River the cost, turmoil, and potentially unsatisfactory results of a Supreme Court decision, Carpenter put forth ideas he had first outlined eight years earlier. To avoid the possibility that population growth and rapid agricultural development in other states sharing the river basin would deprive Colorado, and perhaps other states, of their water rights, he advocated action under little-used Article Six of the US Constitution, which enabled states, once receiving permission from Congress, to negotiate treaties among themselves. Specifically, Carpenter sought to invoke the compact clause of the Constitution—a clause previously used to settle boundary disputes—and apply it to interstate water rights. The result would produce a mechanism whereby complex issues involving interstate water rights could be settled through negotiation without litigating them before the US Supreme Court.[21]

Arizona's congressional delegation—Congressman Carl Hayden, US Senator Ralph Cameron, and US Senator Henry F. Ashurst—appreciated the legal theory that framed Carpenter's proposal. A binding agreement among the states could incorporate both a Lower Basin dam and a provision that would assure slower-developing states that they would receive protection. At this historically significant meeting of the League of the Southwest, delegates adopted a resolution affirming Carpenter's legal concept, calling for a compact to determine present and future water rights of the states who claimed interest in the Colorado River. Additionally, the resolution suggested that the seven states authorize appointment of commissioners for the purpose of entering into an agreement and for the subsequent ratification by the respective state legislatures and Congress.[22]

By spring 1921 all state legislatures had taken a formal first step toward the goal of forming a Colorado River Commission. On May 10, 1921, Arizona governor Thomas Campbell met with the six other basin state governors in Denver to issue a request of President Warren G. Harding that he would publicly support Colorado River Compact negotiations—even though Arizona, at the time, harbored strong state's rights views concerning the river within its borders. Harding responded, and on August 19, 1921, he signed the Mondell bill, sponsored by House majority leader Franklin Mondell (R-Wyoming), who introduced legislation calling for a Colorado River Commission comprising representatives of each of the states plus one federal representative. Harding took another action that surprised the various stakeholders in the river, assigning Secretary of Commerce Herbert Hoover as the federal representative on the commission. Although an astonishing appointment to some observers, Hoover's honesty, integrity, and international reputation were beyond reproach, and in the end he served as chairman of the commission.[23]

Meanwhile, the engineering studies conducted under the aegis of the Department of the Interior approached completion. The Fall-Davis Report, named for Secretary of the Interior Albert Fall and Reclamation Commissioner Arthur Powell Davis, recommended construction of an All-American Canal and a high dam capable of generating power at or near Boulder Canyon. On December 9, 1921, Davis held a public hearing in San Diego and announced: "In the northwestern corner of Arizona there is a profound and very deep narrow canyon, where it would be feasible to build a dam seven feet high."[24] Indeed, the site held other advantages. It was closest to power markets in the areas where year-round irrigated agriculture took place. Moreover, a high dam would create a reservoir that controlled flooding, regulated flow, and allowed silt to settle, eliminating that thorny problem without compromising storage capacity.

A major influence that not only paved the way for the seven-state agreement, but also the two-basin concept was the *Wyoming v. Colorado* Supreme Court decision, rendered on January 5, 1922. The Court reaffirmed the rule of priority over streams shared by states, adhering to the rule of prior appropriation. The Supreme Court thus allowed the application of the priority rule on interstate streams even if the water was transferred out of the basin for use. The decision favored fast-growing California and placed that state in a strong legal position to gain prior rights to Colorado River water. Upper Basin commissioners felt "badly exposed" and now believed that they had to come away from the forthcoming negotiations with a compact.[25]

On January 26, 1922, the Colorado River Commission began its first set of hearings in Washington, DC, After intermittent meetings throughout the West during that year, the commission concluded its negotiations with the signing of the compact at Bishop's Lodge in Santa Fe, New Mexico. The compact's provisions have been widely discussed and analyzed, but its chief innovation—the legal handiwork of Colorado's commissioner, Delph Carpenter—divided the Colorado River watershed into two basins, with the division point at Lee's Ferry, in the rugged canyon lands of northern Arizona near the Utah border. The commissioners, in forging this accord, allotted the Upper Basin—Wyoming, Colorado, Utah, and New Mexico—7.5 million acre-feet of water annually. The Lower Basin, composed of Nevada, Arizona, and California, received 7.5 million acre-feet of mainstream water per year, plus an additional 1 million acre-feet under the vaguely worded and legally controversial Article III (b): "In addition to the apportionment . . . the Lower Basin is hereby given the right to increase its beneficial consumption of such waters by 1 million acre-feet per annum."

Article III (b) reflected Arizona interests in the beneficial consumptive use of the waters of the Gila River, the Colorado River tributary whose watershed lay almost entirely within Arizona. According to one leading scholar of the Colorado River Compact, Arizona's representative William S. Norviel "played a major role in shaping the treaty that was eventually drafted." Norviel may have understated the figure of 1 million acre-feet because Arizona's tributaries, primarily the Salt and Gila systems, according to contemporary accounts, produced between 2 and 3 million acre-feet from annual surface runoff. Norviel later explained that "it was understood though not expressed, that the 1 million acre-feet from the Gila would practically take care of or offset all the water produced" in Arizona. Since the commission apportioned water to basins rather than states, such an understanding, implied or otherwise, could not be expressly written into the compact. Chairman Hoover lent unofficial credence to Norviel's declaration when, after signing the compact, he gave an autographed picture of himself to the Arizona commissioner with the caption: "To the best fighter on the commission. Arizona should erect a monument to you and entitle it 1 million acre-feet!"[26]

Article IV addressed the hydroelectric power issue. Agriculture and water for domestic use took precedence over water for power generation. Importantly, the compact had no provision for a storage dam; the Upper Basin commissioners had scuttled California's attempt to include it in the agreement. This omission meant that flood protection on the lower reaches of the river had to await congressional action. Close observers foresaw future litigation as they grasped the significance of Article VII. That article, in dubious fashion, addressed American Indian rights to the Colorado River. Sarcastically referred to as "Hoover's Wild Indian Article," it was, in fact, a nonarticle. "Nothing in this Compact," it states, "shall be construed as affecting the obligations of the United States of America to Indian tribes." Future generations, unfortunately, have wrestled with this costly equivocation. At the same time, commissioners virtually ignored Mexican rights on the river. If the United States recognized the rights of Mexico on the river, Article III (b) stipulated that both basins must share the obligation in equal proportion.[27]

Despite the Compact's omissions, ambiguities, and limitations, it was soon heralded as the "Law of the River." The misleading label carried the erroneous supposition that the agreement would keep the river out of the courts. Ottamar Hamale, solicitor for the Reclamation Service, Hoover's legal adviser, and a leading voice for federal control of western waters opined: "This settlement was reached within a year, while settlement in the court of the *Wyoming v. Colorado* case required about eleven years and is unsatisfactory to both sides involved. The compact was not intended to be a complete settlement but a big step in the right direction and as big of one that can be made at this time."[28]

While Arizonans prepared to debate the Colorado River Compact in their legislature, no one foresaw that it would take Arizona twenty-two years to ratify it, an unusual and frustrating chapter in Arizona's water history. Still, throughout 1923 the Arizona congressional delegation campaigned for ratification of the compact, spending extended periods of time in Arizona speaking to audiences throughout the state. When Congressman Hayden learned that the state legislature might consider trying to insert reservations or amendments into the agreement, he sent them an especially sharp message: "Acceptance of this compact with reservations is, in fact, no approval at all. Any fair-minded person must conclude that Arizona alone can not undertake the development of the great river without consent of the United States and without an understanding with the other states of the Colorado River Basin, all of which leads to the conclusion that sooner or later the Colorado River Compact must be approved by the State of Arizona."[29]

In supporting the compact, Hayden and Ashurst disavowed that portion of the 1922 state Democratic platform advocating a states' rights approach to Colorado River development. Their opposition more closely resembled the ideas espoused in the Republican platform, which called for "a return to the Republican policy, reclamation, inaugurated under the administration of President Roosevelt." The Republican platform emphasized greater federal involvement in western reclamation in order to facilitate growth and development. At least in the realm of federal reclamation, many Democrats in the Arizona congressional delegation found common ground with Republicans.[30]

The Governor Hunt-led framers of the 1922 Democratic Party plank, however, contended that Arizona's rights to the river were "superior and natural." Benefits derived from river development should be preserved for the people and not for "selfish private interests." Reclamation and flood control would be the primary objectives in any development program. Power, although "tremendously important," was a secondary consideration. The Salt River Valley Water Users Association doubtless influenced the relegation of power production in the majority party's platform. Competition from a power-producing dam at the Boulder Canyon site would eliminate the association's monopoly on power within the Salt River Valley service area. The overall content of the Democratic Party platform of 1922, sprinkled with old-time Populist rhetoric, held great appeal to voters in Arizona in the general elections of 1922. They not only elected Hunt to the fourth of his seven two-year terms, but they also chose an overwhelmingly Democratic state legislature. When the Sixth Legislature convened in Phoenix on January 8, 1923, eighteen Democrats and one Republican comprised the Senate, while forty-one

Democrats and six Republicans sat in the House of Representatives. For all intents and purposes, Arizona was a one-party state. [31]

The former and future governor-elect, George W. P. Hunt, who would again serve from 1923 to 1933 with only one interruption, stood firm in his opposition to the regional cooperation implied in the compact. He promoted "development of Arizona with Arizona resources," echoing the strong southern states' rights strain that permeated much of Arizona's Sixth Legislature. The former mining camp waiter, who paid scant attention to the rules of grammar in his speeches, railed unremittingly against the aggressor-state California. He became the self-proclaimed "savior" of the Colorado River, issuing hundreds of proclamations, messages, and speeches, all of which centered on one theme—opposition to the Colorado River Compact. Governor Hunt summed up his views and those of the "anti-pacters" in one circular issued in 1923: "The Colorado River is our greatest resource, and unless we conserve it and get the maximum benefit from it, we can depend upon becoming a sort of vermiform appendix to Los Angeles," adding, "I am not at all ambitious toward building up Los Angeles with Arizona resources." Hunt knew that Arizona businessmen, workers, miners, and farmers found intolerable the notion of being considered a colonial offshoot of the booming southern California economy. Indeed, the governor parlayed the idea of neocolonial status in the region into a politically attractive and potent issue. [32]

As Arizona's Sixth Legislature met to consider compact ratification, Hunt staked out his divergent position from his fellow Democrats, Congressman Hayden and Senator Ashurst, as well as from Republican Senator Cameron, all of whom supported the compact. While the congressional delegation urged federal development of the Colorado River under the terms of the seven-state agreement—it was a regional resource, they declared—the governor maintained the proprietary position that Arizona possessed the ultimate authority over the river within her borders. The debates within Arizona's legislature over the issue illustrated the philosophical tensions within the Democratic Party and between the major parties, and tested the sensitive strands of American federalism. [33]

As Arizona's elected officials indulged themselves in a desultory years-long debate over states' rights versus federal control of the river, California congressmen and senators redoubled their efforts to instigate federal action, raising new concerns in Arizona and elsewhere. The first of several Swing-Johnson bills—named for California congressman Phil Swing and his upper house colleague, Senator Hiram Johnson—signaled California's intentions. HR 11449, introduced in the House of Representatives on April 25, 1922, contained provisions for storage, power production, and an All-American Canal. Arizona congressman Hayden opposed the bill and used his senior position on the House Committee on the Irrigation of Arid Lands to derail the legislation for the remainder of the session. Hayden informed the committee chairman: "I intend to use any and every legitimate means to prevent action on the bill." Through a variety of maneuvers, the Arizona representative and senators kept the Swing-Johnson bill, in its various guises, from reaching the floor of the House or Senate for nearly five years.[34]

While Arizona's congressional delegation fought California's attempts to pass the Swing-Johnson bill, the state moved into an even more intransigent position thwarting ratification of the compact. Without a seven-state agreement, there could be no legislative action and thus no development of the river's resources. Ultimately, the other states and the federal government attempted an end run around the political logjam created by Arizona. Shortly after Arizona's Sixth Legislature failed to ratify the compact in 1923, Delph Carpenter suggested to Herbert Hoover and the other commissioners that the agreement be allowed to become effective after six states agreed to its provisions. Despite forceful protests from Arizonans, by mid-1925 the legislatures of the upper states and Nevada had approved the six-state blueprint. California, however, threw a last-minute political wrinkle into the plan when it made its ratification contingent upon federal approval of a "storage dam" at or below Boulder Canyon. Upper Basin leaders balked at this notion, but Representative Swing and Senator Johnson devised another way to secure a Boulder Canyon dam. When they reintroduced their bill for the third time on February 26, 1926, it included a provision making the Colorado River Compact effective with six-state ratification. Though Arizona derailed this legislative maneuver, it could not halt the inexorable power of the upper states, California, and the federal government.[35]

Arizona leaders continued to fight what they considered a "force bill," which sought to compel them to conform to the will of the other states, particularly California. Meanwhile, in the elections of 1926, Congressman Carl Hayden, who had been Arizona's lone representative in the House since statehood, ran against incumbent Senator Ralph Cameron, won the election easily, and advanced to the US Senate, where he joined Senator Henry F. Ashurst and would remain until his retirement in 1969. Hayden assumed his new position knowing that passage of the Swing-Johnson bill was inevitable. On January 21, 1927, as he transitioned to the Senate and after he had submitted a "Minority Report" on the Swing-Johnson legislation to the House, he pled for three hours before the House Rules Committee. While the proposed legislation provided not a dollar for reclamation in Arizona, it gave California $31 million for a water delivery system—the All-American Canal—and provided a federally subsidized power plant to generate cheap electricity to stimulate growth in southern California. The requirements of Los Angeles and other southern California cities did not justify passage of legislation at the expense of Arizona. The Swing-Johnson bill, Senator-elect Hayden argued, was "purely and selfishly a California measure."[36]

Hayden also touched upon several potent legal issues in his three-hour presentation. Mexican rights to mainstream water, states' rights, the six-state compact, power royalties, and an equal division of mainstream water between Arizona and California were covered. Enactment of the Swing-Johnson bill, Hayden prophesied, "could only result in protracted litigation over Colorado River rights." "Arizona," he warned Rules Committee members, "would be compelled to file suit in the US Supreme Court to restrain construction of Boulder or Black Canyon Dam until the rights of Arizona are determined."[37]

Ultimately, after a filibuster, threats of fistfights, and an executive session that "deprived the Senate gallery of one of the wildest scenes since war days," the Seventieth Congress began its second session on December 5, 1928, with the express purpose of passing the fourth iteration of the Swing-Johnson bill. Despite Ashurst's and Hayden's vigorous opposition in the Senate, on December 16 it passed the Boulder Canyon Project Act by a vote of 64-11. The House approved the Senate version two days later by a vote of 167-128, and President Calvin Coolidge signed the legislation into law three days after that, ending one phase of the fight for the Colorado but signaling the beginning of another.[38]

Few measures have had greater impact on the region than the congressional action that authorized Hoover Dam and the All-American Canal. It was an action that in 1944 spurred a controversial water treaty between the US and Mexico and precipitated a series of Arizona-generated actions in the Supreme Court that culminated in *Arizona v. California* (1964). Additionally, the legislation repealed the law of prior appropriation as it applied between the Upper and Lower Basins. And, much to the chagrin of Arizonans, the terms of the Colorado River Compact of 1922 became effective with the passage of the Boulder Canyon Project Act of 1928. Pundits wrote that the legislation gave California everything and Arizona nothing.[39]

Concessions to Arizona, however, were evident and testified to the Arizona delegation's resourcefulness. California, in winning support for passage of the Swing-Johnson bill, agreed to limit itself to a specific amount of the 7.5 million acre-feet allocated to the Lower Basin under Article III (a) of the compact. The bill also contained the Pittman Amendment, named for Nevada Senator Key Pittman, who introduced a compromise that outlined an Arizona-supported concept of a Lower Basin tri-state agreement. Under its terms, Nevada would receive 300,000 acre-feet of water annually, California 4.4 million, and Arizona 2.8 million plus "exclusive rights to the Gila River." This aspect of the Pittman Amendment, as interpreted later in *Arizona v. California* (1963), represented a clear victory for Arizona, which had long fought for the exclusion of that river from any mainstream water computation. Finally, under the terms of the Pittman Amendment, Arizona and California would divide any surplus mainstream water.[40]

Other Arizona contentions were included in the law. The construction repayment and maintenance costs of the All-American Canal were separated from those for the dam and power plant. Arizona congressman Hayden had suggested the arrangement as early as 1922 during hearings on the first Swing-Johnson bill, and consistently pressed for their separate financial status. Equally important, the bill recognized the principle of a power royalty or tax on power. Arizona and Nevada were each to receive 18.75 percent of the surplus profits of power revenues. Although Arizona may have lost the first major battle over the Colorado River, she could point to several skirmishes along the way that, in some instances, made California's victory less than complete.[41]

As Norton Jr. found his place in the regional agricultural economy in the late 1920s, he concluded that Arizona and California, though locked in a roiling political and legal dispute over the Colorado River, nevertheless maintained a common regional destiny. As then-Congressman Hayden put it in 1926, during debates on the Boulder Canyon legislation: "The West is an entity and there is a growing homogeneity among its people, an identity of interest which makes the region a reality geographically, commercially, industrially, and economically. Injury could not be done to one western state without affecting the others. If the West is to maintain its identity and become a destined factor in the increased wealth, population, and strength of the nation it will be because the large communities and the states of the West continue to recognize the necessity of mutual consideration of the problems peculiar to the West." Unfortunately for the ideal of regional cooperation, however, the Swing-Johnson formula for developing the Lower Basin's water resources appeared to threaten Arizona's economic future. [42]

By the end of the decade of the 1920s, Norton Jr. knew that Arizona played a distinct, if contentious, role in this early chapter of the Colorado River controversy. In its quest for a share of water and power from the river, Arizona's approach to the issue of the decade demonstrated little philosophical consistency. One of the most ardent proponents of states' rights at the outset of the controversy, when events turned against it Arizona would turn to the central government and the Supreme Court. As the struggle for the Colorado wore on in subsequent decades, the river became a fixation in the Arizona body politic. As one writer described the seemingly irresolvable conflict and its impact on Arizonans: "The river has had a traumatic influence in Arizona history; whole careers, even lives have been shaped by it. It has become Arizona's characteristic obsession."[43]

·6·

DEPRESSION TO EMPIRE

◇◇

"I was born in St. Joseph's Hospital in 1929 and of course I was born in April and that fall the economy fell through the bottom. My dad was flat broke."

JOHN R. NORTON III, AT THE LAW OFFICES OF SNELL & WILMER, PHOENIX, ARIZONA, AUGUST 11, 2009.

"C'mon boys, let's get this done."

US SENATOR CARL HAYDEN (D-ARIZONA) GREETING MEMBERS OF THE ARIZONA INTERSTATE STREAM COMMISSION IN HIS WASHINGTON, DC OFFICE JUST PRIOR TO LAUNCHING THE MONUMENTAL *Arizona v. California* SUPREME COURT CASE CONCERNING WATER RIGHTS TO THE COLORADO RIVER SYSTEM AND ITS TRIBUTARIES, AUGUST 13, 1952.

I n September 1929, as the seemingly endless struggle over Colorado River resources continued, the stock market, which since the last half of 1924 had been on its biggest boom in history, began to behave more erratically than at any time in history. John R. Norton Jr., like the majority of Salt River Valley residents, did not have the resources to invest in the market and hardly noticed this sudden variation. But after each downturn, investors and speculators noted, there was a recovery. The vicissitudes of the fluctuating market alarmed few, if any, knowledgeable financial professionals. Then on October 24, 1929—Black Thursday—the beginning of the end arrived. Immediately after the market opened there was a panic to sell, and as speculators sought to sell rather than buy, stock prices fell. So many shares of stock traded hands that day—a record-setting 12,894,650 shares—the ticker fell hours behind actual activity on the floor of the New York Stock Exchange. Fear and confusion took hold. [1]

Although business leaders tried to staunch the bleeding, the Wall Street rupture escalated. Representatives of four of the largest New York banks met in the offices of Thomas W. Lamont of the J. P. Morgan firm at 1:00 p.m. on Black Thursday, and they agreed to pool some of their resources and buy, thereby assuring the financial community that they—the institutional titans of the banking sector—were not concerned about the sudden downturn. At 1:30 p.m., Richard Whitney, vice president of the Exchange, walked to the post where United States Steel was traded and placed an order for ten thousand shares at $205; the highest bid at the moment was $193.50. To close observers it appeared that the bankers had stepped in to peg the market. At first, the action quieted the jitters, and industrials rebounded enough so that there was only a twelve-point loss for the day. On Friday and Saturday trading remained heavy, yet prices remained steady. When asked to comment on the overall economy, President Hoover stated flatly, "The fundamental business of the country, that is production and distribution of commodities, is on a sound and prosperous basis."[2] But despite occasional gains, the market spiraled downward thereafter. By mid-November 1929, the market value of stocks on the Exchange had declined a jaw-dropping 40 percent, and prices continued to slide. A short list of stock prices on the Dow Jones index in September 1929 and January 1933 illustrated vividly the wreckage: thirty industrials fell from $364.90 to $62.70; twenty public utilities dropped from $141.90 to $28; and twenty railroads declined from $182 to $28.10.[3]

In the Salt River Valley and greater Arizona, the effects of the stock market crash were slow to take hold. At the outset Phoenix suffered markedly less than the urban areas in industrial America. Local newspapers barely mentioned the stock market decline and subsequent events.[4] Yet social challenges, unemployment, and structural economic dislocations were on the not-too-distant horizon. Copper mining, the state's most profitable industry, declined precipitously when consumer demand and purchasing power decreased, and the market became oversaturated. Prices dropped from $0.18 per pound in 1929 to $0.056 per pound in 1932, and between 1931 and 1933 most of Arizona's copper mines shut down or cut back dramatically in production. Put another way, Arizona, the nation's leading copper-producing state, saw the value of its overall mining production drop from $155.7 million in 1929 to $14.7 million in 1932. John Norton Jr. was caught in an economic quagmire as farm production decreased from $41.8 million to $13.8 million during the same three-year period. Additionally, livestock production, a passion shared by Norton Jr. and his father, declined from $25.5 million in 1929 to $14.7 million in 1932.[5]

Industrious and innovative, John Norton Jr. had cleared his inherited debts and begun farming on his lands near Buckeye, Arizona, west of Phoenix. In 1929 he planted cantaloupes, looking forward to a profitable harvest. Then a hailstorm demolished his crop, and like so many other Americans at the time, Norton Jr. found himself broke almost overnight; he lost the family home at 351 West Portland that had been purchased from James Cashion in 1915, and lost title to his lands west of Phoenix. But Cashion, a Canadian immigrant and railroad construction superintendent who had arrived in the Salt River Valley via Texas about the turn of the century, had purchased a section of farmland on Highway 80 west of Phoenix, along the Southern Pacific Railroad tracks. In 1910 he founded the town of Cashion, thirteen miles west of Phoenix, and built a grocery to serve the small group of farmers who worked on the produce farms in the area. Besides naming the community after himself, Cashion brought in section houses and dwellings for the Southern Pacific employees and their families. The little community of Cashion, moreover, was near the old Norton lands. The two men arrived at a fortunate agreement for Norton Jr. and his family: Cashion leased Norton Jr. the entire town and allowed him to run the grocery store, which was located on the main thoroughfare from Phoenix to Buckeye, as well. The family lived in the house directly west and across the street from the grocery. Besides operating the store, Norton Jr. ran the post office and leased out the pool hall, restaurant, auto court, and blacksmith's shop, all of which served the needs not just of the community's seventy-five people, but of laborers in the west valley and travelers who were going to and from Yuma and beyond. Later, when Norton III asked his father why he wanted a grocery store, he responded, "Well, I was so damned broke, I didn't have anything. I thought if I ran the grocery store I could snitch enough food out of it to feed you and your mother."[6] In the 1930s the settlement became known as the "Norton Section," which maintained housing for Southern Pacific workers and their families.

Norton Jr. returned to farming in 1931 thanks to some timely backing from Cashion and a few others who knew him for his honesty, integrity, and hard work. In that and the following year, during the depths of the Depression, he forged an existence, growing vegetables and melons on leased land in western Maricopa County around Tolleson and Glendale and enhancing his reputation as an outstanding grower of all types of produce. Financially, he kept his head above water and provided for his wife and child.[7] Like so many of his friends and neighbors, Norton Jr. had "shortened sail, in good nautical fashion, to meet the gale," and as it lessened, it did not hurt him to find himself wasting less, expecting less, needing less.[8]

In 1932, the Arena Company of California sought to expand their produce growing operations into Arizona to increase the volume of their seasonal harvests. The original A. Arena migrated from Italy to New York during the last decade of the nineteenth century and sold produce from a vegetable cart on the streets of lower Manhattan. He realized that the produce he was selling was grown in California, so in 1910 he moved to southern California and started growing fruits and vegetables. Arena and his son A. T., through acquisitiveness and shrewd analysis of regional markets that were expanding with the unprecedented population boom in southern California in the teens and twenties, soon grew into a corporate agricultural giant with headquarters in Los Angeles and farm operations in the Imperial Valley, Lodi, and Merced. The Arenas used refrigerated railroad cars to ship produce to the East Coast, making conditions right for them to add an Arizona undertaking. They grew fruits and vegetables in every season and viewed Arizona as a new agricultural asset—a seasonal diversification—for their enterprise. The family-owned company sent the youngest son, Sylvester, to Arizona to assess the various possibilities and learned of a bright young farmer in western Maricopa County who had a reputation for hard work and who knew the local market, the labor pool, and the nuances of growing and harvesting produce in central Arizona.[9]

In spite of the fact that Norton Jr. was virtually destitute in 1932, the Arena Company offered him 40 percent of the proposed venture if he oversaw all the growing and ran the entire Arizona farming operation. The new entity in Arizona would be called the Arena-Norton Company.[10] Under the terms of the agreement, the Arenas would arrive during the harvest and market the crop. In a short time the Arena-Norton Company was growing lettuce, cauliflower, carrots, broccoli, cabbage, cantaloupes, and honeydew melons— mixed row crops—on nearly four thousand acres. Five years after its creation, the Arena-Norton Company was the largest vegetable growing operation in Arizona, and by World War II the company branched into the east valley, farming the historic Tremaine Ranch acreage, a landmark property between Mesa and Chandler. Norton Jr. ran that operation as well. Roads remained primitive and, in many instances, unpaved throughout the Salt River Valley, and for the better part of a decade and a half, John Jr. spent hours per day in a car traveling from Tolleson to Chandler—almost two hundred miles per day, round trip. As his son John R. Norton III put it: "He supervised the whole thing, and talk about work! He was a working fool . . . and when World War II came along they just coined money, because anyone who could grow crops in that period made a fortune. The war created shortages of everything, and the farmers here [Salt River Valley] got well." The elder Norton spent long hours running the expanding agricultural domain, but in a few short years he had gone from impoverished fruit and vegetable grower to one of the leading produce farmers in the Salt River Valley. [11]

Compared to the previous decade, the last half of the 1930s and the war years were comparatively prosperous for the Norton family. Their more affluent circumstances were the result both of John Jr.'s partnership with the Arenas operation and of the demand for agricultural products created by the war effort. John III recalled the Phoenix family home on West Lewis Street just west of Fifteenth Avenue—near the Arizona State Fairgrounds. For first and second grade, John III attended Franklin School on Seventeenth Avenue and McDowell (then Christy Road) on the southeast corner across from the fairgrounds. Often, he joined his neighborhood friends Cecil and Duane Miller to climb the fence at the fairgrounds, where they watched rodeos and attended fairs. Cecil was a year older and Dwayne a year younger than John III, and the predictable roughhousing usually ended up being two against one. "But it was always me with one of the Millers," Norton III asserted. "The Millers would never team up against me." One memorable scrap involved the three boys wrestling in a pile of manure for the Miller's winter lawn planting. Mrs. Miller stormed out of the house, took off all their clothes, and blasted them clean with a garden hose. On another occasion John III decided that his school day was over after lunch, so he left campus and began walking north along Seventeenth Avenue. School officials quickly called his mother, who promptly jumped in the car, grabbed the truant, and hauled him back to class. [12]

As the economy sputtered and steadied and the Arena-Norton Company grew into a stable and prosperous enterprise, Norton Jr. yearned to indulge his affinity for livestock. The Arena-Norton Company followed the common practice of growing alfalfa, a less profitable crop than others in most cases, in rotation with produce crops, because alfalfa harbors nitrogen-fixing bacteria that enrich the soil. A three- or four-year cycle of alfalfa plantings enabled them to grow vegetables for another eight to ten years. Norton pastured sheep on the company alfalfa fields for ranchers who took their animals to northern Arizona for summer grazing, then wintered them in the warmer climes of the Salt River Valley. With his experience in sheep and cattle, and specifically having worked the 3V Ranch as a teenager, Norton Jr. was well-prepared for ranching in northern Arizona. In 1933, not long after the partnership between Norton and the Arenas had been perfected, Norton Jr. suggested that the company operate their own sheep enterprise, concentrating on Suffolk sheep. The Arenas, who had no experience with sheep, balked at the proposal, but Norton Jr. persisted, and finally, in early 1935, convinced them to purchase the twenty-five-thousand-acre Hudspeth Ranch, about two miles north of Ash Fork in Yavapai County. Norton III reflected on his childhood experiences on the ranch and recalled, "You could see the house from Ash Fork across the little valley, and my dad took the sheep up there in the summertime and brought them down here [Salt River Valley] in the winter. We sheared the sheep and then sold most of the male lambs for lamb chops, and the female lambs would become ewes." The Ash Fork ranch maintained up to ten thousand head of sheep, and Norton Jr. hired Basque shepherds to maintain the flocks. The sheep operation became a profitable accessory to the large produce-growing operation on the west side of Phoenix. Moreover, Norton Jr. had a ready transportation hub for the sheep, since Ash Fork was the main railhead to Phoenix on the Santa Fe Railroad. In the fall he loaded them into railroad cars, shipped them overnight to the Valley, pastured them for the winter, and then put them back on the Santa Fe to Ash Fork for summer pasture. This made-to-order arrangement put less stress on the animals than driving them back and forth on the old sheep trails, which had been the routine in years past. In a short time Norton Jr. became a leader in the sheep industry and was named to the Arizona Wool Growers Association Board of Directors in 1939.[13]

Norton Jr. also raised his own livestock on Hudspeth Ranch. He stayed in the sheep business for about ten years, but later opined that he was a better cattle rancher. His interest in polled (hornless) Hereford cattle started on the Ash Fork ranch. In 1935 he purchased a small herd from the B. A. Packard estate in Cochise County, but since he was primarily a produce grower this proved to be a dalliance. He fattened the cows and calves on the Hudspeth Ranch pastureland and sold the polled bulls to the nearby Campbell Ranch at Ash Fork.

Several distinct memories from time spent on the Hudspeth Ranch remained with Norton III. In 1939, when he was ten years old, the Big Chino Wash, which held the headwaters of the Verde River and ran one-quarter-mile from the house into the Verde River, swelled to flood stage after a heavy rain. Rushing waters threatened the dam that held the stock pond, and young Norton III joined his father, ranch foreman Evan Harer, and neighbors as they hauled countless railroad ties to the pond to reinforce the dam. They saved it, but it was a harrowing experience for the then ten-year-old. Less frightening but perhaps equally vivid in Norton III's memory was the weekly bath he was forced to take every Saturday at 4:30 p.m. He was dunked into a tub of hot water that had been heated to a near-boil on the wood stove, scrubbed with a steel brush, and was "clean for a week." The ranch house didn't have running water, so rainwater funneled from the sheet metal roof into an underground cistern served as the household water supply. It was not uncommon to find dead rats floating in the cistern. The rodents were removed; the water was boiled on the wood stove and was ready for use.[14]

The shortages in agricultural and livestock products during World War II created opportunities for Arizona farmers and ranchers like Norton Jr. that were well beyond expectations. The company joined other produce operators when it moved into offices at the Security Building on Central Avenue in downtown Phoenix. In 1940 the majority of produce companies also centralized their packing sheds along the railroad tracks there. At least six major packing sheds were built along Grand Avenue. Arena-Norton had maintained an aging shed in Tolleson, where ten-year-old John III sold soda to thirsty workers, but rather than renovate the Tolleson packing shed, they joined the centralization movement in Phoenix and built a new shed near Six Points. Melons from the fields on the west side of the valley as well as from the Deer Valley area were brought into the shed, and Arena-Norton produce led the way in harvesting, packing, and shipping. When he was twelve and thirteen years old, Norton III packed cantaloupes at the new shed, although he was not supposed to work until he was sixteen.[15]

By the end of the decade, Norton Jr. had worked his way into an enviable position in Arizona's farming and ranching community. He had accumulated enough leverage and capital by 1939 to acquire the Aubrey Valley Investment Company (AIC) Ranch in Mohave County. Charles "Charlie" Recker, a Mesa, Arizona produce grower and a close friend of Norton Jr., partnered in the enterprise. Situated on a juniper-laden plateau, the AIC sat at an elevation of five thousand feet, about fifteen miles south of Hyde Park on Old Highway 66 between Seligman and Hackberry. Four years later, Norton-Recker purchased the adjacent X Bar One Ranch in Hualapai Valley and combined it with the AIC, making it a spread of 150,000 acres.[16] In 1941 Norton bought out Recker, leaving the Salt River Valley native as sole owner.[17]

As World War II broke out, Norton Jr. continued to acquire ranch lands in northern Arizona. In 1941 he sold the Hudspeth Ranch near Ash Fork and talked the Arenas into acquiring the Double O Ranch from Phil Tovrea Sr. "I don't know why Tovrea wanted it," Norton III recalled, "but my dad talked the Arenas family into having the Arena-Norton Company buy the ranch. My dad thought he had died and gone to heaven when he got that ranch." Besides the land's outstanding cattle ranching qualities, the house on the Double O was considered one of the best ranch houses in northern Arizona. Tovrea liked to entertain his friends from Phoenix and he spent a lot of money on the showplace, which included a basement furnished with Las Vegas slot machines. Both Norton men enjoyed the Double O. Norton III recalled that as a teenager, "I spent my summers up there; usually I'd work for about three weeks in the cantaloupe harvest in June, then I'd go and spend the rest of the summer up there on the back of a horse. I got to be a pretty decent cowboy because I'd spend a lot of time up there." [18]

John III recalled that his father was a "man's man, a hard-working guy who wanted me to amount to something. He made me work in the summers, and he was a bit of a disciplinarian. I'd get up to the Double O in Seligman and start the roundup. But as a kid growing up we would drive through Prescott to go to the ranch, and I knew Prescott had the big Fourth of July rodeo weekend—the famous Prescott Frontier Days Rodeo and Celebration. My dad never let me go; I've never been to the Fourth of July in Prescott." He wondered aloud if his father's discipline in this regard was necessary. That missed right-of-passage notwithstanding, the Norton family affinity for this ranch was palpable and was reflected in the popular Double O brand produce box labels and advertising for J. R. Norton, Company[19]

In 1935 Norton Jr. invested his earnings in residential real estate and built and moved into a beautiful two-story home at 2233 North Ninth Avenue, in the heart of the well-heeled Encanto neighborhood. The two-story colonial-style white brick house with green shutters, designed by prominent Phoenix architect Lester Byron, was constructed for $11,000 in 1936 and sat across fields of cotton bisected by a canal with cottonwood trees clinging to

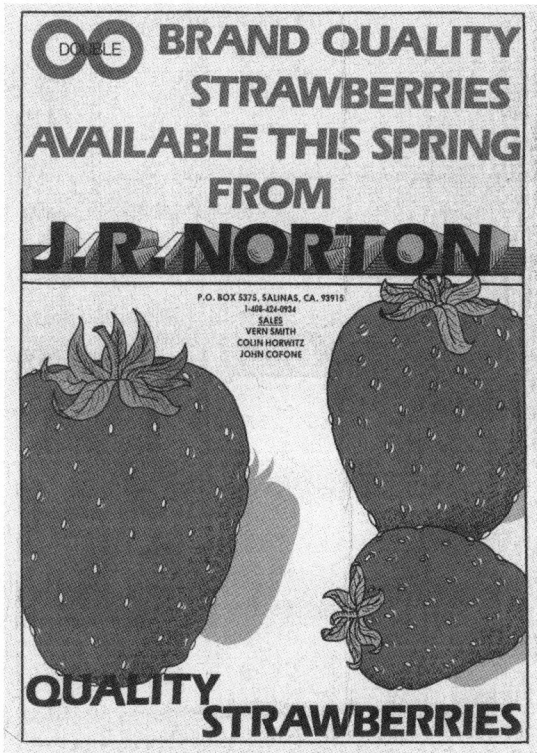

A typical advertisement used in trade journals and newspapers for Double O strawberries.

its banks. As Norton Jr. made more money, he also bought his wife a LaSalle automobile. "It was the last word," according to Norton III, "and my mother thought it was great." This move prompted a change in grammar schools—from Franklin to Kenilworth—which brought John III into contact with the children of the most prominent Phoenix residents. The Encanto neighborhood was convenient too; most shopping was done at a large supermarket at Seventh and McDowell, with a Rexall Drug Store and a gas station and barbershop at the same intersection. As Norton III remembered it, "You could get anything you wanted at that intersection there. I'd ride my bike to Kenilworth School, go down Seventh Avenue and we'd have lunch at Rex's Malt Shop on Third Avenue and Roosevelt." When the proprietor's daughter worked the counter and her father was absent, she made Norton III and his friends free malts and shakes in order to gain favor with the neighborhood boys.

John III kept and raced homing pigeons during these grade school years, and his bedroom window looked out upon the 220 acres of cropland. Between 1936 and 1938 these were transformed into one of Phoenix's signature attractions—Encanto Park. William Hantranft, a millionaire philanthropist and president of the first Phoenix Parks and Recreation Board, had visited California and wanted Phoenix to have a stunning park that would serve as a showpiece for the valley, like Balboa Park in San Diego or Golden Gate Park in San Francisco. He persuaded the Parks and Recreation board to convince the City of Phoenix to purchase the farmland across from the Norton home, which at the time was just outside the city limits. "I saw them turn over the first shovel of dirt," Norton III stated in 2010, at Encanto Park's seventy-fifth anniversary, and "the clubhouse, lagoon, boat dock, band shell, playground, tennis courts, and eighteen-hole golf course took three years to complete."[20]

As Norton Jr.'s fortunes rose and he grew active in various farm and livestock boards in the late 1930s and early 1940s, he also became involved in politics. Like his father, Norton Jr. was a registered Democrat, as were the majority of Arizonans at the time, but Arizona's economic and political environment proved fertile for a conservative brand of Democrat that had earlier become known as a "Pinto Democrat." Conservative approaches to public policy issues were not anathema to "Pintos," and Norton Jr. found that wing of the party more to his liking than the New Dealers and their wearisome meddling in the market economy. In fact, friends urged Norton Jr. to enter politics as a gubernatorial candidate, but he preferred focusing his energies on his farming and ranching enterprises. He exercised his service to community in ways other than seeking elective office, becoming an active spokesman for Arizona agriculture and serving with numerous trade associations. He was president of the Maricopa County Association Farmers in the 1930s and a director of the Western Growers Association, representing the fresh fruit and vegetable industry in Arizona and California. Norton Jr.'s affinity for horse racing, like that of his father, led to his role as one of the incorporators of Turf Paradise. He was a working member of the Arizona Cattleman's Association, the Maricopa County Sheriff's Posse, and the Pioneer Historical Stockmen's Association. Though these activities drew him into the orbit of politics, he never made the final leap into seeking elective office as his father had at the turn of the century.

Like other "business-type farmers," Norton Jr. was moved to support the Republican nominee for president in 1940, Wendell Wilkie, reflecting FDR fatigue and a growing impatience with the seemingly never-ending New Deal. In fact, the thirty-nine-year-old produce grower-activist became state chairman of "Democrats for Wilkie," in the utility executive's upstart campaign against incumbent President Roosevelt, who was seeking an unprecedented third term. For most of the campaign Wilkie was hindered by his basic agreement with Roosevelt's policies. He centered his critiques on the failure of the New Deal to solve the unemployment problem and charged that national defense was not adequate. Wilkie also tried to gain traction with the third term issue, but none of his criticisms evoked a marked response from the voting public. "My dad was full of piss and vinegar," Norton III recollected, "and when Wilkie came to speak in Phoenix during the campaign, he had dinner at our house [that] I will never forget.... We had the big house at Ninth Avenue and Encanto and a very nice formal dining room and a long dining room table, and Wilkie [was] sitting at the head of the table and my dad [was] sitting at the other end. So they told me I should come in and meet Mr. Wilkie, who's going to be the next president."[21] On November 5, 1940, Roosevelt received 27,243,466 popular votes to Wilkie's 22,304,755. FDR won a huge majority in the Electoral College, 449–82. Nevertheless, the plurality was less than any since the extremely tight 1916 election, when Woodrow Wilson squeaked out a victory over Charles Evans Hughes.[22]

Meanwhile, Norton Jr.'s marriage of sixteen years began to unravel. Norton Jr., who ran in what his son described as "that fast produce crowd" and a rancher by nature, grew apart from his more cerebral and contemplative wife. Norton III was twelve years old and remembered that his mother was devastated. "We were in that big house at Ninth Avenue and Encanto," he recalled, "and I remember my dad was in their bedroom and he was packing his suitcase and I wanted to know where he was going." He said he was going on a trip and John III asked how long a trip. Norton Jr. responded, "It's going to be a pretty long trip, son." "I knew what was going on. I was crying," Norton III remembered, "sitting there on the floor while he was packing his bags." By early 1942 Norton Jr.'s marriage to Earlene had ended. Among other things, she received the house in the divorce settlement, but quickly sold it and ultimately moved to a more modest residence at 1545 West Wilshire with her son. Norton III spent much of his early adolescence calming and reassuring his mother and, in his "stumbling teenaged way," took on the emotional re-

"The Arizona Produce Gang." Left to right, Joseph Martori, Ned Leonard, Kemper Marley, Chet Johns, and John R. Norton Jr. (1950).

sponsibilities of an adult in the aftermath of the divorce. As he put it, "This sort of thing matures you because you have to be the steadying influence to a distraught parent; it grew me up a little bit." [23] By early 1943, Norton Jr. had remarried. He wed Texan-born Mary Nell Gambrel, who loved ranching, horses, and cooking for large groups of cowboys and visitors, and shared the love of the cattleman's lifestyle with Norton Jr.

Despite the domestic disruption and John Jr.'s remarriage, the Arena-Norton Company continued to flourish as demand for foodstuffs increased dramatically during World War II. By 1941, Norton III recalled, "my dad is growing four thousand acres of row crops. Now that is a nightmare: lettuce, cauliflower, broccoli, cantaloupes, and honeydews." The main operation remained on the west side of the Salt River Valley in Tolleson and Glendale, but in 1943 the company branched over to the east valley and leased a landmark property, the Tremaine Ranch, between Mesa and Chandler. Norton Jr. ran the entire operation, driving back and forth two hundred miles a day on primitive roads, working twenty-hour days. As a result of this prodigious work ethic the company nearly cornered the market. Norton III claimed his father was a "working fool," but Arena-Norton and other farmers responded to the market forces in admirable fashion.[24]

The transformation from wartime mobilization to postwar prosperity scarcely altered the ascending arc of the Norton-Arena agribusiness enterprise. Norton Jr. worked longer hours and covered more miles than ever as he sought to keep pace with the emerging postwar boom in population and its concomitant demand for fruits and vegetables. But by 1947 Norton Jr. grew weary of the twenty-hour days and he and the Arenas, in amiable fashion, parted ways. They split the company, Norton assuming control of the Double O Ranch while the Arena family maintained proprietorship of the produce operation in Arizona. Norton became the sole owner of the Double O and, as his son claimed, "He loved it." He ranched there into the mid-1950s when his cattle raising interests changed and he focused his energies on Polled Hereford cattle. As a result he sold the Double O Ranch to a group headed by Rex C. Ellsworth, the Safford, Arizona-born owner of the 1955 Kentucky Derby winner, Swaps. As the *Western Livestock Journal* noted, "The Chino, California-based horseman [Ellsworth], his brother Reed, Jockey Willie Shoemaker, and Trainer M. A. Tenney negotiated the purchase of the 100,000–acre Double O." Ellsworth and his partners merged the Double O with the adjoining G. E. "Blondie" Hall Ranch, and the publication reported that the two ranches together were valued at $1,250,000.[25]

His hunger for ranching unabated, in 1956, the same year he sold the Double O, Norton Jr. purchased both the Del Rio Ranch at the north end of Chino Valley in Yavapai County and the JV Ranch in western Maricopa County near Litchfield. His return to farming reflected pragmatic business sensibility. He formed a partnership with Robert "Bob" McElroy and operated as Norton and McElroy until McElroy retired in the late 1960s, leaving Norton Jr. as the sole operator under the auspices of his J. R. Norton Company.

The Chino Valley Ranch property operated under the name of Norton Polled Hereford Ranch.[26] Shortly after Norton Jr. purchased it, an industry publication announced: "John R. Norton, perhaps better known as one of the Salt River Valley's large vegetable operators, has dabbled on and off with Polled Herefords for a long time, though for over a decade his Polled operations have been one of his major interests, with the operation of the herd directed toward a top-deck show herd as well as a fountainhead supply source

135

of Polled bulls for commercial operators. But in 1935 he bought the Polled Hereford cattle of the B. A. Packard estate. Packard was a well-known, even famed Cochise pioneer." At this earlier interval, Norton Jr. was not, as he put it, "primarily a cattleman," so he sold them off. But in 1946 he revisited Polled Herefords, purchasing bulls from Kansas.

At the new Del Rio Ranch, once owned by the Fred Harvey Company—the historic ranch had produced water for Ash Fork and the Santa Fe Railroad, mules for the Grand Canyon, and produce for the chain of Fred Harvey restaurants along the railway—Norton Jr. finally indulged his passion for Polled Herefords. The Del Rio Ranch was comparatively small—approximately thirty-five hundred acres—but one thousand of them were irrigated. In 1960 he showed the top Polled female at the Arizona National Livestock Show. She subsequently was the Grand Champion female at Denver. Norton and his foreman, Carl Taussig, sold Polled bulls that were purchased by many well-known area ranchers, as well as by out-of-state buyers. During this period Norton Jr. also helped start the Polled Hereford Association of Arizona.

According to Lee TePoel, longtime coordinator for the annual Arizona Livestock Show, "John had a good eye and even better judgment when it came to cattle and people. He loved to ride and was always on top of things himself, even though he had a good foreman. His great personality and sense of humor helped a lot of ranchers and farmers through tough times. John had a great understanding of human nature and he could negotiate ranching deals with ease. He would buy an ailing ranch, make a lot of improvements, and sell it as a profitable operation."[27]

Like many of his fellow Arizona agriculturalists, Norton Jr. recognized the possibility of a water shortage in Arizona. The seemingly endless battle royal between Arizona and California over the use and distribution of the Colorado River worried many Arizonans that they could end up on the dry side of the river. According to one leading agricultural pioneer, "We were expanding agriculture for the cotton boom in those years and doing a lot of drilling and pumping; the groundwater tables were declining."[28] Depleting groundwater supplies, the grim outlook for a much-discussed central Arizona diversion project from the Colorado River, and another historic bout with drought and consequent water shortages prompted Norton Jr. to begin looking for agricultural opportunities elsewhere.[29] In effect, he needed to hedge his bets. He and his longtime friend, Chet Johns, concluded that Arizona could not outmaneuver California for an equitable share of Colorado River water.

They grew increasingly intrigued by the Palo Verde Valley around Blythe, California. Significantly, the Palo Verde Irrigation District had the longest-standing rights to water from the Colorado River and thus, at the time, maintained the most secure rights of any place in the American Southwest. In 1950, Johns and Norton Jr. acquired some farmland on the California side of the river, formed the J. R. Norton Company near Blythe, and commenced a relatively small farming operation that, within two decades, developed into a diversified cotton and produce venture with feedlots and a cotton gin. At Blythe, Norton also began the process of reclaiming several thousand acres of raw land in the Palo Verde Valley and at one time fed ten thousand head of cattle at this fertile and productive operation.

During this period of expansion Norton Jr. devoted a significant amount of his energies to lettuce growing. He established a year-round operation growing and harvesting lettuce in several locations: in Marana, where they could plant earlier in the fall and later in the spring; in California's Imperial Valley, where they grew midwinter lettuce; in Blythe in December and March; in Willcox and Hatch, New Mexico, in September and early June; and in Salinas Valley, California, for the summer harvest. J. R. Norton Company used an airplane to facilitate the management of these areas. With over twenty thousand acres in cultivation, Norton Jr. had created a substantial regional agribusiness empire.

The elder Norton also realized the potential for growing lettuce at Aguila, Arizona, twenty miles west of Wickenburg. Fields reclaimed from the desert at an altitude of 2,500 feet were ideal for fall and spring seasons. In 1955, after a series of discussions with fellow grower Keith Walden, he worked out a farming agreement for his recently formed Norton and McElroy Produce Company to level fields in the Aguila area and put in dikes and diversion canals to protect their crops against sudden flooding from occasional desert rains. Eastern markets responded enthusiastically to the high quality of lettuce harvested at the Aguila operation. Moreover, Norton and McElroy harvested more than six hundred cartons of lettuce per acre at Aguila which was then trucked to the partners' westside shed in Glendale for cooling and loading. After that process it was shipped to the coasts. Indeed, by the end of the 1950s, Norton Jr., who had tried to wean himself away from the grind of the produce business, was more successful than ever. Eight other lettuce growers followed Norton and McElroy to Aguila, and two years after they embarked on their venture in the new "Aguila District," close to two thousand

An advertisement for Wilcox lettuce, grown by Norton & McElroy Produce.

acres had been reclaimed from the desert and were under cultivation. Norton and McElroy were lauded for their "procedures, techniques, and [just] plain confidence."[30] During the late 1950s and early 1960s Norton and McElroy launched a clever advertising campaign to draw attention to their Double O, J-B, and Pace Setter brands. The ads had the appearance of newspaper articles with catchy titles such as "Rooms filled with Lettuce Crates Found in Home of Aged Recluse," "New World's Heavyweight Champ Says He Won by Eating Lettuce," and "King of Gambles Turns to New Career as Market Produce Buyer." The advertising garnered much attention and praise among growers.

In the 1960s, Norton Jr. purchased a ranch near Meeker, Colorado, where he again raised Herefords. And at the end of the decade, in 1969, he added the 130,000–acre Buenos Aires Ranch at Sasabe, Arizona, on the Mexican border, to his string of agricultural and ranching properties. He exhibited at the Arizona National Livestock Show, where, in 1970, he had the grand champion pen of Arizona-fed steers. He never tired of this lifestyle, and he spent as much time as possible on his ranches. He raised quarter horses and Santa Gertrudis at Sasabe until the mid-1970s, when, his health beginning to fail, he sold the ranch to Peter Wray and the Victorio Land and Cattle Company. In 1985, the US Fish and Wildlife Service purchased the Buenos Aires Ranch and turned it into a National Wildlife Refuge for the masked bobwhite quail, with plans to bring back the area's native pronghorn. [31]

Throughout his business career, Norton Jr. gave back to the community that provided so much to him. He especially supported ambitious young people, and hired student workers during the acute labor shortage generated by World War II. Over the years he employed hundreds of young men on his ranches, farms, and produce sheds, enabling some to earn college degrees. He donated produce to various organizations like the City of Hope and St. Vincent de Paul Society. He received a service award from the Arizona Boys Ranch, and over the years was recognized by various growers and cattlemen's organizations. He was especially proud of the service award he received from the Boys Clubs of America in 1979. Occasionally he joined forces with other Valley growers, brokers, cooling companies, and railroad and truck lines to donate the proceeds from carloads of lettuce to charities like Easter Seals. He furnished financial backing for Joseph Wood, a young man who wanted a start as a produce shipper and who later became very successful as the J. A. Wood Company in Tolleson. For years he looked forward to the prestigious invitation-only "Rancheros Visitadores" spring trail ride in Santa Barbara, California, which he participated in from 1951-1965. In these ways and many others, Norton Jr. symbolized the best in civic engagement. [32]

In 1963 Norton Jr.'s personal life took another turn when his two–decade marriage to Mary Nell ended. After a two–year hiatus, he married Bonita Kent, widow of Dr. Leo J. Kent of Tucson, in 1965. Then in 1972, at the age of seventy-one, John Jr. was diagnosed with Parkinson's disease. He withdrew to his home in Paradise Valley, where he lived for fifteen years, gradually taking leave of his businesses and placing them in the hands of his son, John III, who since the mid-1950s had been an integral factor in most of his father's affairs. Norton III, who had been living in Blythe, California, with his young family, returned to Phoenix to assist in handling Norton Jr.'s business and related issues. Norton Jr. passed away in Phoenix just short of his eighty-sixth birthday, on May 17, 1987.

John Norton Jr. was thrust prematurely into his agricultural career by the unexpected death of his father in 1923. With ingenuity, sound management, and hard work, his farming and ranching enterprises grew exponentially during the middle decades of the twentieth century. He expanded his enterprises from Arizona into New Mexico and California, and his workforce grew from a handful in the beginning to more than two thousand employees. By the time he passed the company on to his son, John III, it was an agribusiness of the first order. Norton Jr.'s life paralleled that of his father in many ways. Both started with little, and at times lost crops. They persevered and they prospered. They recognized the importance of water for the future of agriculture in the Southwest, and they capitalized on that knowledge and insight. They served their communities well in organizations relating to their vocation. And their characters were above reproach. From the dust of the Depression John R. Norton Jr. had created a sprawling agricultural empire and left a remarkable legacy for his only child, John R. Norton III, to build upon.

·7·

TO THE OTHER SIDE
OF THE RIVER

"*The Colorado River is our greatest resource, and unless we conserve it and get the maximum benefit from it, we can depend upon becoming a sort of vermiform appendix of Los Angeles.*"

ARIZONA GOVERNOR GEORGE W. P. HUNT IN A POLITICAL CIRCULAR ISSUED IN 1923.

"*Why stay here [Arizona]? The place is going to dry up.*"

JOHN R. NORTON III, SEPTEMBER 28, 2010, REFLECTING ON THE ARIZONA WATER SITUATION IN THE LATE 1940S AND EARLY 1950S.

John R. Norton III watched, listened, and learned from his father's life experience. As Norton III recalled in 2003, "He was a good father. . . . And he never shortchanged me on attention or advice or counsel or coaching. He showed me the way. He paved the road for me. . . . He would give me 'holy hell' when he thought I was wrong or wasn't working hard enough. He was a hard worker and he demanded hard work." By the time John III was six or seven years old, prior to his parents' divorce, he spent summers on his father's sheep or cattle ranches in northern Arizona and "just loved it. . . . I thought going to that [sheep] ranch when I was a little kid was like dying and going to heaven." After the divorce he continued to spend summers with Norton Jr. on the X Bar One or Double O. Earlene supported her son's time with his father, and she knew young John was treated well. "He was good to me, he loved me, and he challenged me a lot," Norton III recalled, "and I wouldn't have had the wonderful opportunities as a youth growing up if it

Branding on the Double O, 1948. At left, John R. Norton Jr.; center, with his back to the camera, is John R. Norton III; at right is Boyd Clements.

hadn't been for my time with him on the ranches . . . because I really learned about life. And I wanted to work. I was a kid that liked to work."

When he was fourteen and working on the X Bar One Ranch in the summer of 1943, Norton III was stationed with an old cowboy at a camp six to eight miles south of headquarters. The cowboy worked John III from sun-up to sundown, as the mentor and his student scoured the lower deserts for cattle around Hackberry to take to the high country for summer pasture. "So I spent about two months in that camp with that cowboy, just me," Norton III remembered. "And when you are fourteen years old, you learn to ride and you learn to work, because there wasn't anything else to do, and that cowboy wanted help; he wasn't going to let me get away with anything." In ensuing years, when Norton III worked on the Double O Ranch in the summers, he knew that he and his father had to be the first two out of bed in the morning because they did not want the cowboys standing around waiting on them. And at night, the father-son pair corralled and fed the horses and completed side work until every cowboy was ready for dinner. Such experiences, Norton III noted, "gave me the strength to deal with life." Later, he placed this ethic in succinct terms: "My dad always thought that the boss should work harder than anybody else there, as an example. He should be first out in the morning and last in at night. He was old school in that you couldn't expect men to work unless you set an example. And so that's how I was raised. And boy, I mean, when we were going to be out on horseback at daybreak, I was saddling my horse well before daybreak."[1]

Formal education at North Phoenix High School, where he graduated in 1946, and then college formed another critical part of Norton III's character. His mother preached intellectual rigor and achievement so that John III maintained good grades throughout high school. Earlier, Earlene guided her son to the Phoenix Library at least once a week, a ritual she rarely departed from during his youth. In high school, besides the core subjects, she encouraged John III to take typing and public speaking. Sixteen-year-old Norton III was selected to represent North High in a valley-wide extemporaneous speech contest held at Glendale High School's auditorium in the spring of 1946. There were several topics to consider for this formal speech, so the students had to conceptualize their talking points in the broadest terms. As it turned out, World War II and its aftermath, including discussion of a United Nations, was a front-burner public issue. Woodrow Wilson and the League of Nations failed in the days after World War I, and many pundits and academics argued that refashioning the failed concept was a waste of time. Norton III, during this period, had read an article in *Reader's Digest* about the pros and cons of forming a United Nations. At the contest, each of the extemporaneous speakers was given a slip of paper with a title and subject and allowed to speak for ten minutes. His slip read "United Nations," and he gave a talk that "knocked them cold." Norton III won the contest and took home first prize.

Upon his graduation from North High, he entered Stanford University, his mother's alma mater, for his freshman year. Yet his father's lifestyle and his summer experiences on the ranches lured him away from Stanford, which, at the time, did not offer a major in agriculture. Norton III was compelled to look elsewhere for his college education. In 1947, at the outset of his sophomore year, he transferred to the University of Arizona (UA), and for the next three years studied agriculture, especially animal husbandry, and indulged in a host of elective courses and extracurricular pursuits.

Academically, Norton excelled at Stanford and at UA. He worked with his professors in the College of Agriculture at UA and established a close relationship with Dr. E. B. Stanley, head of the Department of Animal Husband-

John R. Norton III in his high school football uniform, 1946.

ry. He also took several classes from Dr. W. J. Pistor, who taught courses in veterinary science. In his senior year, Norton III served as a grader for Dr. Pistor.

Dr. Bartley Pratt Cardon also played a significant role in his education as well as in his later professional life in agribusiness. John III "enjoyed [Cardon's] animal nutrition classes immensely," learned a great deal from him, thought Cardon was a "fine professor."[2] Cardon developed a reputation as a youthful, well-liked and charismatic professor "who was able to motivate people to do more than they ever thought they could and make them enjoy it." Professor Cardon even provided wise counsel to Norton III's concerned father, who was troubled when his nineteen-year-old son's vocational interests turned less than practical. By this time John III was on the UA Rodeo Team and expressed to his father that he wanted to pursue a life on the rodeo circuit. The increasingly distressed Norton Jr. sought out Dr. Cardon, who stated simply, "Be patient, he'll grow out of it."[3]

At UA Cardon emerged as one of Arizona's leaders in higher education. He became the longtime Dean of the College of Agriculture and Life Sciences, and for a good portion of the mid-twentieth century he devoted his considerable energies to pioneering work in agriculture and resource econom-

The University of Arizona rodeo team, 1950. John R. Norton III is second from right.

ics. He counseled countless students like Norton III as well as government officials and businessmen throughout his varied and distinguished career, instilling the idea that understanding the history, law, and politics surrounding the immutable relationship between water and food production was key to understanding how natural resources relate to families, food supply, markets, imports, and exports.

Norton III, wisely, took elective courses in business, accounting, and corporate finance in the business college, and as he put it, "I'm glad I took those courses; they have helped my career."[4] The course in corporate finance proved especially challenging; the professor "poured it on," but Norton III remembered it as one of the best courses he took in his three years at UA. And in addition to his core courses and electives, two extracurricular activities played important roles for Norton III at UA: rodeo competition and ROTC.

Norton III was a member of the rodeo team at UA and became captain in 1950, his senior year. The training that gave rise to his leadership on the team actually began much earlier than college, on the ranch lands of northern Arizona and during his years of participating in rodeo competitions during the mid-1940s. Norton III took one of his favorite horses, Tracy, down to

John R. Norton III at the Peter Martinez Ranch north of Tucson; below, John on his favorite horse, Tracy (1950).

Tucson and boarded him at Peter Martinez's spread in north Tucson. In the afternoons he practiced and roped calves. Prior to his successful rodeo victory in Flagstaff, Norton III and another collegiate rodeo competitor at UA, Gila Bend's Ralph Narramore, pulled their horses and competed in a rodeo at Hardin-Simmons University in the panhandle region in Texas as well as one at Sul Ross State University in West Texas.[5] On May 9, 1949, during his junior year, John III led UA to the team trophy award in the two-day second annual intercollegiate rodeo at Arizona State College at Flagstaff.[6] According to the *Arizona Republic*, Norton III led the six-man team for the trophy. He placed first in team roping and second in bull riding, and won the saddle and title of all-around champion cowboy.[7] Norton III's rodeo accomplishments thrust him into the limelight; he appeared in Tucson area newspaper advertisements for the Arizona Shirt Company. A dashing photo of him with the tagline "I always buy my shirts at the Arizona Shirt Company because they feature the finest in custom tailoring and stock sizes" appeared in the *Arizona Daily Star* and *Tucson Citizen*.[8] As captain in 1950, Norton III was Rodeo Boss and was responsible for the tenth annual UA rodeo that year. It was a great success, and he again won the trophy saddle as best all-around cowboy at the UA event. Not only did Norton III garner accolades for his work as Rodeo Boss, but he was also awarded a UA Service Blanket for outstanding service to the student body. The UA board of control and student council honored him for his work on the incorporation of the Associated Students of the University of Arizona (ASUA) and for work on a proposed student gas station. He and other campus leaders received the awards just before Easter vacation in 1950.[9]

One other elective activity at UA shaped Norton III's immediate future: the United States Air Force Reserve Officers' Training Corps (ROTC) program. ROTC was designed as a college elective program to focus students on leadership development, problem solving, strategic planning, and professional ethics. For Norton III, at least, these values struck a chord and played a role in his unexpected and brief military career after he graduated with a bachelor of science in agriculture—with distinction—on May 31, 1950.[10]

The 1950 UA graduation ceremonies also honored newly minted ROTC officers like Norton III at the football stadium, one day before diplomas were issued to the entire graduating class. Norton III was one of fifty graduates who were to receive their Second Lieutenant's commissions that day. Arizona Governor Dan Garvey walked down the line of ROTC graduates and con-

Above, bareback riding; below, bull riding; both at the intercollegiate rodeo, 1950.

gratulated them. After the ceremony, Governor Garvey forgot that he had already congratulated the ROTC honorees and proceeded to repeat the act.[11]

On June 25, 1950, a world away and three weeks after Norton III's college graduation, North Korea invaded South Korea, signaling the first military engagement of what historians consider the Cold War. The conflict between South Korea—supported by the United States and its United Nations allies, and North Korea—backed by the Peoples Republic of China with military aid from the Soviet Union, was a proxy war resulting from the division of Korea by World War II Allies at the conclusion of hostilities in the Pacific theater. Japan had ruled the Korean peninsula from 1910 until the end of World War II, when victorious American administrators divided the peninsula along the 38th Parallel, with US troops occupying the southern end and Soviet troops the northern. The failure to hold free elections in the Korean peninsula in 1948 created sharp fissures between the two former allies, and the North established a Communist government. Unification negotiations carried on, but tension intensified. Cross-border skirmishes and raids at the 38th Parallel became commonplace. The situation escalated into open warfare when North Korean forces invaded South Korea in late June 1950. In a short time, the Korean War would summon Norton III to the other side of the world.

Consciously or not, Norton III prepared in other ways for the possibility of military service in Southeast Asia. In the late 1940s, one of his father's produce and cattle ranching partners, Charlie Recker, purchased a private airport south of Tucson. The facility had been used as a military training field during World War II, and after the cessation of hostilities, the US government sold it at auction in 1947. Recker, ever the entrepreneur, turned the airfield into a flying school called the Airborne Incorporated Flying School. World War II veterans, taking advantage of the broad flexibility of the GI Bill, took flying lessons. Norton III, in his junior year at UA, took flying lessons at Recker's flight school as well, and at the end of 1949 he boasted a pilot's license. In 1951, the federal government suspended the portion of the GI bill that paid for flying lessons, so Recker was forced to put the company into bankruptcy and began selling his airplane inventory.[12]

Norton III, realizing that Recker needed to liquidate, bargained for a two-seat, sixty-five horsepower Luscombe trainer that he had trained in two years earlier. He purchased the light, fabric-winged plane for $300. Since it had sat idle for several months, the fabric had begun to peel off the wings.

CLIMATE
plus CITRUS
equals
BEAUTY!

Doris Mae Schaefer, Arizona Citrus Queen, 1950.

John III expressed trepidation about flying his first private plane back to Phoenix, so he contacted a friend with a Cessna from Phoenix and hitched a ride to Tucson. Norton III fired up the plane, took off, and had his friend tail him a bit above and behind. The two flew just over one hundred feet off the ground, and John's friend remained close enough to see if the fabric began to tear. If the wings started to lose their fabric, the spotter would zoom past Norton III and alert him to trouble. The plane, fortunately, remained airworthy, and Norton III flew his first private plane around the state for the next several years. [13]

Immediately after graduation, even as war broke out on the Korean peninsula, Norton III decided to spend the summer and fall of 1950 working as a yard boy for the Henderson Livestock Commission Company in the Los Angeles Union Stockyards. The Union Stockyards, occupying thirty-five acres at 4500 Downey Road, was the largest in the western states. It maintained pens for cattle, sheep, and hogs, all of which Norton III had experience with. From June until December 1950, Norton III moved cattle from one pen to another on their way to the scales, but he learned a great deal during these months in Los Angeles. He recalled, "I worked hard all summer and through the fall trying to learn more about cattle and the livestock business."[14]

As Norton III worked through the transition from graduating college student, ROTC Second Lieutenant, young airplane pilot, and entry-level stockyard laborer, he also married. He met Sioux Falls, South Dakota native Doris Schaefer at the 1948 Valentine's Day dance at Bear Down Gymnasium on the UA campus. She was standing in the Coca–Cola line with a Sigma Chi fraternity brother, and Norton III, who asked to be introduced to the young woman in front of him, did not know that Doris was the reigning Arizona Citrus Queen. She had spent a year traveling and promoting Arizona's citrus industry, visiting Hollywood and, at one juncture in her reign, having lunch with Frank Sinatra. John III was smitten with Doris and later claimed he called her the next week, although in her recollection he called weeks afterward, at the beginning of Easter vacation. As she recounted it, "He said this is John Norton, and I said, who?" Despite the less than auspicious initial foray into courtship, Norton III began dating Doris, who was fascinated with the fact that the native Arizonan was a cowboy and rode and roped in such an effortless and elegant fashion. Doris, a horse lover, also liked the fact that he kept horses in Tucson and that he would "go to any little rodeo that was around there." In fact, when John III practiced with his horse or roped calves, Doris would sit on the fence and watch him. Dating a cowboy was unheard of, even exotic for the South Dakota native, and her mother let her know of her disapproval.[15]

Despite her mother's misgivings, on November 17, 1950, the couple married in Phoenix, and Doris moved into a little one-room apartment in Los Angeles with her new husband. Norton III was making $300 a month. He kept thinking that he would be called into the Air Force for service, and his father believed the same. Nevertheless in December Norton Jr. asked Norton III and Doris to return to Arizona to take over daily operations of his father's vast ranching operation in northern Arizona. For the next two years, John III ran the 101,000-acre Double O Ranch near Seligman, where he had spent so

John and Doris's wedding day, November 4, 1950.

many summers during his youth. Doris recalled, "We had a little house there and I thought it was a wonderful life experience." They had friends; Doris enjoyed taking lunch to the cowboys, and she remembered many of the wonderful characters she met during the roundups.[16] As Norton III explained, "To get a foreman it's usually a guy with a wife, not just a cowboy. So my dad built a little house, just a cracker-box house . . . and the bunkhouse was nearby, and [there was] an orchard, and my dad's house was the big house about a hundred yards away."[17] After two years at the Double O, on January 15, 1953, Norton III was finally called to active duty. He knew he had to serve two years. In many ways Norton III was biding time at the Double O, though it was a valuable experience in many ways. As he explained his call to duty in 2003, "Well, most of the kids were a little luckier than I was and they got called in the first six or eight months after graduation and they got their two years in and over with. I sat out for two-and-a-half years waiting to be called. I got a lot of experience on the ranch and that was fine."[18] Doris and John III left the Double O Ranch in early January 1953, and on the fifteenth of that

Bulldogging at the Williams's, Labor Day 1952.

month he reported to Floyd Bennett Field in Brooklyn, New York.

Later, Norton III paused and laughed at the notion of an Arizona cowboy thrust into Brooklyn, New York at a US Air Force base. It was actually located on an old navy base on the southern tip of Long Island, abutting Long Island Sound and across from the working class community of Rockaway Beach. The young couple rented a small upstairs apartment from a woman named Blanche Colton, and they settled in for Norton III's tour of duty. On his first day at Floyd Bennett, Norton III learned that he was to be the provost marshal—in effect the supply officer and mess officer. He knew nothing about the job description: "I didn't know anything about being a provost marshal," he said. "I didn't know anything about any of it!" [19] Within a week, however, he and a "nice Master Sergeant" who reported to Norton III were directed to move his office from one location to another. Norton III thought nothing of the small task. But a Major came up to him and snapped, "Lieutenant Norton! Attention!"

"Yes, sir," Norton replied, standing at attention.

Calf roping.

"What are you doing?" the Major inquired.

"We are carrying a desk from this office to that office," Norton III explained.

The Major shot back, "Do you realize you are an officer of the United States Air Force? You do not carry desks. We have enlisted men carry desks. I never want to see you do anything like that again. You are a disgrace to the ranks of the Air Force!"

"Yes sir," Norton III responded, in spite of the fact that it was "not the way [he] was raised."

He was indeed taken aback, and Norton III later admitted that for several years after this incident, he had trouble dealing with the military's hierarchical protocol. "I had a great deal of trouble telling people to do things when it would have been just as easy to do it myself, or to tell them to do something difficult and not help them when I could easily have helped them to do it," he reflected. "I think that officer went beyond the pale," he allowed, "but that is part of the military culture."[20] The memory of that incident remained with him throughout his life.

From January to late April 1953, Norton III reported to one side of the base and Doris, who quickly found work as a secretary, went to her job on the other side of the base. In their spare time, the young couple drove their car to Flatbush Avenue, picked up the subway at the "Flatbush Line," and ultimately navigated to Times Square where they would see Broadway shows like "Hello Dolly." Then they would return to Flatbush, pick up their car, and head back to Rockaway Beach. "We really saw New York," John III said later, "and had an enjoyable time." [21]

In late April 1953 Norton III received his orders: he was to report to Travis Air Force Base north of Sacramento, the main staging point for Air Force personnel heading to the Far East.[22] He and Doris loaded the car, drove from New York to Phoenix, and dropped off Doris and their personal belongings at her parents' house there. As Doris remembered this period, "It was difficult when John was in Korea. Living at home again after being married and up on the ranch with such an entirely different life; that was hard. So, we just had our correspondence, our letters going back and forth."[23] At Travis, John III soon received his orders and was sent to US Air Force headquarters, Tachikawa Air Force Base, Japan, just a few miles outside Tokyo. When he assessed his placement, John III assumed he was heading to Korea. But upon arriving at Tachikawa, he was ordered to Iwo Jima, one of the Japanese islands that became infamous during World War II. Norton III was fascinated with the chain of islands in the western Pacific, six hundred to seven hundred miles south of Tokyo, that included Monami Jima, Kita Jima, and Iwo Jima, his station. What was curious to the young Second Lieutenant was the fact that they were part of an active volcanic chain that was gradually rising up from the ocean depths. One time, flying back from Japan, the pilot of his transport plane grew excited and called his military passengers to the front of the plane, "Look," he said pointing at a tiny piece of land jutting up from the ocean—it was not much bigger than a building—"an island is being born."[24]

Norton III arrived at Iwo Jima on the first of May 1953, and according to protocol, was to take over the supply account. Part of the process was that as the incoming officer, he was entitled to take a complete inventory of everything at the base. After signing for them, he would be held accountable for all supplies. He quickly concluded that shortages existed, probably substantial shortages. He was replacing a Lieutenant Robert Hobbie, from the small hamlet of Red Level, Covington County, Alabama, a hard-drinking military man intent on returning to the states since Norton III had arrived to take over for him. With only twenty-two officers and two hundred enlisted men on the island, Norton III would be in charge of supplies for a relatively small contingent of American military men. He speculated that his predecessor, Hobbie, paid scant attention to supply matters on the base, but didn't believe that the Alabaman was the root cause of the problems. Hobbie inherited these supply inconsistencies because he took over just a year prior to Norton III and had signed for the account without bothering to do an inventory, something that often happened. Norton III intended to avoid Hobbie's dilemma, and his careful inventory turned up the shortages he suspected.

The Quonset hut that housed the Officers' Club towered above the rest of the base on a hill that overlooked the enlisted men's huts and other military facilities that lay toward the beach. One night early in his assignment, Lieutenant Norton took a break in the Officers' Club after taking inventory of the steel cots in one of the Quonset huts below the club. He counted eight hundred cots, but that was half of what was supposed to be on the base. He asked where the rest were located, and the response alerted him to potentially nefarious activity: "Well, we will get the rest of them tomorrow." Not comfortable with the equivocal answer, he sauntered down to the group of Quonset huts near the beach and saw men carrying cots from one hut to another. He kept the observation to himself; then the next morning he took inventory again, and the men took him to a second Quonset hut to count the rest of the cots. So Norton III began counting the second set of cots, then said, "I made a mistake in my notes here; I want to look at the Quonset hut we counted yesterday."

"Oh, we can't do that," the enlisted men replied, "We can't go over there, they are doing some maintenance on the building."

"Oh, I don't mind," Norton responded. "I'm heading over there anyway." When he entered the first Quonset hut there were no cots to be found; they had been moved to the second hut so they could be counted twice.[25]

In the Air Force the supply officer is "accountable" and the commanding officer is "responsible" for supplies. Shortages reflect on the commanding officer. When Lieutenant Norton completed the inventory he stated simply: "I am not willing to sign for the count until we make sure shortages are accounted for." This upset everyone, because the Lieutenant and the Colonel in charge could not leave Iwo Jima and head home until the matter was settled. The Air Force sent a team from Tachikawa Air Force Base to conduct an inventory. The Colonel and Lieutenant Hobbie spent five days distracting the inventory team, keeping them intoxicated at the Officers' Club. The team then returned to Japan and covered for the discrepancies.

The Colonel then called Norton III into his office and, as he put it: "Read me the riot act for not signing the account." He said, "Do you realize that if you don't sign for this account, Lieutenant Hobbie can't go back to the States and I'll be here until this thing is cleared up? Do you realize what you are doing?" He proceeded to inform Lieutenant Norton that he would make life unpleasant for him on the island. Norton III was in a tight spot, but he would not budge. "Well, sir," he responded, "I hope we can get the matter straightened up but I can't sign for the account." The Colonel sneered, "You'd better rethink that, Lieutenant Norton." Norton III then wrote a report to Tachikawa Air Force Base headquarters stating that the inventory was not adequate, and in response, another Colonel arrived on Iwo Jima, took a careful inventory, and corroborated Norton III's findings. They found $480,000 in shortages. The ruse had finally been exposed, and a full-scale inquiry and hearings took place at Tachikawa Air Force Base. [26]

The hearings revealed that Lieutenant Norton had initiated correct and appropriate action. A notation was placed in his records that stated he had performed in an exemplary fashion to protect the assets of the US Air Force. The hearings, which took place in July and August 1953, only three months after his arrival in Iwo Jima, corresponded with President Dwight D. Eisenhower's signing of the armistice that ended the war on July 27, 1953. Immediately after the armistice, yet while the hearings on the inventory shortfall continued in Japan, an order came down that there was to be a Reduction in Force (RIF) that would result in 10 percent of the Iwo Jima contingent—two officers and twenty enlisted men—being sent back to the United States. In his short time there, Norton III had befriended the personnel officer, Luther J. Cramer from Pennsylvania. Norton III approached his friend and said, "Luther, this RIF means that two officers are going home and I know one of them is you." Cramer laughed and said, "You got it!" Norton III, not missing a beat, said, "But I want to know who the other is and it damned better be me!" Cramer said, "Okay, Norton, you're on." [27] A few days later, in early September 1953, the orders arrived that Lieutenant Norton and Lieutenant Cramer were rotating back to the United States.

A few days after Norton III, Cramer, and the twenty enlisted men secured their orders, a large DC-3 personnel plane landed at Iwo Jima and the group packed their gear to board. Everyone on the base stood watching, including the Colonel and Lieutenant Hobbie, as the group boarded the aircraft that would take them home. "And if looks could kill," Norton III told an interviewer in 2003, "they would have killed me on the spot, because I came in May and I'm going home in September after all this mess. . . . And Hobbie was supposed to have rotated home in June, right after I took over, and he was still there when I left. . . . I was not a military hero, but I did my duty; I found the supply shortage." [28]

During his time in the Far East Norton witnessed some oddities, observed an impoverished but resilient Japanese people, and began a photography hobby turned into a passionate lifetime commitment. Iwo Jima, for example, was scarcely seven years removed from the horrendous World War II battle that took place in the spring of 1945. Lieutenant Norton claimed, "there was so much ordnance all over the small two-mile-long, one-mile-wide island; tanks, jeeps, planes." The island was covered with the leftover military reminders of the struggle. An entrepreneurial American businessman from the United States bid to clear the scrap metal from the island, and during his four-and-one-half months at Iwo Jima, Norton III observed this private sector initiative. The businessman constructed a camp with three dormitories and mess hall and hired inexpensive Japanese laborers. They loaded the metal on a huge press that flattened the material. The scrap was then loaded on a barge and taken to the Japanese mainland and sold. The process, Norton III noted, ran around the clock, twenty-four hours a day. Incredibly, the American businessman bid $70,000 for the scrap and obviously reaped huge profits for his ingenuity and enterprise.

Iwo Jima was sulfuric and was geologically and seismically active. In the American occupation of the island at the end of World War II, the soldiers dropped grenades into caves where Japanese soldiers hid. With all the sulfur that resulted from steady seismic activity, the bodies in the caves never decayed. While Norton III was stationed there, the medical officer, a captain who was a recent medical school graduate, paid some of the enlisted men to open the caves and provide him with preserved cadavers for surgical practice. It was a rather ghoulish benefit of their unique situation on Iwo Jima.

Norton III also experienced a typhoon. He was told to "get your mattress and lie down on the floor of your Quonset hut." The typhoon hit and he was not prepared for its force and fury. The hut filled with water, so Norton III, along with some of the enlisted men, left their huts and sat outside on the leeward side of the island, though he allowed, "It was not the smartest thing to do." In the end nobody was hurt, and the cleanup took place in uneventful fashion.

He also toured Tokyo during the supply scandal hearings that took place at Tachikawa Air Force Base. He took a few sojourns and was struck by the abject poverty of postwar Japan, where farmers used human waste as fertilizer. He witnessed a vivid example of the negative consequences of eating raw fish, and thereafter never ate sushi. The yen was pegged at 325 to the dollar, so he purchased a set of Noritake china, paying the equivalent $5.50 for eight place settings, plus an $8 charge to ship it back to the United States. When he learned of a first-rate furrier in Tokyo, he bought a fur cape for Doris for $300, and when it arrived, Doris had it appraised. It was worth ten times that.

Inspired by Iwo Jima's natural environment—oceans framed by azure skies—Norton also secured a brownie camera and began photographing his surroundings. He enjoyed the process, the results, and the artistic endeavor. Though his comparatively primitive camera yielded less-than-compelling images, a few decades later Norton III would return to photography with a passion.

Norton III's return trip to the states proved unforgettable. After the plane left Iwo Jima for Tachikawa Air Force Base, he waited for a few days and then was ordered to the USS *Corregidor*, an aircraft carrier stationed in Yokohama. The craft was one of the minicarriers that Henry Kaiser constructed during World War II, and the military was now using it to ship troops home after the Korean War. Norton III boarded with fifteen hundred Air Force personnel in late September 1953, and as everyone settled in, Norton III noticed that the ship listed slightly. One slender, pale-looking Air Force officer became seasick before they left the dock. For the entire fifteen-day trip from Yokohama to San Francisco, Norton III was witness to the retching of the unfortunate officer. When the *Corregidor* sailed through a storm in the northern Pacific, it moved at a glacial one knot per hour for two days. Soon everyone on board joined the seasick officer in his suffering. In spite of the challenging return trip—maybe in part because of it—Norton III recalled that looking at the bottom of the Golden Gate Bridge was "one of the most thrilling moments of my entire life."

Doris drove from Phoenix to San Francisco and greeted John III. They celebrated, returned to Phoenix, and Norton III received welcome orders: he was assigned to Luke Air Force Base and made the property accounting officer, just below the base supply officer in seniority. John III and Doris rented a duplex in Phoenix, and he drove to Luke Air Force Base each day. "I was a supply officer, just like I had been an executive of some company, and it wasn't a very military-type lifestyle." Though the Korean War had ended, Luke hummed with activity through 1954. "We were in full training mode, and it was pretty educational for me to have to . . . account for all the property at such an active Air Force base," he explained in 2003. "We had jet engines that had to be overhauled all the time, and we had a constant stream of trucks . . . flatbeds. . . . And one of the things that stuck in my memory was the fact that the summer was so hot that we'd only get four or five landings out of a set of tires [before] they would burn up." Another odd aspect of Luke Air Force Base during Norton III's tenure was the fact that it did not have a jail. They had to use a regular barracks building that was not fortified, so it was not difficult to escape. To make sure that no one fled for freedom, the prisoners had to wake up every night on the hour. They had to go outside, count, then head back to bed.[29]

Norton III learned a great deal during his stateside service, but when he left the Air Force in 1955 he allowed that he wanted to return to agriculture. He did not want to return to the Double O Ranch, however, because working as a foreman on the ranch offered little hope for long-term advancement. He weighed the romance of the range and the $400 a month salary against the prospect of supporting his wife and his first-born son, Michael Robert Norton, born at St. Joseph's Hospital in Phoenix on August 3, 1954. John III arrived at a different conclusion than he had a few years earlier. The notion of cattle ranching as a vocation and a main source of income seemed impractical: "I saw it as kind of a dead-end job."

Meanwhile he earned wages with the spring 1955 lettuce harvest and listened to his father and Chet Johns, both of whom had grown increasingly attracted to the Palo Verde Valley, on the California side of the Colorado River in Blythe. For decades, the Colorado River Basin states had wrangled over the use and distribution of the waters of that interstate and international stream, but one issue was certain: the lands near and around Blythe had maintained first priority in California's rights to water supplied by the river, while Arizona's rights were yet to be determined. For most of his boyhood and teenage years, Norton III heard his father and the produce-growing community express fear, even hysteria, about the water situation in Arizona. Cotton growers in Pinal County, especially, voiced this concern as they pumped more and more groundwater each year. As Norton III recalled the ongoing discussion in the late 1940s and early 1950s, the concensus was "Why stay here? The place is going to dry up. . . . There was no Central Arizona Project—it was a pipe dream—and the huge lawsuit with California didn't start until 1952. And California is taking more and more water and those of us in Arizona said, 'We will never get any of that water.' The place was drying up, the wells were going down, there was no future here for agriculturalists. So, what did I do? I went to Blythe, which had the first rights."

Blythe lies in the Palo Verde Valley, an agricultural area along the west bank of the Colorado River, 110 miles west of Phoenix and 120 miles east of San Bernardino. The history of the Palo Verde Valley and the Palo Verde Irrigation District is interwoven with the history of the Colorado River. The boundary between Arizona and California, the Colorado River also forms the eastern and southern boundaries of the Palo Verde Irrigation District. The valley is relatively level; nine miles wide, thirty miles long, and ranging in elevation from about 300 feet above sea level at the northern end to 220 feet above at the southern end. Thanks to years of Colorado River flooding, the soils are alluvial and range in texture from grainy clays to silty loams to light sandy soil, with the predominant soil being sandy loam. Moreover, the entire valley is undergirded with permeable sand at shallow depths. Finally, its long, hot growing season and mild winters are ideal for year-round farming, and permit the growing of crops not suitable for production in other areas. [30]

When Norton III moved to the district, the community of Blythe had a population of roughtly five thousand. It was named after Thomas Blythe, a gold prospector who was responsible for establishing primary water rights in the region. [31] Blythe, a native of England born under the name Thomas Mold in 1832, was drawn to the gold strikes of mid-nineteenth century California, settled in San Francisco, and grew wealthy through real estate investments. In turn, he invested in agricultural and mining schemes. He acquired 40,000 acres in and around what is now the city center of Blythe and made the first of several filings for Colorado River water in California when he recorded a request for 95,000 miners inches (1.905 cubic feet per second) in what was then the county seat at San Diego on July 17, 1877. The ultimate request, which totaled 190,000 miner's inches, was filed for "agricultural, mining, manufacturing, domestic, and commercial purposes." These longstanding rights, Norton III and his father knew, held unquestioned priority in the effort to acquire Colorado River water to put to beneficial use. [32]

Blythe hired a manager, George S. Irish, and an engineer, Oliver P. Calloway, to assist in developing the area. Calloway, an engineer who had carved out the first stage road from San Diego to Yuma and who played a small role in the development of San Diego Harbor, saw the potential of the Palo Verde Valley, and looked for a financial angel. He found him in Blythe. Irish and Calloway proceeded to develop the first irrigation structure—a slough—from a swamp area in the northern part of the valley, and gave it the vainglorious name Oliver Lake. It irrigated some pasture lands and small agricultural plots. Shortly thereafter a canal was built from the river at Block Point, a location one mile north of present-day Diversion Dam, to the slough, but it cost a great deal of money—$82,000. Blythe made only two visits to his development—once in 1875, then again in November 1882—before his death of a heart attack in San Francisco in 1883. Upon his death, Blythe's creditors froze his assets. Irish and Miller sold off the inventory, turned the funds over to the estate, and left the area. Blythe, a bachelor, left no will, and his estate was tied up in court for decades. No agricultural development took place in the Palo Verde Valley until the early twentieth century. [33]

Arizonans Frank Murphy, a Prescott, Arizona entrepreneur involved in the cattle industry, and his partner Ed Williams visited the valley in 1904 and saw its potential for cattle raising and agriculture. They convinced W. A. Hobson from Ventura County to join forces with them and form the Palo Verde Land and Water Company. This entity, which became the parent company of the Mutual Water Company, purchased the Blythe estate. In payment for the intake, headworks, and related infrastructure, the land company assumed the right to sell the water stock which was issued for the entire valley. They also hired W. F. Holt, who had experience developing the nearby Imperial Irrigation District, as their general manager. On August 8, 1916, the California Southern Railroad laid down a track to Blythe from a desert station called Blythe Junction, which was later renamed Rice Junction to honor G. W. Rice, the superintendent of the railroad.

The first engineer of the new water company, C. K. Clark, built the entity's first intake structure and located the canals, most of which remain today. The Colorado River inflicted almost predictable annual flood damage. This yearly problem required the formation of the Palo Verde Joint Levee District, which was organized in 1917. The new entity sold bonds to build a levee to protect the valley against floodwaters. In 1921, when it became apparent that a drainage system was needed, the Palo Verde Drainage District was organized and sold bonds for construction of those facilities. At this time valley water users realized that it was more practical—perhaps even necessary—to have one entity administer irrigation and drainage functions. They petitioned the California State Legislature to take action, and in 1923 the Palo Verde Irrigation District Act was passed.[34] The District was then organized and commenced operation in 1925, taking over the assets of the three predecessor organizations: the Palo Verde Mutual Water Company, the Palo Verde Joint Levee Committee, and the Palo Verde Drainage District.

Three decades after the official formation of the Palo Verde Irrigation District, in the spring of 1955, Norton III and his father visited Blythe and located what they considered an attractive parcel of land with water rights. A Los Angeles area homebuilder owned the acreage, which supported a well-developed farm with good housing and a sound infrastructure. With bank financing, some creative planning, and some financial help from his father, Norton III scraped together a down payment and bought the farm, taking up residence in the Palo Verde Valley on July 1, 1955. Doris, now pregnant with their second child, remained in Phoenix during this period, while Norton III camped out in a local hotel-apartment facility during the period of transition. John Phelps Norton, his second son, was born on November 6, 1955, and soon the family, with the help of a GI veterans' loan, moved into a 1,100-square-foot stucco tract house in the small Rodeo Gardens subdivision slightly north of downtown Blythe. The $11,000 purchase price worked out to a monthly mortgage payment of $101, and by the end of 1955, Norton III and his young, growing family were residents of Blythe, California. The three-bedroom home had air conditioning, and Norton III had never lived in an air-conditioned home prior to this time. It was a move that altered his personal and professional life. He described his reasoning in moving to this comparatively isolated agricultural community in the simplest terms: "I was attracted by that water." From Phoenix to Korea to Blythe, Norton III had, in a remarkably action-packed decade, attended and graduated college, served his country in the Korean conflict, married and fathered two children, and taken the bold step of embarking on an agricultural career on the western banks of the most litigated river in the world. Though an Arizona cowboy by heart, he was, in the fall of 1955, a California farmer.

·8·

AN EXPATRIATE'S
◇◇
DILEMMA:
◇◇◇◇◇◇◇◇◇◇◇◇◇◇◇◇◇◇◇◇◇◇◇◇◇◇◇

Arizona v. California

Arizona had a long history of political and legal setbacks as California sought to entrench their lion's share of the Colorado River's resources. Norton III and other Arizona agriculturalists were hedging
their bets when they ventured into such areas of California as the Palo Verde
Valley and the Imperial Irrigation District, which held prior rights to Colorado River water. In many ways the Colorado, and the legal issues surrounding
it, shaped the nature and direction of Norton III's agribusiness. In taking
steps to enter the world of irrigated agriculture, Norton faced several contra-
dictory and challenging situations for a native-born Arizonan. For one thing,
new Palo Verde Irrigation District (PVID) members like Norton III and his
Arizona confreres found themselves having to send money to the California
Colorado River Board—the entity that supported California's water interests
against Arizona and other states. Of course the PVID supported California
in the decades-long water rights struggles, but Norton III and his fellow Arizona expatriates silently rooted for out-manned and out-financed Arizona.
John R. Norton Farms began operations as a partnership between Norton
Jr. and Norton III, and as the latter expressed it, "I farmed aggressively in the
Blythe area, and over a period of years I acquired more farmland and was very
active in producing lettuce, melons, cotton, alfalfa, wheat, and barley in what
I considered a typical irrigated farming operation."[1]

For nearly three decades prior to Norton III's move, Arizona fulminated
against California "water greed" in a situation that produced endless frustration for those farmers and ranchers involved in irrigated agriculture in Arizona. Norton III had settled in Blythe after careful observation of the water sit-

Advertisement for John R. Norton Farms in Blythe, California.

uation in Arizona: he and other Arizonans who took like action had formed strong and unwavering opinions about Arizona's future water supplies and believed the prognosis was dismal. A host of factors—including political demagoguery, lack of a coherent and focused water policy among Arizona's elected officials, and, on the other side, a unified California leadership—hindered Arizona at every turn, and eventually Arizonans like Norton III justifiably moved to the water. Norton relocated to Blythe just as the two states, after three decades of struggle, took up their arguments before the United States Supreme Court.

Politics and the Courts: Toward *Arizona v. California*

Indeed, from the time of his birth in 1929 until his fateful move to PVID in 1955, Norton III lived mostly in Arizona and experienced the years of fruitless efforts, or, as he called it "a pipe dream" for the development of some kind of reclamation project that would transfer water from the Colorado to central Arizona. His father witnessed California and the federal government derail Arizona's hapless state's rights arguments in the 1920s, and the end of that decade, which saw the Boulder Canyon Project Act passed, brought more pain to Arizona's water litigants and agriculturalists. On January 15, 1929, just three weeks after President Calvin Coolidge signed the Boulder Canyon Project Act into law, Arizona's Colorado River Commission authorized At-

165

torney General K. Berry Peterson to file suit in the US Supreme Court seeking to halt its implementation. On October 30, 1930, the State of Arizona formally submitted its petition. Arizona contended that the Colorado River Compact (1922), the Boulder Canyon Project Act (1928), and subsequent water and power contracts attendant to federal agreements with California agencies were unconstitutional. Strangely, Arizona looked outside the state for legal expertise and hired Idaho attorney John Pinkham Gray, reputedly "one of the outstanding lawyers in the West," to assist its stable of attorneys. Attorney Gray avowed, "California and the federal government were not going to override [sic] roughshod Arizona sovereign rights."[2] The states' rights approach proved legally anemic, and on May 13, 1931, in an eight-to-one decision, the Court rejected Arizona's suit without prejudice, asserting that the Boulder Canyon Project Act "represented a valid exercise of congressional power." The Court, per Justice Louis Brandeis, held, *inter alia*, that the Compact and the Project Act were constitutional, that the river was a navigable stream, and that the Secretary could construct the dam authorized by Section 1 of the Project Act. One month later, workers began excavating the diversion tunnels for the dam at Black Canyon: after much congressional wrangling, the dam was named "Hoover Dam."[3]

A host of unresolved problems frustrated the state's leaders, and the confusion and factionalism that characterized Arizona water policy during the 1920s continued through the 1930s and the New Deal era. Several powerful political and agricultural organizations continued to excoriate California's water and power greed while at the same time reaffirming their commitment to state-owned and -operated reclamation programs. States-rights Democrats, for the most part, dominated Arizona's political affairs during the period, promising to "save" the Colorado for Arizona, yet offering little new or progressive in water resource development policy. Aging Governor George W. P. Hunt, who served his last term between 1931 and 1933, and his successors in the thirties—Dr. Benjamin B. Moeur, (1933-1937), Rawleigh C. Stanford (1937-1939), and Robert T. Jones (1939-1941)—all Democrats—maintained proprietary views concerning the Colorado, and opposed Arizona entering into the Colorado River Compact. In effect, Arizona experienced the expansion and benefits of federal reclamation during the New Deal in theory, but not in practice.[4]

During these years California began enjoying the benefits of the Boulder Canyon Project Act. Although slowed by the Depression, construction of

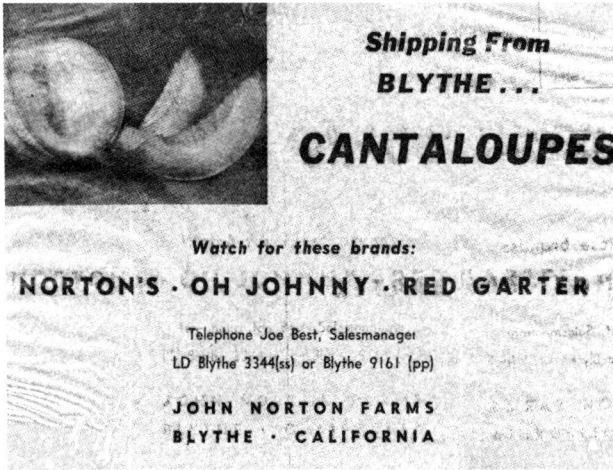

The Norton farms in Blythe produced some of the region's best cantaloupes.

Hoover Dam was completed in 1935. The following year hydroelectric power reached the southern California coastal plain. Beginning in 1939, the Metropolitan Water District (MWD) of Southern California, much to the alarm of Arizonans, commenced delivery of Colorado River water to its customers. In 1942 Imperial Valley residents saw water delivered by the All-American Canal. As a result of these developments, southern California and the Imperial Valley prospered, attracting millions of newcomers in subsequent decades. Not surprisingly, Arizonans looked warily to their west and vented their frustrations in a variety of ways—including another interstate suit. [5]

On February 14, 1934, in its second petition to the US Supreme Court, Arizona Governor Benjamin Baker "B. B." Moeur and his administration asked to perpetuate testimony intended for use in future legal action. In this unusual legal maneuver, the Arizonans, once again, hoped to gain what they had lost in their opposition to the Colorado River Compact and the Boulder Canyon Project Act. State officials wanted a specific amount of mainstream water for future use, and exclusive rights to Gila River water. Ironically, Arizona based its longstanding claims on testimony given during negotiations surrounding the drafting of the compact. Arizona's junior senator, Carl Hayden, predicted accurately the outcome of the suit: since Arizona had not ratified the Colorado River Compact, it could not base its claims on that document. Indeed, in a unanimous ruling on May 21, 1934, the US Supreme Court rejected Arizona's claim, ruling that the introduction of the proposed testimony was not relevant because, among other things, Arizona had refused to ratify the compact.[6] As the Supreme Court considered its second *Arizona v. California* case, Arizona leaders took dramatic steps to express their extreme displeasure with California and the federal government.[7] In 1932, the Department of Interior announced that it intended to contract with the Metropolitan Water District of Southern California (MWD) for the construction of a water storage and power dam near Parker, Arizona. When news reached Arizona, Governor Moeur informed federal officials that he opposed further development on the river until Arizona's rights were clearly defined. His opposition was based upon three fundamental issues: though Arizona was not a signatory, he wanted to protect Arizona's water rights as promulgated in the compact; he had been advised that construction of Parker Dam would not commence until a satisfactory resolution to Arizona's water claims was resolved; and finally he believed that construction of Parker Dam was illegal. In February 1933 Moeur wrote outgoing Secretary of the Interior Ray Lyman Wilbur, stating flatly that the proposed diversion dam could not be placed on the river without Arizona's consent, and advised that the state would take action in opposition to construction.[8]

Despite this warning and others, the Interior Department entered into a contract with MWD for the construction of Parker Dam. Late in the fall of 1934, as MWD employees began construction of a bridge between Arizona and California, Governor Moeur acted on his threat "to repel any invasion or threatened invasion of the sovereignty of the State of Arizona." On November 10, 1934, Moeur declared marshal law on the construction site and ordered a unit of the National Guard to occupy the area. In addition to the 101 men sent to Parker, an "Arizona Navy," composed of the riverboats *Julia B* and *Nellie Jo*, joined the guardsmen in the "war" with California. Offering to diffuse the highly charged situation, an unhappy Secretary of the Interior Harold Ickes ordered work suspended until the dispute had been resolved. After months of fruitless negotiation, the federal government, on June 14, 1935, filed suit in the Supreme Court to enjoin Arizona from interfering with the construction of Parker Dam. The Court, however, ruled against the federal government and held that the consent of Congress was necessary before the dam could be constructed. In effect, the Court asserted that the US government had failed to show that construction had been authorized, and therefore no grounds existed for granting the injunction. Arizona leaders, however, had little time to rejoice: shortly after the ruling, the Seventy-fourth Congress, by Act of August 30, 1935, authorized construction of Parker Dam.[9]

In November 1935, Arizona filed another bill of complaint against California, Colorado, Nevada, New Mexico, Utah, and Wyoming asking for a judicial apportionment of the unappropriated water of the Colorado River. The Court, per Justice Harlan Stone, denied the petition on the grounds that the United States was an indispensable party. Specifically left undecided was the question of whether an equitable division of the unappropriated water of the Colorado River could be decreed even if the United States *were* a party in the suit. Between 1930 and 1936 Arizona had been a party in three Supreme Court cases as they pertained to the Colorado River.

Arizona's congressional delegation, meanwhile, found itself at odds with state leaders. They questioned a dogmatic adherence to a "states' rights" water policy, and disavowed Governor Moeur's histrionics concerning the calling out of the National Guard. Such tactics, though colorful, helped Arizona congressmen little in gaining favorable federal compromises. Moreover, in the context of the nation's economic crisis, it seemed foolhardy to expend state funds for the purpose of halting construction of Parker Dam. While the celebrated miniature war with California made exciting press, sold newspapers, and focused attention on their perceived aggressors, the days of "saving" the Colorado for Arizona were over, and alternative approaches needed to be explored.[10]

With the election of Sidney P. Osborn to the governorship in 1940, Arizona underwent a water policy revolution at the state level. Once a states' rights advocate and ardent opponent of the regionalism inherent in the Colorado River Compact, Osborn informed the electorate that he now favored the interstate agreement. He told Arizona lawmakers that with the passage of the Boulder Canyon Project Act of 1928, the era of philosophizing and theorizing about the river had ended: "Whatever our previous opinions about the best place or plan, we can only recognize that decisions have been made and the dam constructed."

Moreover, the series of legal calamities in the 1930s combined with a massive drought in the 1940s to create a water and power crisis in the state. A pending water treaty with Mexico, and California's announced plans to increase annual use of mainstream water by 2 million acre-feet, posed serious new threats to future Arizona water supplies. Thus on February 9, 1944, in a special session called specifically to deal with the Colorado River, the legislature passed a bill authorizing a water delivery contract with the secretary of the interior that provided the annual delivery of 2.8 million acre feet of mainstream water, plus one-half of "any excess or surplus . . . for use in Arizona . . . under the compact." Soon thereafter, on February 24, 1944, Osborn signed the bill that ratified the compact, thereby ending over two decades of controversy within Arizona. These actions, importantly, enabled Arizona to fight its reclamation battles within rather than outside of federally approved guidelines.

California planned to increase its annual use by 2 million acre-feet through a proposed project at Pilot Knob. Even more disturbing for Arizonans, the Pilot Knob Project included selling additional water to Mexico at $1 per acre-foot. Additionally, the drought prompted Arizona farmers to pump groundwater in unprecedented quantities. By 1940 most of the state's water storage reservoirs were nearly empty. In fact, hydroelectric power generation dwindled to the point that, in 1939, state leaders faced a severe power shortage. In response to the pleas of Governor Jones and other lawmakers, Senator Hayden contacted the interior department and the Bureau of Reclamation officials, who hastily constructed a transmission line on wood poles from Parker Dam. Thus Hoover Dam power, through a connecting link at Parker Dam, brought much-needed electricity to the Salt River Valley in 1940-1941. The irony of the situation was not lost on close observers, who realized that the decision to accept Hoover power represented a sharp reversal of Arizona policy. The drought and its immediate consequences—overdrafting of groundwater supplies and power shortages—forced state leaders to distance themselves from states' rights policies.

Other forces drove Arizona leaders to reevaluate their positions. An unprecedented population influx taxed water supplies and forced officials to take a more scientific approach to solving the problems of an arid environment. In the two decades preceding 1940, Arizona's population had grown 67 percent, to approximately 500,000. By 1945, 200,000 more people had moved to the state. Significantly, most of this growth was concentrated in and around Phoenix and Tucson, prefiguring Arizona's urban growth in subsequent decades. [11]

A final—yet major—factor convinced Arizonans that they must chart an alternative course. The issue of Mexican rights to the Colorado River had been one that Arizona leaders hoped they could address after they resolved their differences with neighboring states. On February 3, 1944, however, State Department officials completed arduous, complex negotiations with their Mexican counterparts culminating in an agreement which promised annual delivery to Mexico of 1.5 million acre-feet of water from the Colorado River. For once, it seemed Arizona and California agreed on an issue: both states believed the amount excessive. Nevertheless US negotiators, motivated by the Good Neighbor Policy and the recent close cooperation between the two countries in the global conflict, favored the generous allocation. And in view of the pressing international situation, the Roosevelt administration hesitated to force Mexico to accept a lesser amount. [12] The implications of the Mexican Treaty were abundantly clear. Unless Arizona took steps to put mainstream water to beneficial use within the state, California and Mexico could claim a prior right to Arizona's claimed share. During hearings on the Mexican treaty, Bureau of Reclamation engineers underscored the point when they reported that water supply figures were significantly lower than previously believed.[13]

In October 1943, with all these factors in mind and in his typically discreet fashion, Arizona's senior Senator Carl Hayden secured Congressional appropriations for the US Bureau of Reclamation to conduct an inventory. The inventory would identify irrigation and multiple-use projects in the lower Colorado River Basin that could be made ready for construction when servicemen returned home at the end of World War II. Capitol Hill observers, preoccupied with winning the war, hardly noticed this innocuous legislation. After all, Hayden had spent his long career in Washington supporting the bureau's affairs, and few lawmakers questioned his judgment in reclamation matters.[14]

To close observers, however, Arizona's tacit cooperation with the Bureau in conducting a survey punctuated the dramatic shift in the state's stance regarding the Colorado River. Beyond signaling Arizona's apparent rapprochement with the federal government and the other basin states who had signed the Colorado River Compact of 1922, the inventory held profound significance for the development of the Southwest during the second half of the twentieth century and the early years of the twenty-first. It marked the first meaningful step in Arizona's long quest to divert its claimed share of Colorado River water to the burgeoning heart of the state.

No wonder long-time Senator Hayden felt a great sense of accomplishment in 1944, when Bureau of Reclamation officials announced the results of their inventory. Paradise Valley, north of Phoenix, and the Gila River Valley, east of Yuma, were likely candidates for postwar projects. Bureau engineers also recommended building a water storage and power dam at Bridge Canyon, above Hoover Dam and near the southern boundary of Grand Canyon National Park. Viewed as the key structure in the bureau's plans for Arizona, the proposed Bridge Canyon Dam would regulate silt build-up and provide 75,000 kilowatts of additional electric power. Although cast in the most general terms, this "working paper," as reclamation officials called it, gave hope to those who dreamed of finally bringing water to central Arizona.[15]

Hayden quickly took advantage of the new political climate, convincing the Senate Irrigation and Reclamation committee to make a complete study of the need for irrigation and electric power development in his home state. On July 31, 1944, a subcommittee convened five days of hearings in Arizona to discuss the bureau's recently completed preliminary reports on importing Colorado River water and to survey the state's various irrigation needs.

Arizona Governor Osborn meanwhile continued his efforts at water policy reform and institutional reorganization. He created two new agencies of government: the Arizona Power Authority (APA) and the Arizona Interstate Stream Commission (AISC). The APA became the agency with the authority to bargain for, take, and receive electric power from the waters of the Colorado River mainstream. With the advice and consent of the Senate, the governor appointed its five members. With the ratification of the Colorado River Compact, the Arizona Colorado River Commission passed out of existence, and in its place, Osborn formed AISC to protect the state's interest and prosecute its claims to Colorado River water before Congress and the courts. Again with the advice and consent of the Senate, the governor appointed its seven members. Significantly, from the date of its formation, AISC focused its energies on gaining approval of the Central Arizona Project.

In the context of the effort to secure CAP authorization, local business leaders formed a private, nonprofit association, the Central Arizona Project Association (CAPA) to support the efforts of the state's congressmen. Organized on July 1, 1946, CAPA was composed of agricultural, business, professional, and industrial people who saw diversion of Colorado River water as fundamental to the future of Arizona's economy. The association raised money, provided research and legal assistance, lobbied for, and publicized the project. CAPA also worked closely with AISC, lending its services and personnel. Finally, CAPA provided links among federal government, state government, and private sector individuals involved in issues related to CAP. [16]

During the first set of Senate hearings on the CAP concept, E. B. Debler, the Bureau of Reclamation's director of planning, presented various plans for what he called the "Central Arizona Diversion." Debler explained that the investigations commissioned by Senator Hayden's committee on postwar planning had located three possible routes for the proposed diversion of approximately 2 million acre-feet of water. The Marble Gorge Plan, the most expensive and technically complex project, would require seven years to construct at an estimated cost of $487 million. The Bridge Canyon Dam Plan, the most widely discussed proposal, would take six years to construct at a cost of $325 million. The Parker Pump Plan, the least expensive of the three alternatives, could be completed in three years at a comparatively modest cost of $134 million. [17]

A noteworthy feature of the hearings was the introduction of Tucson and Pima County into the emerging CAP concept. During the third day of testimony, Tucson City Manager Phil Martin announced that the city and county were interested in obtaining a share of Arizona's allotment of Colorado River water. He presented the senators with petitions from the City of Tucson and Cortaro Farms Company requesting an allocation of 70,000 acre-feet annually for domestic and agricultural uses. [18]

Shortly after the conclusion of the Arizona hearings, Hayden directed funds to the Bureau of Reclamation for the purpose of commencing full-scale feasibility studies for a central Arizona water delivery system. In the context of postwar planning and reconversion to a peacetime economy, he quietly funneled funds to the bureau for the Central Arizona Project.

The earliest CAP planning studies focused on selecting the proper route for diverting Colorado River water into central Arizona, and by February 1947, with surprisingly little controversy, federal and state officials arrived at a general agreement on the Parker Pump Plan. Secretary of the Interior J. A. Krug signed the final feasibility report on February 8, 1948, and justified construction of CAP because it was essentially a rescue project designed to avert serious disruptions in Arizona's predominantly agricultural economy. Beyond that, Tucson, which obtained its water from an overdeveloped and rapidly shrinking groundwater basin, desperately needed additional supplies of surface water so that it could survive and develop. Krug also cited the unprecedented postwar migration of people to the Southwest, which imposed an urgent need for additional hydroelectric power in Arizona, southern California, southern Utah, and southern Nevada. CAP would transform the Colorado River into a truly multipurpose natural resource that would replace depleted groundwater, create hydroelectric power for a developing region, provide supplemental water to lands currently in production but not adequately irrigated, and increase Tucson's domestic water supply. Citing the varied and pressing need for additional water in Arizona, Secretary Krug recommended that "the Central Arizona Project be authorized for construction, operation, and maintenance by the Secretary of the Interior under the general plan set forth herein."[19]

Three issues needed resolution before any further action took place. First of all, the lingering water allocation question between Arizona and California required a final answer. California contended that the annual flow of the Gila River, estimated at one million acre-feet, should be included in Arizona's allotment of Colorado River water. Arizona, of course, countered that the Gila, including water put to beneficial use under the San Carlos Reclamation Project and Coolidge Dam, was exempt from allocation formulas. As proof, Arizona cited the Colorado River Compact and the Boulder Canyon Project Act of 1928, neither of which deducted the Gila River flow. Krug noted that if the controversy were resolved in favor of Arizona, the Interior Department could move quickly to implement CAP plans. At the same time, he hinted that if California prevailed, there might not be enough water allocated to Arizona to justify construction of CAP.[20] Arizona also needed to adopt a groundwater control law that effectively limited the average annual withdrawal from groundwater basins within and reasonably tributary to the areas served by CAP. Finally, the state had to organize an improvement district to help repay construction costs and oversee local management of the project.[21]

Meanwhile, Hayden and Ernest McFarland initiated congressional action on CAP. On June 18, 1946, McFarland introduced SR 2346—the Bridge Canyon Project Bill—to the Seventy-Ninth Congress. Because McFarland's legislation had been drafted and introduced prior to completion of the comprehensive feasibility study, Congress refused to take action on the early version of a CAP bill, citing the lack of sufficient technical data to conduct hearings. Hayden and McFarland introduced an identical bill (SR 433) at the opening of the Eightieth Congress on January 19, 1947. It met the same fate as its predecessor. Despite their failure, these early drafts of CAP bills signaled Arizona's legislative intentions.[22] Hearings, votes, political maneuverings, public relations campaigns, and struggle against an obstinate and formidable California commenced anew. No small number of parliamentary maneuvers could override California's superiority in numbers, especially in the House of Representatives. Arizona senators Hayden and McFarland campaigned within Arizona for CAP. They declared that the project was "the future of Arizona," and that California was the obstacle to the enactment of desired legislation. In a 1950 radio address, Senator Hayden discounted California's repeated claims that CAP was economically infeasible, asserting, "By compact and by contract there is sufficient water in the Colorado River belonging to Arizona to provide an adequate supply for Arizona's agricultural and domestic needs." Also, this phase of the controversy differed in distinct ways from the earlier debates over the Swing-Johnson bills and the Boulder Canyon Project Act of 1928. During the earlier confrontation, the people of Arizona were divided and expended much energy fighting each other; the CAP effort saw the state united.[23]

Local and state politicians, business leaders, and the general public actively supported the legislation. Governor Dan Garvey proclaimed a "CAP Week" while the CAPA and AISC churned out press releases and literature geared to educating the public. And, on June 2, 1949, the Senate Interior and Insular Affairs Committee reported out S 75, yet another CAP bill, with a "Do Pass" recommendation. While Arizona's senators had to wait for the second session of the Eighty-first Congress to convene, this bill held promise. After much more astute political maneuvering in the Senate Rules and Appropriations Committees, Hayden and McFarland watched confidently as the Senate passed S 75 on February 21, 1950, by a vote of 55-28.[24]

The political battle shifted to the House of Representatives, and California prepared to respond to the Arizona victory in the Senate. Their congressmen trained their sights on the able Arizona congressman, John Murdock, a former history professor at the Arizona State Teachers College (later renamed Arizona State University), who ranked second in seniority on the Public Lands Committee. California banked on the strategy of delaying action on the bill for several years, until the House delegation from California increased. Preliminary reapportionment figures indicated that in the 1952 federal elections California's representations in the House would grow from twenty-three to thirty members. It would then have exactly as many representatives as all other reclamation states combined, prompting the two most outspoken critics of CAP, California Congressmen Claire Engle and Norris Poulson, a former mayor of Los Angeles, to suggest that California could then exercise veto power in important water matters.[25]

Murdock knew he faced an uphill battle in the House. He was Chairman of the Subcommittee on Irrigation and Reclamation of the Public Lands Committee, but three Californians—Poulson, Engle, and Richard Welch— sat on the subcommittee and were all outspoken opponents of CAP. The *Phoenix Gazette* lamented, "With no Hayden in the House to steer the project, and a solid block of California representatives to stop it, the CAP had practically no chance of success." During the Eighty-First Congress, California introduced twenty-three separate Colorado River bills and referred the entire question of water rights to the Judiciary Committee, thus outflanking the Arizona delegation of Murdock and Harold "Porky" Patten.

Close observers knew the delay tactic was masterminded by Northcutt "Mike" Ely, the determined California lawyer who had spent virtually his entire legal career on issues related to water law. Ironically, Ely's father, Sims Ely, was a former editor and publisher of the *Arizona Republic* and wrote a widely read book on Arizona's colorful past, *The Lost Dutchman Mine*.[26] The younger Ely, who was born in Phoenix on September 14, 1903, was in every sense a contemporary of Norton Jr. He attended Stanford University and Stanford Law School, graduating from the latter in 1926. When fellow Stanford graduate Herbert Hoover assumed the presidency of the United States in 1929, Ely was appointed executive assistant to the Secretary of the Interior, Dr. Ray Lyman Wilbur, whose remarkable overlapping careers also included that of physician, Dean of the Stanford Medical School, President of the American Medical Association, and President of Stanford University. During his four-year stint at Interior, Ely became intimately involved in reclamation policy, negotiating Hoover Dam water and power contracts. He maintained a ringside seat during the all-important debates surrounding the Boulder Canyon Project legislation. During the 1930s and 1940s Ely became one of the nation's most prominent experts on water issues and mineral rights and became the leading legal adviser and litigator for California in Colorado River matters.[27]

Despite Arizona Congressman Murdock's efforts, the House counterpart to S 75 never emerged from the Public Lands Committee. As the second session of the Eighty-First Congress adjourned in December 1950, CAP legislation had passed the Senate, languished in the Public Lands Committee of the House of Representatives, and brought forth again the unresolved question of rights to the Colorado River.[28]

In January 1951, the Eighty-Second Congress convened, and Arizona's congressional delegation reintroduced CAP bills in both houses of Congress. House legislation was introduced by Congressmen Murdock and Patten, HR 1500 and HR 1501, identical CAP bills. Close observers voiced mild optimism because Murdock had risen to Chairman of the House Interior and Insular Affairs Committee. That apparent advantage notwithstanding, California and its supporters still outnumbered Murdock and his allies on the committee. On February 21, 1951, Murdock called to order the first of twenty-three sessions on CAP legislation. The committee rehashed previous testimony, introduced new reasons to support their respective positions, but made little progress. Always, the issue of water rights remained. Finally, California's representatives moved to break the stalemate and in mid-April, Congressmen Sam Yorty, Clair Engle, and Norris Poulson asked fellow member John Saylor of Pennsylvania to offer a preferential motion to postpone further consideration of the bill until the water rights issue had been adjudicated in the Supreme Court or the ever-elusive agreement among the lower basin states had been made. [30]

On the morning of April 18, 1951, Chairman Murdock, sensing he was losing control of his committee, reluctantly recognized Congressman Saylor. The Pennsylvanian said: "I move you, Mr. Chairman, that HR 1500 and HR 1501 be postponed until such time as the use of the water of the lower Colorado River Basin is either adjudicated, or a binding mutual agreement as to the use of the water is reached by the states of the lower Colorado River Basin." The motion was quickly seconded and by a vote of 16-8, the committee shelved the bill. A stunned Arizona Governor Howard Pyle, who was waiting in an anteroom preparing to address the committee in behalf of CAP, nevertheless requested to make a statement. His brief, impromptu fulmination reflected the sentiments of CAP supporters. "I think this is one of the most depressing moments of my life. . . ." he said. "A delaying action on the part of California has been the thing they have aspired to most of all." [31]

Representative Engle of California termed the developments "a signal victory for California," yet Arizona senators Hayden and McFarland refused to abandon their cause in the Eighty-Second Congress. They reintroduced S 75 in the Senate in a last-ditch effort, though this version, significantly, carried an article that provided for the adjudication of the water rights issue in the Supreme Court. On May 29, 1951, the Senate, for the second time within a year, held a floor debate on a CAP bill. On that day, Senator Hayden delivered one of the longest addresses of his career. He explained and interpreted for the Senate the recent events that transpired in the House of Representatives and questioned California's "unjustified objection" to what was due water-poor Arizona. He reminded California's two senators, William F. Knowland and Richard Nixon, how he had helped in securing funding for their state's Central Valley Project. This exceedingly rare floor speech from the so-called "Silent Senator" prefigured future developments.

Section 12 of the 1951 CAP Senate bill contained a provision for a Supreme Court determination of water rights. Senators Joseph C. O'Mahoney of Wyoming and Eugene Milliken of Colorado, distinguished lawyers and Upper Basin supporters of the bill, drafted that section. Echoing Governor Pyle's lament, Hayden told the Senate that California wanted "delay, and more delay; delay for many years to come in hope that in the meantime more people will go to southern California and thereby a greater need for the water will be built up." Section 12 of S 75, therefore, addressed directly the intent of the Saylor motion in the House, which had the effect of withholding any appropriation for CAP until the Supreme Court decided that enough water existed for the project. If California succeeded in killing CAP, Hayden warned, it could conceivably apply the same tactics to hinder development in the upper states. On June 5, 1951, the Senate, despite persuasive opposition arguments presented by senators Nixon and Knowland, voted 50-28 to authorize the $788 million for CAP. This gave the bill new life in the House, where six weeks earlier it had been left for dead.[32] Unfortunately for Arizona and Congressman Murdock, whose political career hung in the balance, hopes that the House would reconsider the legislation never materialized. The House adjourned without further deliberation, and Arizona appeared to have lost yet another major legislative battle with California over Colorado River water.

Twice then, in 1950 and 1951, Arizona secured passage of a CAP bill in the Senate only to see it derailed in the House. California Congressman Norris Poulson aptly described his state's perspective on this politically charged phase of the fight over the Colorado: "The only sure way to be reelected in California is to oppose the Central Arizona Project."[33] Indeed Arizona had lost the overall congressional battle, but Section 12 of S 75, which called for a new effort in the Supreme Court, held promise for Arizona as political leaders calculated their next move. Ultimately, the legislative battles over CAP provided a catalyst for addressing the sharp and longstanding differences between Arizona and California. Unable to forge a compromise acceptable to both parties, in 1952 Arizona leaders surprised many when they took assertive action and put forward the mosaic of contested issues to the Supreme Court. As several students of *Arizona v. California* (1963) have noted, the case was one of the most complex and fiercely contested in the history of the high court.

Arizona Governor Howard Pyle, whose background as a war correspondent in World War II and owner of KFAD Radio (later KTAR) in Phoenix, knew how to utilize the print and electronic media to full advantage and on Wednesday, August 13, 1952, his flair for the dramatic and his considerable political communications skills were on full display. On that day Arizona filed suit against California in what Pyle called "a new chapter in a fight for life by the whole state" and he added that Arizona asked the Supreme Court to take jurisdiction in the battle with California on use of Colorado River water. J. H. "Hub" Moeur, Chief Counsel for the Arizona Interstate Stream Commission (AISC) and son of the former Arizona governor who declared war on California, filed a fifty-four-page motion and complaint with the high tribunal, charging that failure to settle the longstanding dispute threatened the state's economy with destruction. The action was prepared by Moeur, Perry M. Ling, Special Counsel to AISC, Burr Sutter, Special Counsel to AISC, Alexander B. Baker, Chief Assistant Attorney General, and Fred O. Wilson, Arizona Attorney General.[34]

The filing addressed the primary reason for Norton III's bold move to Blythe in 1955. It declared Arizona's groundwater supply could no longer support its 725,000 irrigated and ranch acres, and unless additional water was obtained, over 30 percent of that cultivated acreage would be lost. The only source of water to prevent such a catastrophe, the complaint stated, was the mainstream of the Colorado River, and it asked that the Supreme Court take jurisdiction in the dispute. It requested further that the Court hold that Arizona was entitled to 2,800,000 acre-feet of mainstream Colorado River water annually, as set forth in the Boulder Canyon Project Act of 1928, and that California was limited in perpetuity to 4.4 million acre-feet of water annually as provided in the California Limitation Act of 1929. In Washington, where the filing took place at 2:50 p.m. that day, Arizona Senator Hayden welcomed a small group of somber-faced Arizonans to his office. After a brief exchange of pleasantries, he stuck a battered white straw hat on his bald head, strode to the door and beckoned, "Come on boys, let's get this done." Together, the group walked to the US Supreme Court building, and Moeur filed the bill of complaint against California. After witnessing the filing, the then-seventy-four-year-old Hayden issued a short statement to the press. "I believe this action," he told those gathered at the steps of the Supreme Court, "will make possible the settlement of a most serious controversy which is delaying the development of the Colorado River basin. . . . If the Californians are sincere in their oft-repeated demands for court action," he added, "then they will welcome the opportunity to present their side of the case." With that, Arizona launched the monumental *Arizona v. California* Supreme Court case.

California's response to what newspapers described as a "surprise action" was immediate and negative. Their representatives in Washington, unable to secure a copy of Arizona's motion before the close of the clerk's office on that Wednesday, nevertheless blasted the suit. John Terrell, lobbyist for the Colorado River Board of California, called the filing "part of a plot to get Senator Ernest McFarland and Representative John R. Murdock reelected to Congress." He stated further, "I think . . . it was part of a program designed by Interior Secretary Oscar Chapman. . . . McFarland and Murdock are both in serious political trouble in Arizona. The administration is making a serious effort to rescue them." Though Terrell's political instincts regarding the fate of McFarland and Murdock were on the mark—both lost their reelection bids to Barry M. Goldwater and John J. Rhodes respectively—his analysis of the motivation for the suit was not.

The comments and actions of Arizona's junior senator, Ernest McFarland, an attorney, former county judge, and longtime friend of Mark Wilmer's, accurately framed the suit filing and infuriated California.Indeed California's water leaders had singled out McFarland as the nefarious broker from a God-forsaken state who had worked a deal with the Truman administration in furthering Arizona's interest. An August 3, 1952 *Los Angeles Times* editorial stated that the purported court filing was "cooked up . . . in a backroom of the Democratic National Convention. It was designed to aid the reelection of Senator McFarland of Arizona. It was put into execution by Secretary of the Interior Oscar Chapman with all the finesse of a butcher boy pole axing a steer in the Chicago stockyards, the aroma of which it appropriately bears."

McFarland, the Democratic Senate Majority leader, was indeed in a pitched political battle for survival with former department store scion and Phoenix City Councilman Barry Goldwater, the Republican nominee for the Senate. Senator McFarland took advantage of the situation to deliver his message. "This was not the first time California interests have made charges against me," McFarland allowed, "but I want the people of Arizona to know that this action was carefully planned months in advance—not in Chicago but in my office with Senator Hayden, Congressmen Murdock and [Harold] Patten, and representatives of the Arizona Interstate Stream Commission. . . . I went over our complaint with our attorneys. It is carefully drawn and in my judgment states a cause of action. This is the reason Representatives Engle and Poulson are protesting so vigorously."

And, in a prophetic afterthought, he added that "There is no question but what California, from their statements made now and earlier, will fight our securing funds for the Central Arizona Project, even after our rights are adjudicated. Even Senator Nixon has said in the past that he would make an all-out fight to defeat what he calls our 'steal' of Colorado River water. This suit is just one step: we will have to take other action in Congress to build this project." [35]

As she had done in Congress, California commenced a campaign of judicial delay. A blizzard of motions and filings postponed the start of the proceedings, while the number and complexity of the issues raised prompted the Court to appoint a special master to hear arguments. After the first few years of judicial jousting, but prior to formal hearings, the case entered an important preliminary phase where California desired to have the four upper basin states, New Mexico, Utah, Wyoming, and Colorado, enjoined in the suit, reasoning that these states, too, had a pertinent interest in the outcome. As Arizonans knew by this time, this procedure could threaten to extend the case by many years. California could continue drawing water from the mainstream and Arizona could not move forward on CAP. [36]

At this stage, incredibly, the sitting governor of Arizona officially entered the proceedings. Those opposed to the joinder motion prevailed upon Governor Ernest McFarland to become involved in the pending hearings because of his "position, prestige, and experience." Former Senator McFarland, who two years after his upset loss to Barry Goldwater had revitalized his political career with a gubernatorial election victory over incumbent Howard Pyle in 1954, needed little persuasion to become directly involved in the lawsuit. Earlier, representatives from the upper basin states and Arizona held a strategy session at the old Hay-Adams hotel in Washington, where they determined that McFarland should make a final summary and a "personal" plea to the Supreme Court. [37]

McFarland traveled to Washington with then-*Phoenix Gazette* reporter Ben Avery; the two had struck up a cordial relationship and often caravanned together. In fact, the governor would customarily request permission from publisher Eugene Pulliam to have Avery accompany him on trips. Pulliam, the inveterate Republican promoter, doubtlessly obtained inside "Democratic" information from Avery, who, it turns out, was a longtime "conservative" Democrat. [38]

Avery related McFarland's careful preparation for his appearance before the Supreme Court. He stayed up until 4:00 a.m. pacing the room the night before, meticulously organizing, checking, and rechecking his thoughts before committing them to paper and then, to memory, and much to the horror of future archivists and historians, casting crumpled pages in the wastebasket. McFarland felt it was important to appear without notes or brief.

In his argument Governor McFarland noted that three years had already been spent on this issue and predicted that another fifteen would be added if the four additional states became involved, during which time "the economy of Arizona would be ruined." He then related the legislative history of CAP, highlighting California's litany of delaying tactics, and suggested that this "was more of the same." [39]

The hearing was held on Thursday, December 8, 1955, and all expressed great surprise with the speed of the Supreme Court's decision. The following Monday, December 12, 1955, the Court ruled in a 5-3 decision (Earl Warren abstaining) in favor of Arizona. "The motion of California to join States of Colorado and Wyoming as parties to this cause is denied," the decision read in part, and "the motion to join Utah and New Mexico as parties is granted only to the extent of their interest in Lower Basin waters." This unprecedented appearance by a sitting governor and the victory suggested to some a turning point, as well as a newly configured political and legal approach to the case. California's governor, Goodwin Knight, expressed great bitterness and uttered ominously that the "tea leaves had turned," though much more lay ahead. [40]

At this stage of the case, two years into an eleven-year Supreme Court odyssey, Norton III was engaged, intellectually, at least, in the legal and political battle over the rights to regional water resources. Also, the twenty-six-year-old co-owner of John Norton Farms had evolved politically, and this evolution mirrored developments in Arizona more precisely than those in California. In effect, he represented a three-generation transformation that

Doris Schaefer Norton, Etta Wright Norton, and John R. Norton Jr. (mid-1950s).

in many ways reflected the political history of the American Southwest. This transformation swept Democrats Ernest McFarland and John Murdock from the US Senate and House, respectively, and heralded the rise of Senator Barry Goldwater and Congressman John J. Rhodes.[41] Though the dizzying machinations and complex legal and political posturing in the struggle for legal rights to the Colorado River may have played a small part in Norton III's political departure from Democratic family tradition, more was afoot.[42] It was manifested in the Republican surprises in Arizona in 1952 as well as in President Dwight D. Eisenhower's sweeping victory over Adlai Stevenson: in Arizona, at least, there was a tectonic shift in political ideology. Norton III would, in the future, see his increasingly conservative views play a more direct and relevant role in the growth and development of his home state of Arizona, and in time he found ways to influence the state's changing political culture.[43]

When Norton III announced to his family that he had registered as a Republican on his twenty-first birthday, members of his mother's family were aghast, wondering if he had suffered some kind of physical or emotional trauma. But in the Norton trilogy as it moved from John R. Norton Sr., the states' rights Democrat; to John R. Norton Jr., a Pinto Democrat (a conservative Democrat with probusiness leanings) who supported Wendell Wilkie in the 1940 presidential election; to John R. Norton III, a young, conservative Republican when he registered to vote in 1950, the transformation mirrored the state of Arizona's political evolution from its territorial period to mid-century and beyond. In a broader sense, this fundamental shift—the increasingly conservative posture in the West overall and in Arizona in particular—contrasted markedly with the late-nineteenth century and early twentieth-century political culture in the world of the first John R. Norton, who lived in a political environment of radical politics, unionism, and socialism.[44] When Congressman Carl Hayden, a Democratic political rival of John R. Norton, entered the House of Representatives in 1912, Arizona embraced unorthodox political bromides, Populist insurgency, and radical political ideas like women's suffrage and allowing voters to legislate directly. The first Norton supported these progressive reforms. According to political historian Paul Kleppner, Arizona reflected broader trends and changing patterns of electoral behavior in the American West. He characterized this transformation as "the radical west becomes conservative." Certainly, by the early 1950s, both Norton III and the majority of his fellow Arizonans had begun identifying with conservative Republicanism, entrepreneurial capitalism, and traditional values. Goldwater versus McFarland and Rhodes versus Murdock, as well as Governor Howard Pyle's victory over Democrat Anna Frohmiller in the 1950 governor's race—all captured vividly the radical to conservative drift that had taken place in the region's electoral politics and political culture. Norton III, as he demonstrated with his move to Blythe, was also ahead of the curve in his assessment of Arizona's predicament. At the same time he continued to support Arizona, in spirit, in its half-century quest for its rightful share of the Colorado River. [45]

·9·

WE MUST NOT BE
INDECISIVE LEST
WE BE INEFFECTIVE

"Going to the Western Growers Association meetings, well, they started promoting him as a candidate to be Deputy Secretary of Agriculture . . . and it appealed to him right away."

DORIS SCHAEFER NORTON, JUNE 6, 2011, PARADISE VALLEY, ARIZONA.

I n Blythe Norton III commenced a routine of hard work, long hours, and endless travel that extended through one decade and into the next until he relocated his center of operations to Phoenix in 1972. It was a period of exponential expansion, both in land use and crop diversification. From the original eight-hundred-acre John Norton Farms in PVID he quickly added to his land base, whether through purchase or leasing agreements, and at the same time diversified the variety of crops from his benchmark work in lettuce growing, harvesting, and shipping. As he often said, "My business philosophy is based on diversification," and though lettuce "kept him hopping" during his first four years in Blythe (his father focused on most of the Arizona-based holdings during this period) he eventually broadened the circle of crops he grew and even started a cattle feeding operation in California, and later, in Arizona. He described the process in Blythe: "Over a period of years I acquired more farmland, and was very active in producing lettuce, melons, cotton, alfalfa, wheat, and barley, and a variety of other crops typical of a desert, irrigated farming operation."

By the time Norton III took over his father's agricultural interests in 1972, he and his growing number of employees harvested crops or raised and sold livestock in Palo Verde Valley, Imperial Valley (Brawley), San Joaquin Valley, various locations in Arizona, southern New Mexico (Hatch), Colorado, and Nevada.[1] After 1972 all products from these were marketed under the J. R. Norton Company and its wholly-owned subsidiaries like Garin Company, Superior Berries, and San Tan Ranches, among others. In 1982 the company celebrated its first big strawberry harvest in Salinas, California, where Norton III had, over twenty years earlier, purchased acreage to harvest lettuce in his expanding agribusiness. It now reached throughout the Pacific Southwest.[2]

In order to keep abreast of his region-wide interests, in 1958 he purchased a Cessna single-engine plane, because, he noted, "We were harvesting lettuce in Blythe, and also the Imperial Valley, and Aguila, west of Wickenburg, and Willcox, and in the Phoenix area, and my dad was anxious for me to help him with some of the operations in these other areas."[3] He added that "My father had always been in the lettuce production and harvesting business, and so when I started this farming operation in Blythe, it was only natural that we'd grow lettuce there, so we had a full desert operation." Norton Jr. grew lettuce in Phoenix and Aguila, which is at 2,500 feet above sea level and a little bit cooler than Phoenix, so the crop could be harvested earlier in the fall and later in the spring. And in Imperial Valley, which was ready for harvest in mid-winter—January and February—with Blythe following in March, the Nortons were never idle. When they expanded to Willcox, which lay at 4,000 feet, they could start in September and go until early June. Being able to fly his own small plane enabled Norton III to gain essential intimate knowledge of these disparate growing and harvesting operations. Though his base of operations remained in Blythe for seventeen years, he built hours of flight time in the routine visits to other operations. Meanwhile he kept growing the business and buying more land in Blythe as he could afford it, and in a colorful description of his business acumen he stated, "I kept myself broke because if I'd make a dollar, I'd borrow ten against it and spend twenty to buy a piece of land."[4]

Alfalfa, which had played a central role in his grandfather's and father's agricultural pursuits, played a significant role in the Blythe operation. Norton III grew alfalfa for the same reason his progenitors did—to rotate with vegetables and reinvigorate the soil with nitrogen-affixing bacteria. Norton III practiced a three- or four-year alfalfa cycle at John Norton Farms, and in Blythe, he discovered that he could grow vegetables for another eight to ten years. But alfalfa was problematic. "If you pinch your pennies," Norton III avowed, "you can make a dime or two on alfalfa, but not much." In the late 1950s and through the 1960s John Norton Farms harvested a lot of alfalfa, baled it, and shipped it west to the Los Angeles basin in Covina, Riverside, and up to Chino. "When I was farming in Blythe, it was solid dairies from Pasadena to the pass that goes into the desert from that point into Los Angeles.... It was all dairies, though I think most of them have moved up into the southern San Joaquin; so we shipped a lot of alfalfa out of Blythe when I was there."[5]

He purchased the Cessna in time for the delivery of his third child, Melanie, who was born February 20, 1959. Norton III hoped that Doris could somehow detect Melanie's arrival in advance so she could be born, as he and his two sons were, at St. Joseph's Hospital in Phoenix. On cue, Doris woke up in the middle of the night on February 19, and said, "Uh-oh, I think we are going to be having a baby pretty soon." So Norton III packed up essentials, drove to the field where the Cessna sat ready to depart, and flew Doris to Phoenix. It was a beautiful moonlit night, the couple had a safe, uneventful trip, and the next day, Melanie Norton entered the world.[6]

Melanie and her brothers, Michael and John, attended elementary and part of high school in the Blythe public school system. All participated in 4H Club, with the boys raising steers and Melanie, lambs. They experienced great success in 4H and, of course, as had their predecessors, they became accomplished riders. Melanie, especially, took to the riding and won numerous competitions. Blythe, with its agrarian lifestyle, reminded Norton III of his upbringing in Arizona in the 1930s.

Seasonal farm labor formed another significant challenge for John Norton Farms; Norton III employed numerous Mexican nationals under the well-known Bracero Program, and Public Law 78. Under that law, actually a corollary to the original agreement between the United States and Mexico, the employer contracted for the number of workers needed for the harvest season. The Bracero Program named for the Spanish term that meant "strong-arm" (literally, "one who works with his arms"), was actually a series of laws and agreements originally initiated by an exchange of diplomatic notes between the United States and Mexico in August 1942. The content of the program, promulgated when President Franklin Roosevelt met with Mexican President Manuel Avila Camacho in Monterey, centered on the importation

The Norton family in 1964:
from left, Johnny, John, Melanie, Doris, and Michael.

of temporary contract laborers from Mexico to the United States. During World War II, Congress responded to growers' worries about a shortage of agricultural workers by approving the temporary entry of migrants from impoverished rural areas in Mexico. The Bracero Program became the largest guest worker program in US history, employing more than four million Mexican workers over its twenty-two year history (1942-1964). The program was controversial; some argued that the low wages at which migrants were willing to work threatened the jobs of domestic farm workers. But the "Good Neighbor Policy" that characterized US-Mexico relations during that era suited the wartime labor needs of the United States. After the expiration of the initial agreement in 1947, the agricultural program was continued under a variety of agreements and corollaries based on past practices. Although it benefitted residents on both sides of the border in several substantive ways, the program was discontinued in 1964, when Lyndon Baines Johnson was elected president and kept promises made to US labor unions—especially George Meaney and the AFL-CIO.[7] Norton III, under the terms of Public Law 78, housed and fed the workers and passed each US Department of Labor inspection. The program worked well for both Norton III and the braceros and in later years many of the same people worked harvests at John Norton Farms.

In 1957 Norton III was asked to serve on the PVID Board of Directors. Only twenty-eight years old at the time of entry to the board, he was the youngest among a group of agriculturalists dominated by men in their fifties and sixties. His service lasted until he left the area in 1972. Aware of the sensitivities surrounding the *Arizona v. California* Supreme Court case, which at the time was in the oral argument stage in San Francisco, Norton III walked a fine line during the early years of his tenure on the board. PVID, by law, had to support its financial share of the lawsuit to defend California against the plaintiff, Arizona. As he put it in 2004, service on the PVID board was "a tremendous education in what the water law is about and what problems face the region, and where we [needed to] go in the future." When Norton III moved to Phoenix in 1972, his general manager in Blythe, Robert Micalizio, took his place at the monthly board meetings. Micalizio continues to represent the company's interests on this important irrigation district board.

Despite the move, the heart of this regional agribusiness enterprise remained in Blythe for the next fifteen years. Norton III oversaw all aspects of the operation, including advertising various crops to wholesalers and retailers with the help of his sales manager, Joe Best. In Blythe, for example, John Norton Farms, under the Norton's Supreme and Red Garter brands, touted their cantaloupes. Under the tagline, "The Nortons Say We Have Gone All Out," the copy read: "We have determined to give you absolutely the finest in Blythe cantaloupes. To accomplish this objective, we have gone all out everywhere along the line from beginning to end. . . . First, middle of the bed planting has made sounder, more attractive melons. Second, Likens mechanical loading machines . . . minimize bruising and handling damage. Finally, double selective sorting in our packing plant, one of the most up-to-date and fully equipped in the business—all packing is under the supervision of Clarence Lane—assures you of getting the product in top condition." The ad ended with a sign-off by John Norton Jr. and John Norton III.[8]

Similar advertising campaigns linked the various John Norton Farms operations in the 1960s, drawing attention to the Double O, J-B, and Pace Setter brands, among others. Norton & McElroy Produce, Inc., one of John

John R. Norton Jr., 1960.

Norton Farms Arizona-based subsidiaries, declared the superiority of their lettuce: "On the march with better lettuce from Outstanding Aguila and Phoenix Districts of Arizona and Blythe, California." Always, the ads touted the "top quality of growers, packers, and shippers" that formed the John Norton Farms network.[9]

In the first month of 1960, Norton III's meteoric rise in the area's agribusiness world was recognized by the Blythe Jaycees. After a six-week search conducted by the Blythe Junior Chamber of Commerce, in which eleven nominees were considered, John R. Norton III was named "Outstanding Farmer" in the region. On its front page, the local *Palo Verde Valley Times* ran a headline "John Norton Named Outstanding Farmer" and remarked, "Norton has been farming in Palo Verde Valley since 1955, when he came here from Phoenix. He is the local head of John Norton Farms, owned in partnership with his father." The article went on to describe Norton's four years in the area: "He started on 800 acres of land owned by the partnership, which has increased to 1,100 acres owned, 900 acres leased, and a 2,500-acre reclamation project which Norton is managing for the Estes Investment Company, making a total of 4,500 acres under direct control of Norton." The paper added that Norton III grew cotton, lettuce, alfalfa, cabbage, melons, grain, and onions and pastured a large herd of cattle on a rotation system on coastal Bermuda grass. By the time of the awards dinner, on January 6, 1960, Norton III served not only on the PVID board of directors, but also the California Lettuce Marketing Board and the Blythe Lions Club. Following the announcement, Norton III issued a statement: "What I know about farming I learned here. I was a pretty green kid when I came to Blythe. We have nice people here who have helped me very much. The attitude and cooperation [are] good. We are a small community, and we must work together for mutual success."[10]

Norton III, who admired his father's accomplishments in cattle raising and his expertise in Polled Herefords, started a cattle feeding operation in Blythe around the time he won the Outstanding Farmer award in 1960. Determined to engage in pursuits that his father enjoyed, he maintained an interest in the cattle industry. The feedlot in Blythe at the John Norton Farms, which handled upwards of 15,000 cattle, proved profitable, and he was asked to serve on the board of directors of the California Cattle Feeders, which he did for several years. The large-scale operation reflected, yet again, Norton III's philosophy of agribusiness diversification. In 1968, Norton III and his father, who never tired of the cattle business, purchased the Buenos Aires Ranch in Santa Cruz County from Clifford Dobson of Mesa. The 130,000-acre ranch, which abutted the Mexican border just east of Sasabe, ran along the border for about fifteen miles, then extended north to the Altar Valley near Three Points. Norton Jr. loved the ranch, in many ways reliving his days at the Double O, X Bar One, and Del Rio ranches.

The elder Norton spent a lot of time at the Buenos Aires, which had a long and celebrated history. In the nineteenth century the Altar Valley was an open grassland teeming with pronghorn and masked bobwhite quail. In the 1850s, shortly after the region became part of the United States after Congressional approval of the Gadsden Purchase (1854), Pedro Aguirre Jr. began a stagecoach and freight line operation between Tucson and the mining communities of Arivaca in Arizona Territory, and Altar, Sonora. Aguirre added a homestead in 1864 and named it Buenos Aires, translated "good air," a reference to the constant gentle winds that blew through the area. He also drilled the first well in Altar Valley, built earthen dams near the homestead, and thus created Aguirre Lake on the property. Further well drilling revealed a stable water supply for the cattle that soon grazed throughout the valley. Between the turn of the last century and the time the Nortons purchased the ranch in 1968, Buenos Aires Ranch changed ownership several times. Shortly after the Buenos Aires purchase, Norton Jr.'s Polled Hereford received the grand champion trophy for Arizona-fed steers from none other than Dr. Bart Cardon at the Denver National Livestock Show. The Hereford was fed by Spur Industries, Inc., which ultimately attracted Norton III, who declared in 2001, "I got involved in the cattle operation over here [Arizona] too and wound up owning the Spur Feeding Company" headquartered in Glendale.[11] As he did on California's cattle board, and a result of his activities with Spur Feeding, he served on the board of directors of the Arizona Cattle Feeder's Association.

Immediately after the Denver Livestock Show, Norton III's flying skills came into play in harrowing fashion as Norton Jr. and Norton III flew back to Arizona from the event. They boarded Norton Jr.'s Aero Commander and flew to Springerville, Arizona, to look at some bulls on a ranch located on the New Mexico side of the border. A hired pilot flew the Aero Commander, a sophisticated and technologically advanced aircraft. The party landed at Springerville, drove to the ranch to look at the bulls, then returned to the northeastern Arizona community at nightfall. In the morning they braced for the winter cold and boarded the plane at 6:30 a.m. As they took off, the pilot, heavy-set and not in the best of physical shape, moaned and slumped to the side. Norton III uttered, "What's the matter?" He reacted quickly to the pilot's apparent heart attack, though he recalled "climbing uphill in that plane was hard, like grabbing a chain and trying to pull yourself to the top of a building."[12] The plane began to "wing over," and when Norton III finally reached the pilot, he could tell that the stricken man was entirely incapacitated. Norton III gained control of the plane, grabbed the radio, and tried to contact Springerville, but there was no radio at what was essentially a primitive landing strip. He then tried Show Low, which had a radio, and told the air traffic controller there: "I've got a sick pilot here. I'm a licensed pilot but I'm not familiar with this aircraft, and I need to make an emergency landing at your airport." "But our runway is covered with snow," the obviously nervous controller responded. Norton shot back, "I can't help it, I've got to bring it in," to which the controller said, "You get lined up with the fence posts that are sticking out of the snow on each side, [and] the runway is dead center in the middle of those fence lines." Norton III landed the plane and slowed it, but the snow pulled the craft to the left and it slid off the runway into the mud. An ambulance arrived quickly, and the ailing pilot had to be flown to Phoenix; he survived what was described as a massive heart attack and Norton III, thanks to his piloting skills, saved three lives.[13]

But after Norton Jr. contracted Parkinson's in 1972, the family realized spending time at Buenos Aires Ranch would be a hardship for him. So in 1975, the Nortons sold the Buenos Aires to Peter Wray, husband to Gay Firestone Wray, heiress to the Firestone fortune. During the seven years of cattle raising at the Buenos Aires, the Nortons made money, and they sold the ranch at a marginal profit. In 1985, the US Fish and Wildlife Service purchased the Buenos Aires Ranch, and it became a National Wildlife Refuge. Conservation strategies are now restoring the grassland and bringing back the native pronghorns and masked bobwhite quail.[14] In many ways, the Buenos Aires venture served as a kind of capstone to Norton Jr.'s career. "Over a period of time in the early 1970s," Norton III confided to a friend who was on the UA Agriculture faculty, "I acquired all his interests, so at the time of his retiring in 1972 I took over the operations, both here and in California. That's why I moved back here in 1972, because it seemed like a logical thing to do."[15]

Another factor that altered the agribusiness landscape for Norton III—and perhaps influenced his return to his land of birth—was the legal reversal of fortune that Arizona's outmanned legal team had orchestrated in the US Supreme Court in *Arizona v. California*. In 1957 the Arizona Interstate Stream Commission and Governor Ernest McFarland agreed that a change in legal counsel was necessary, because close observers of the case determined that the Arizona legal team was playing into the hands of their California opposition. As a result attorney Mark Wilmer, who at the time was considered "the best litigator in the state," began arguing the case, changed Arizona's legal theory, and convinced the Special Master of the wisdom of his arguments. An exasperated California legal team, headed by the legendary Northcutt Ely, could only watch in wonderment and despair—while Norton III and his expatriate Arizona brethren farming in California could scarcely disguise their delight and surprise at the outcome. Stunning changes took place between 1957, when Wilmer took over, and 1963, when the Supreme Court decided the complex case.

The *Arizona Republic* announced on its June 4, 1963 front page, "The end of the eleven-year legal battle with California will shape the future of Arizona for all time to come and ranks in importance with the state's admission to the union. It will open the way for the long sought Central Arizona Project to bring water from the river into the populous centers of the state. [Senator] Hayden has already prepared legislation to authorize construction of the $1 billion project and expects to introduce it very soon." Characteristically, Wilmer, a soft-spoken litigator, took the victory in stride, expressing gratification and accepting congratulations from all who knew him, yet remaining low key and sticking close to family. The Court, Wilmer told those around him, reasoned correctly that Congress "intended to and did create its own comprehensive scheme for . . . apportionment." In addition, Congress had authorized the Secretary of the Interior to utilize his contract power to implement a lower basin agreement. In fact, the Court gave Arizona what it had wanted since 1922, and *Arizona Republic* reporter Cole called the decision "a personal triumph for Carl Hayden because the decision referred back twenty-five years to the December 12, 1928 debate in which Hayden pointed out that the Boulder Canyon Dam bill and its allocation formula settled the dispute over the lower basin waters."

California, stung by the outcome, reacted with apprehension. They charged the Court with misreading the intent of Congress, and eroding the rights of the states, and argued that the ruling represented the first time that the Court had interpreted an act of Congress as apportioning water rights to interstate streams. They believed it to be an untoward judicial prerogative that threatened California, and Arizona leaders knew well that California would try to regain in the political arena what they had lost in the judicial decision. Roy Elson, Senator Hayden's administrative aide, described the situation on the heels of Arizona's Supreme Court victory: "We knew that California and Northcutt Ely would try some way to stop this through the legislative process, even though they had lost. . . . What they couldn't accomplish in court they would try in the field of politics."[16] Indeed, Elson's musings were on target as the Associated Press, on June 4, 1963, quoted California Attorney General Stanley Mosk: "California's thirty-eight-member House delegation still could give Arizona substantial opposition in authorizing new Colorado River projects."[17] In short, California would find new ways to oppose CAP.

Arizona's congressional delegation were keenly aware that Wilmer's legal handiwork had, by 1960, created an atmosphere for an Arizona victory in the case. It had already begun laying the political groundwork for another, and hopefully final, legislative push for CAP. The Arizona delegation, led by Senator Hayden, let it be known that it had supported several big reclamation packages for the upper basin states, including the Colorado River Storage Act of 1956, that led to the construction of Glen Canyon Dam. Also, Arizona backed numerous individual state projects like New Mexico's San Juan Project, which passed Congress in 1962. In light of these palpable examples of support for regional development, Arizona's congressional leaders believed they deserved the same kind of consideration for CAP.

Yet between 1960 and 1963 the anticipation of a Supreme Court decision favorable to Arizona prompted federal administrators and representatives in the basin to begin formulating a regional plan for the entire basin—not just Arizona. In January 1962, Secretary of the Interior Stewart Udall, a former Arizona congressman, encouraged Congressman Wayne Aspinall (D-Colorado), chairman of the influential House Interior and Insular Affairs Committee, to request a comprehensive study of water development in preparation for the authorization of individual state projects—including CAP—as soon as a decision in *Arizona v. California* was rendered.[18]

The Court's final decision departed from past rulings. The Secretary of the Interior would allocate future surpluses and shortages among and within states. This dimension of the opinion marked, as one expert on the Colorado River has written, "an especially sharp break with tradition." Moreover, the Court majority ruled that Congress could invoke the navigation clause of the US Constitution as well as the "general welfare" clause to divide the waters of nonnavigable and navigable streams. This dimension of the ruling, as Justice William O. Douglas wrote in his tart dissent, increased drastically federal control over the nation's rivers. He wrote, in part: "Much is written these days about judicial lawmaking, and every scholar knows that judges who construe statutes must of necessity legislate interstitially . . . the present case is different. It will, I think, be marked as the baldest attempt by judges in modern times to spin their own philosophy into the fabric of law in derogation to the will of the legislature. The present decision as Mr. Justice [John] Harlan shows, grants the federal bureaucracy a power and command over water rights in the seventeen western states that it has never had, that it has always wanted, that it could never persuade Congress to grant, and that this court up to now has consistently refused to recognize." In spite of the scathing dissent and charged rhetoric, the ruling seemed to clear the way for Arizona to press forward legislatively for CAP.

Beyond the positive implications for Arizona and CAP, *Arizona v. California* gave American Indians cause for hope. When Arizona filed suit in 1952, the federal government intervened not only to protect its interests in the river, but also to defend the rights of American Indians living on the twenty-five reservations within the lower basin. Predictably, US attorneys petitioned for water for all practically irrigable lands on Indian reservations as well as for national parks, forests, recreation areas, and other federal lands. As the water situation between Arizona and California had been altered dramatically and the former now maintained a credible chance of diverting mainstream waters into central Arizona, Norton III continued to purchase land south and west of Blythe in the PVID. By the early 1970s John Norton Farms held tens of thousands of irrigated acreage and the valuable water rights.

In 1972, San Diego Gas & Electric (SDG&E) began planning a new nuclear energy facility for southern California. In that same year, advance agents surreptitiously arrived in Blythe to learn more about local land owners and the assignation of water rights to privately held lands. The utility's proposed Sundesert Nuclear Power Plant would comprise two 974-megawatt Westinghouse pressurized water reactors, and the proposed facility would be built sixteen miles southwest of Blythe and ten miles from the Colorado River. Norton III learned quickly of the outsiders and their intentions, and like other area farmers, he preached caution to his PVID neighbors about SDG&E's intentions. He knew that any nuclear plant needed a lot of water for the cooling towers, and as he said at the time, "They had the Colorado River there but they didn't have any water rights." So in a series of negotiations that lasted into 1973, SDG&E's attorneys negotiated with Norton III to purchase approximately 6,500 acres. They paid a premium price for the land, which, of course, held essential water rights. Norton III reaped huge financial rewards with this one transaction; he not only garnered $13 million in the land sale, but also "had his cake and ate it too."[19] As it turned out, because of political and cultural challenges to nuclear energy in California, SDG&E faced years of delay before ultimately being disappointed: the nuclear plant was never built, so the utility leased the lands back to John Norton Farms, and he kept farming the acreage. "The land and water rights I acquired around Blythe," Norton III declared, "changed my life in the most profound way."[20]

By 2001 the land was sold again, this time to the Metropolitan Water District of Southern California (MWD), which purchased it for $42.5 million. Approximately 30 percent of the 16,000 acres involved was immediately removed from cultivation, and the water rights appropriated for domestic and urban uses along the southern California coastal plain. Although 9,704 acres were leased back to PVID farmers under multiyear contracts, and are still being farmed, no one knows when MWD will take all of it away from the agriculturalists in PVID. [21] The events of 1972 represented what Norton III called "the gradual drying up and attrition of desert agriculture to satisfy the water needs for this enormous population growth in both Arizona and California."[22]

At forty years young, John R. Norton III had built what began as an eight hundred-acre farm in California's Palo Verde Valley into a virtual agricultural empire, with operations in Arizona, California, New Mexico, and Nevada that boasted gross annual sales of $50 million. What began as a modest produce and cotton farm developed into a growing and shipping juggernaut—a corporation that was producing, among many other crops, more than 7 percent of the nation's iceberg lettuce. During the 1970s J. R. Norton Company shipped its various products from coast to coast and to Canada, Europe, and the Far East. Norton III also pioneered new concepts in agriculture as his business prospered. In the late 1950s, for example, he introduced the use of field-loading machines in the harvesting of honeydews and cantaloupes, which became the industry standard. In short, Norton III combined agricultural expertise, business acumen, product integrity, and process innovation to become one of the West's—and the nation's—leading agriculturalists. [23]

He also entered into a new and, in the long term, profitable limited partnership, with Edward Robson and another business partner, James "Bud" Smith, in 1976. He invested in a real estate development and land venture in the southeast valley of Phoenix called Sun Lakes Properties, Inc. Robson, the principal owner in Sun Lakes, had fallen short of expectations in selling lots, and Norton III and Smith saw the possibility of a profitable farming operation on what had been known as the "undeveloped" Hanna/Lewis farmland, which was part of Robson's vast holdings. Norton and Smith formed S&N Farms, and for an initial investment of $100,000 became 22 percent partners in Sun Lakes. This venture, which prefigured the nature and direction of Norton III's increasingly diverse investment portfolio in ensuing decades, profited from the fact that the farming rights were worth more than their investment. Robson benefitted in two important ways: it helped his struggling company make it through the difficult summer of 1976, and he was now associated with Norton III, which, as Robson acknowledged in his recent autobiography

Outrageous Good Fortune, "gave our company instant recognition as a company of substance."[24]

At this juncture, Norton III became interested in the overall welfare of all agribusiness industries. He broadened his conceptual focus to include the produce, livestock, and food producing entities in their entirety. This interest corresponded with his return to Phoenix, Arizona, and from there, he undertook, consciously or not, a new direction—into national agribusiness leadership. He first joined the board of directors of the Western Growers Association (WGA), the organization that represented row crop growers, in 1975, and the following year, the United Fresh Fruit and Vegetables Association, which was the trade association for the nation's produce industry. Throughout the 1970s his involvement in these two associations represented a new vector in his professional life. Along with his engagement in various water and power boards and other professional organizations, these associations transformed him into something more than a successful agribusiness titan. In the 1970s and 1980s he became an industrial leader and inspirational figure of national significance. At the same time he began to flex his muscle in the realm of Republican party politics, which was fortuitous in an increasingly conservative state like Arizona. For someone like Norton III, with strong and definitive views on tax policy, free markets, and the overall role of government in private affairs, political activism was a natural extension of his public engagement.[25]

WGA, founded in 1926, organized as a result of a major issue between Imperial Valley farmers and the railroads. By the time Norton III joined the board in 1970, it had evolved into an association that addressed issues ranging from health care to government relations, human resources, financial planning, international trade, legal services, science and technology, and importantly for Norton III, trade practices and labor.

The United Fresh Fruit and Vegetable Association, founded in 1904, has been the industry's leading trade association committed to driving the growth and success of produce companies and their partners. In 2006 it merged with the International Fresh-Cut Produce Association, and the new entity has been rebranded the United Fresh Produce Association. In 1980 Norton III became chairman of WGA; two years later he became chairman of the United Fresh Fruit and Vegetable Association, which he described as "national and a very prestigious position." Norton III's timely involvement with both organizations resulted in another career advancement that reconfigured his professional career and placed him, unexpectedly, on a track for national public service.[26]

One day in 1982, when Norton III was serving as chairman of the United, he decided to clear off his cluttered desk and picked up a bulletin from the National Cattleman's Association, in which, of course, Norton III maintained a membership. He read a short column about the Packers and Stockyards Act of 1921. According to Norton III's reading of the law, if a cattleman sent his steers to a meat packer, and that packer declared bankruptcy, the bank or any lender could take possession of the cattle on premises as collateral. But there were abuses; for example, the lender would enter the plant and seize all the assets of the meat packing plant even though these cattlemen had not yet been paid.[27] The Packers and Stockyards Act addressed this and other abuses that had victimized livestock growers and shippers. World War I triggered reform; the war created a sharp increase in living costs for Americans, and President Woodrow Wilson directed the Federal Trade Commission (FTC) to investigate the industry from cattle range to dinner table to determine whether, in the parlance of the day, there were any manipulations, controls, or restraints in variance with the law or the public interest. In 1919 the FTC issued a damning report: there had been a history of abuses in the seizure of assets, as well as monopolistic practices and the manipulation of markets, restricting the flow of food and controlling the price of processed meats. The FTC went so far as to recommend government ownership of stockyards and related facilities.[28]

Norton III recalled, "This little article I happened to read—instead of tossing it in the waste basket—changed my life." The fact that the trust provisions of this 1921 federal legislation protected those cattlemen who had shipped their stock to the meat packer, and that they were paid instead of losing their rightful compensation, stirred Norton III to consider broader applications of the act. Perhaps its principles could be applied to the produce business? If a carload of Arizona lettuce arrived at a receiver—a wholesaler in New York City, for example—the shipper did not necessarily know the financial condition of the company. On occasion, one of those companies declared bankruptcy, and prior to 1982 many would accept a load of lettuce or some other vegetable or fruit shipment, even though they knew they were going under. The next day, the bank would place a yellow tag on the door and claim the entire company for their debt. The farmers received nothing for their product, losing the assets to the bank. Norton III concluded, "We ought to have the provisions included in the Packers and Stockyards Act integrated into the produce industry. If they have a federal law that protects the livestock shipper, why can't we have something similar to protect the produce shipper?" Since he was in both businesses and now served on powerful boards representing these agricultural industries, he knew the parallels.[29]

As chairman of United Fresh Fruit and Vegetables, Norton III traveled around the United States. speaking to various regional organizations. The Florida Shippers, Michigan Cherry Growers, South Texas Vegetable Association, and a host of others welcomed Norton III and listened to his canned speech: "All of you fellows ought to be with United in addition to the Michigan Cherry Growers, because United helps you at the national level." The various regional groups conducted their quarterly or annual meetings and told Norton III of their concerns and challenges. Bernie Imming, who ran the daily operations of United Fresh Fruit and Vegetable, often accompanied Norton III on these jaunts.[30]

In June 1982, at one of these United events in McCall, Idaho, Norton III addressed the annual meeting of the Oregon-Idaho Onion and Potato Growers Association. In an impulsive moment as he stood before this group of growers, Norton III decided to jettison his canned speech and began, "I want to talk to you guys about a concept I have. I want to run this by you and see what kind of a reaction I get." He proceeded, "I don't know how many of you are in the livestock business or have ever heard of the Packers and Stockyards Act," and one listener shouted, "Oh yeah!" Norton III continued, "Well under the Packers and Stockyards Act you have a trust provision," and he continued to outline the details for the Oregon-Idaho group. He concluded his extemporaneous talk with the exhortation, "We ought to try this with the produce business." Norton's traveling companion, Imming, who was not one to depart from established protocol, appeared apoplectic in response to Norton III's departure from the usual script. But when Norton III finished his talk, "the whole damned room" raced toward the podium and they "pulled him and tugged at him like he were some kind of presidential candidate."

Significantly, a reporter for the *Packer*, described as the *Wall Street Journal* of the produce industry, was in attendance, and covered the speech. The next week he wrote a piece about the "Norton Proposal" arguing for an amendment to the Perishable Agricultural Commodities Act of 1930 (PACA) to make it easier to collect money when a customer files bankruptcy. "Well anyway," Norton III recalled, "you couldn't stop it. From then on every place we went, everybody wanted to hear about the 'Norton Proposal,' and there was a groundswell of support." By the time he finished his tour of the country and his chairmanship in February 1983, the industry decided to submit a bill to Congress and, not surprisingly, United had the support of the Packers and Livestock administrators. Though some of his produce constituents advised him to take a more prudent course on this issue, Norton III responded: "We must not be indecisive, lest we risk being ineffectual." Norton III visited with administrators in the Department of Agriculture—the Perishable Agriculturalists Commodity Administration (also known as PACA)—and they were on board. Norton took the lead and in short order a bill was drafted and submitted to Congress in the spring of 1983.

These developments took place with remarkable speed and the *Packer*, on March 5, 1983, in an interpretive profile of Norton's yearlong leadership at United, declared that "Probably no chairman had more impact on the produce industry than John Norton III, who stepped down from his post at the recent convention in downtown Anaheim." In attending industry association conventions throughout the year, the *Packer* columnist asserted, "Norton advocated what later became known as the 'Norton Proposal.'" His efforts were so successful, the article continued, that twenty-six industry associations, mostly shipper groups, endorsed the idea. United's annual Congress of Committees in Washington, DC, supported the proposal in September 1982, as did United's board of directors. "Norton," the *Packer* concluded, "has made a sincere and honest effort to find a way to help the entire produce industry when bankruptcy proceedings occur."[31]

As his United chairmanship ended, Norton III "carried the water" for the produce industry as the House Agricultural Committee held hearings on the "Norton Proposal" to amend the PACA. The president of the American Bankers Association waited in the anteroom to testify after Norton III, and before he entered the hearing room, the banker told him, "I hope you realizeyou've bitten off a little more than you can chew on this one; we are not going to let you get away with this type of thing because this is detrimental to the interests of the American Banking Association and we think you are wasting your time." Norton III listened to his intimidating language, entered the hearing room, and testified. The banker testified after Norton III. The bill passed out of committee unanimously and sailed through Congress. No longer could bankers claim all assets from a receiving house if it underwent financial catastrophe.[32] Thus Congress, in 1984, amended PACA to impose a statutory trust in favor of an unpaid supplier of produce held by a commodity seller. For Norton III, this was a huge accomplishment, and for suppliers in the produce industry, a tectonic shift in their abilities to recoup money owed them.

These developments placed Norton III in the center of the nation's agriculture world and his national reputation grew in exponential fashion. The *Packer* mentioned him almost daily in 1984, and awards, previously unthinkable, came his way. United named him "Produce Man of the Year" in 1983, and the Western Growers Association gave him their "Award of Honor" the following year.[33] According to his wife Doris, these accomplishments, in turn, led to his promotion for deputy secretary of agriculture by the memberships of both the WGA and United. All of the aforementioned developments in 1983 and 1984, moreover, took shape against the backdrop of the end of President Ronald Reagan's first term in office. Richard "Dick" Lyng, a Californian from Modesto, had served as deputy secretary under John "Jack" Block during Reagan's first term. Lyng had experienced health problems and expressed to the administration that he wanted to step down. Many groups and individuals, along with stalwarts in Arizona's increasingly influential Republican party, lobbied for a Norton III appointment to the department of agriculture. Then, on Friday, December 5, 1984, Norton III received a phone call at his office in Phoenix. "Can you be in the White House on Monday morning, December 8, for an interview?" "Okay," Norton III responded happily, thinking to himself, "Why not try something like this?" It meant leaving his business temporarily, if he were selected, but, as he put it, "I decided I'd take a whirl at it."[34] On Monday, December 8, 1984, Norton III boarded a plane at Phoenix Sky Harbor Airport and flew to Washington, DC.

·10·

THE DEPUTY SECRETARY
OF AGRICULTURE

"The committee will come to order. As it is well known, President Reagan has nominated Honorable John R. Norton, of Arizona, to be Deputy Secretary of Agriculture."

CHAIRMAN OF THE COMMITTEE ON AGRICULTURE, NUTRITION, AND FORESTRY, SENATOR JESSE HELMS (R-NORTH CAROLINA), ROOM SR-328-A, RUSSELL SENATE OFFICE BUILDING, WASHINGTON, DC, MARCH 27, 1985.

"Subsidies . . . have made for a generation of farmers dependent not on the market for their profits but on the government."

JOHN R. NORTON III, SIX MONTHS AFTER HIS TENURE AS DEPUTY SECRETARY OF AGRICULTURE, PHOENIX, ARIZONA, NOVEMBER 10, 1986.

O n the morning of December 8, 1985 John Norton III entered the west wing of the White House and was greeted by President Ronald Reagan's Director of Personnel, Robert "Bob" Tuttle, along with three members of his staff. Would he be interested in the position of Deputy Secretary of Agriculture? The honor posed a potential financial risk: he would be absenting himself from the daily operations of the far flung J. R. Norton Company and its subsidiaries, but he decided to allow his name to be placed before the Senate Committee on Agriculture, Nutrition, and Forestry. It would vote on a recommendation to confirm or reject his nomination. Two weeks later, after President Ronald Reagan was sworn in for his second term, the administration formally placed Norton III"s name in nomination, and John and Doris made arrangements to move to Washington, DC.[1]

One of the first challenges Norton III faced in the confirmation gauntlet was full disclosure of his holdings. "I don't think anyone fully enjoys disclosing all for public viewing, but it is a requirement of the law," he told one reporter covering his nomination.[2] At an informal meeting of the Senate Committee that oversaw his confirmation, he stated flatly, "There are some who have questioned my ability to serve as Deputy Secretary of Agriculture because of my apparent financial success and the size of my farming operation. For the record, I make no apologies for whatever success I have been fortunate enough to attain. I regard this success with a combination of humility, pride, and gratitude." Norton III added that his financial and professional success in western agriculture came with hardship, sacrifice, and considerable risk. "Without equivocation I can assure you that I am intimately familiar with the many agonies and cruelties associated with the business of agriculture," he informed the senators. "As a young man I witnessed my father liquidate his entire cattle herd twice, due to drought. On many occasions I have raised crops to the point of maturity but have been able to harvest none or only a small portion of the crop due to markets so low that the value of the crop would not even cover the cost of harvesting." He experienced the sudden and severe drop in the value of his wheat crop when the US government imposed an embargo on grain sales to the Soviet Union, and he witnessed the value of his feed yard full of cattle drop more than $200,000 overnight when President Lyndon Baines Johnson placed an embargo on the shipment of American hides in the 1960s.[3]

Deputy Secretary of Agriculture-Designate John R. Norton III awaited his confirmation hearings while Arizona's entire congressional delegation rallied around their native son. Besides senior Senator Barry Goldwater's (R-Arizona) support—"Both Arizona and California have benefitted from John's willingness to serve his community"—junior Senator Dennis DeConcini (D-Arizona) hosted a reception at his McLean, Virginia home, where he introduced Norton III and Doris to Democratic senators on the committee. DeConcini's wife's family, the Hurleys, were long acquainted with the Nortons through their farming enterprises and their political affiliation with Arizona's growing and increasingly influential Republican party. Though DeConcini was a Democrat, love trumped party affiliation, and he had married into a powerful Republican family who wholeheartedly supported Norton III's nomination.[4] From the House of Representatives, Morris Udall (D-Arizona), John McCain (R-Arizona), Eldon Rudd (R-Arizona), and Jim

Doris and John R. Norton III with Senator Orrin Hatch and then-Congressman John McCain, 1985.

Kolbe (R-Arizona) prepared statements in Norton III's behalf. Representative Udall, in characteristic fashion, offered an amusing comment but left no doubt about his support for Norton III: "John is an unusually well-qualified person, and he is superior to almost everyone I know. I do not know why anyone in his right mind would want to take on this job . . . but the country is lucky that he is, and you will make no mistake if you confirm his nomination." As Senator DeConcini informed committee members: "We have an opportunity to have absolutely the best that there is to offer from the private sector to come in and serve in a very difficult position at a very difficult time."[5] Indeed, from the outset, Norton III maintained bipartisan support, and he specifically recalled how gracious and helpful DeConcini, a member of the opposite party, was throughout the confirmation process.[6]

California Senator Pete Wilson (R-California) offered a lengthy and detailed statement of support that was noteworthy, given the longstanding antipathy between Arizona and California over the rights to Colorado River water. But with agricultural interests in both states, Norton III served as a bridge between the two. He tilted toward Arizona in the longstanding Colorado River controversy, but his agribusiness practices in California, and his leadership role in various agricultural groups, prompted Wilson to attest that the nominee was "a long-time California farmer." Senator Wilson added, "I have a great interest, as do all Californians who have any passing familiarity with agriculture, in this nomination of John Norton, because in addition to his long and happy experience with agriculture and philanthropy in the state of Arizona, he has long and very deep California roots." Wilson continued, "He is a man who is deservedly admired and has been recognized for his achievements.Since 1955 when he founded his own farming operation in Blythe, California, until his more recent position as the president of the J. R. Norton Company, he has overseen the growth and development of a highly respected and diversified agricultural operation. Its headquarters are in Phoenix, but he has branch offices in the California cities of Blythe, Brawley, and Salinas."[7] The California senator made sure to point out to his fellow committee members that Norton III had also been a member of the presidential agricultural task force in 1980 and 1981. As he ended his presentation, his comments took a slightly more personal turn: "I value not only the experience and knowledge, but also the character of the man, his candor and reasonableness, and the fact that he has given so generously of himself in so many efforts that have made agriculture better in Arizona and California."[8]

The confirmation hearings in 1985 shed light on the considerable accomplishments of Norton III—and revealed the extent of his considerable holdings. If confirmed, he would have to take steps to avoid actual or apparent conflict of interest. At the time of his nomination he was president, chairman of the board of directors, chief executive officer and majority stockholder of the J. R. Norton Company, an Arizona Corporation headquartered in Phoenix. The company, at the time of his nomination, was engaged in farming operations on approximately seventeen thousand irrigated acres in Fresno, Imperial, and Riverside counties in California, and Pinal and Maricopa counties in Arizona. The crops produced on these lands included cotton, wheat, alfalfa, iceberg lettuce, and a variety of vegetable and row crops.[9] In addition, the J. R. Norton Company owned 100 percent of the stock of the Garin Company, a California corporation producing iceberg lettuce, celery, asparagus, and cauliflower on approximately fifty-five hundred acres in California's Monterey, Fresno, and Riverside counties, as well as in La Paz County, Arizona. J. R. Norton Company owned 100 percent of the stock of Supreme Berries, Incorporated, an Arizona corporation that produced strawberries on approximately two hundred forty acres in Monterey, California. The company also owned about an 80 percent interest in San Tan Ranches, an Arizona joint venture, and operated the ranch, which produced cotton, wheat, and iceberg lettuce on approximately four thousand acres of irrigated land leased from the Gila River Indian Community in Pinal County, Arizona.

In order to avoid a conflict of interest in connection with these holdings, Deputy Secretary-Designate Norton agreed to resign several positions: as president, chairman, and chief executive officer of the J. R. Norton Company, as a director of the Garin Company, and as a director of Supreme Berries, Incorporated. Moreover, Norton III agreed to abstain from any management role in connection with any of the farming enterprises, and assented to forego salary or compensation from the enterprises during his tenure as deputy secretary. Finally, he pledged that none of the farming entities or operations described above would participate in any USDA farm price support or lending program during his government service.[10]

There was more to Norton III's seemingly limitless agribusiness empire in the Pacific Southwest. The J. R. Norton Company also owned 25 percent interest in Mohave Farms, an Arizona farming partnership that produced cotton and grain on approximately four thousand acres of leased land from the Mohave Indian Community on the eastern banks of the Colorado River in Mohave County, Arizona. In years past, Mohave Farms had participated in USDA's deficiency payment and price support programs for cotton and grain. Norton III assured the committee that beginning with his tenure, the J. R. Norton Company would not participate in any part of the deficiency payment or in any other USDA price support program with respect to its 25 percent share of the Mohave Farms operation.

Norton III owned a 50 percent interest in the Spence and Norton partnership, which operated a cattle ranch of approximately eighty-six hundred acres in Merced County, California. The partnership owned and pastured three thousand head of cattle on the ranch and provided pasture for other cattle growers. Beyond the California ranch lands, Spence and Norton leased pasture for another thousand head on the Denny Ranch, located in parts of Mohave, Yavapai, and Coconino Counties, Arizona.[11] Norton III pledged to resign his positions and withdraw from all management responsibilities in connection with all of the farming and ranching operations and other businesses with which he was involved. Since he continued to hold ownership interests in the described operations, he also pledged to recuse himself from any and all particular matters that arose within the USDA concerning farming or ranching in the California and Arizona counties where the operations were located.[12]

Norton III's business ties didn't end there. He served as a director in a number of organizations, joining the board of the Western Growers Association (WGA), headquartered in Irvine, in 1972, and the board of Calcot, Ltd., the California-based cotton marketing cooperative, in 1976. A past chairman of WGA, he was serving as vice chairman of Calcot at the time of his nomination. Beginning in 1978 he served as a director of the Alexandria, Virginia-based United Fresh Fruit and Vegetable Association, becoming the chairman in 1983. Upon his confirmation, Norton III pledged to resign from the positions he currently held with these organizations.[13]

Even more, Norton III served on a number of significant Arizona-based business entities that shaped Arizona's economy outside the realm of irrigated agriculture and ranching. He was a director of Ramada Inns, Incorporated; the Arizona Public Service Company; Turf Paradise; and United Bancorp of Arizona. He was also an officer or director of a number of other business, civic, and social organizations—all of which were listed on Schedule D of his lengthy Financial Disclosure Report. He agreed to resign from all of these positions when he was confirmed. The only directorship he continued to maintain was on the Chief Executives Organization, an association of business executives that had no business relationship with the USDA.[14]

The confirmation process took some unexpected turns and an inordinate amount of time. Thanks to Norton III's service on the Board of Directors of Ramada Inns since 1978, what was to be their temporary home for a few weeks in one of the Ramada Mini-Suites, located between Georgetown and the Capitol, turned into a two-month-plus stay. Senator John Melcher (D-Montana), a veterinary medicine practitioner and second term senator, held up Norton III's nomination, apparently for political reasons related to his displeasure with the Reagan administration. Specifically, Senator Melcher sought to raise questions about Norton III's participation in the Payment-in-Kind Program (PIK), both in his role as chairman and CEO of J. R. Norton Company and in his position as vice president of Calcot, Ltd., the California cooperative that marketed cotton harvested in the Golden State. Norton III reflected on the process: "What happens is that some senator gets crossways with the administration over some little issue, and he puts a hold on one of the nominees just to be ornery and cause trouble."[15]

But prior to the bulk of the question and answer session, Norton III made his official statement to the committee at his second, formal hearing, on March 27, 1985:[16] "To have been nominated for Deputy Secretary of Agriculture by the President of the United States is a very high honor and one which represents a pinnacle achievement to someone like myself, who has devoted his entire life to the profession of agricultural production," he began. "My grandfather, my father, myself, and my son represent generations in ranching and farming in California and Arizona, and we have participated in the operation of cattle ranches, cattle feeding operations, raising sheep, and the production of a variety of crops on irrigated land for over a century [i.e. from 1881 to 1985]." Chairman Jesse Helms (R-North Carolina) and Senator David Pryor (D-Arkansas) posed the hypothetical questions designed to showcase Norton III's balanced approach and intimate knowledge of the intersection between federal policy and private sector challenges for the nation's farmers and ranchers. But committee member Melcher honed in on Norton III and the benefits derived from the PIK program under the statutory authority granted by the Agricultural Act of 1949 as amended and the Commodity Credit Corporation Charter Act. [17]

Notably, the PIK program was designed and implemented within existing statutory authority but did not receive specific congressional authorization. It offered commodities to farmers who agreed to reduce their planting of wheat, feed grains (primarily corn and grain sorghum), rice, and upland cotton beyond what was called for in the 1983 programs for those crops. Historically, the USDA had used a number of adjustment mechanisms to take cropland out of production. These mechanisms were part of a group of farm programs designed to stabilize and enhance commodity prices and farm incomes. The Agriculture and Food Act of 1981 authorized cropland reduction programs for the 1981-1985 crops of wheat, rice, cotton, and feed grains.

These products fell within Norton III's agribusiness wheelhouse. Prior to his nomination, like many other growers throughout the country, Norton III participated in the PIK program, which was essential in keeping agriculture profitable in the early 1980s. In his statement to the Senate Committee, on March 27, 1985, Norton III fired a preemptive strike and addressed Melcher's hostile questions with candor. "There is a question that has received some attention in the press recently concerning my participation in the Payment-in-Kind program administered by the US Department of Agriculture in 1983," Norton declared, and "to avoid the need for probing by your committee, I would like to submit for the record the following information. I did not participate in the PIK program on a personal basis, although my company, J. R. Norton Company, participated for the 1983 crop year and received payment-in-kind cotton worth $1,676,898 and payment-in-kind wheat worth $296,977, for a total of $1,973,875." In addition to this disclosure at the outset of his hearings, Norton III informed the committee, "J. R. Norton Company is a 50 percent partner in San Tan Ranches, and that entity received payment-in-kind cotton worth $979,015, and payment-in-kind wheat worth $36,670, for a total of $1,015,685. Further, J. R. Norton Company was a 25 percent partner in an entity known as Mohave Farms, which received payment-in-kind cotton worth $647,440. . . . I offer no apologies to anyone participating in the Payment-in-Kind Program. It has been my policy and the policy of the J. R. Norton Co. to maintain rigid adherence to every detail of the law concerning all programs pertaining to our operations, including but not limited to those of the USDA. Had I not participated in the Payment-in-Kind Program, members of this committee might well be questioning my management ability. Besides showing poor business judgment, it would have constituted a failure to participate in a government-sponsored program designed to rid the nation of price-depressing agricultural surpluses."[18]

Though Norton III's one detractor on the committee, Melcher, tried to find gaps and inconsistencies in Norton's record concerning cotton exports to the Soviet Union, and evidence of favoritism or inaccurate or incomplete reporting of the amounts received from the PIK program, the Montana Democrat failed to derail the nomination. Norton III accounted himself well in the thrust and parry that characterizes confirmation hearings. Melcher's arguments were soon subsumed and even dismantled as Norton III and the committee participated in an informed discussion concerning agricultural export policies as well as the subtleties and complexities that marked the roles and relationships among the various growing regions in the country. As the hearings progressed, Norton III's distinguished private sector career and accomplishments, his community service, his mastery of national agricultural issues and federal agricultural policy, and his intelligent and cogent responses impressed the vast majority of the committee. His performance, combined with glowing testimonials from Arizona's delegation and a host of private-sector individuals and entities that had dealt directly with Norton over his thirty-year career, served the nominee well. Senators got a clear picture of Norton III's body of work, skill sets, and fitness to serve as Deputy Secretary of Agriculture, and the committee recommended him to the full Senate.[19] He was finally confirmed in May, five months after his arrival in Washington, DC.[20] At the ceremony on May 15, 1985, recently appointed Associate Justice of

United States Supreme Court Justice Sandra Day O'Connor swears in John R. Norton III as Deputy Secretary of Agriculture; John's wife Doris holds the Bible (1985).

the US Supreme Court and longtime friend Sandra Day O'Connor came to the Department of Agriculture and swore in Norton III as Deputy Secretary while his family, including his father and his wife looked on. Norton III's recollection of his swearing-in ceremony was vivid: "So, the family came back for the swearing in. My dad, his wife . . . and our children, our grown children and their spouses. In one case, Michael had his girlfriend. But they all came back to visit Doris and me for the swearing in. And afterwards we took them to Mount Vernon and saw George Washington's home. . . . And they were there for a couple days. That was really the last time I was able to have a real good association with my father, because I spent the rest of that year on into March of the next year in Washington. His health was declining pretty badly by then. His Parkinson's had him. And he was pretty well bed-ridden when I got back here [Phoenix] in March of 1986. And he died the next year."[21] Norton III had now entered the world of federal agricultural policy, administration, and bureaucratic challenge.[22]

According to Ward Sinclair of the *Washington Post*, who covered Norton III's entry into his first government job, people were "shocked by his candor and frankness; he is a quick study, but he's walked into an administrative nightmare at USDA." Sinclair lauded Norton during the hearings, when "the

Doris Schaefer Norton.

Western agribusiness titan "easily derailed" the contrary Senator Melcher, who had been "poised to pounce at his confirmation hearing." Sinclair quoted one lobbyist: "You can disagree with Norton on philosophy or principle, but you have to respect his views because he is honest." And both the *Washington Post* and hometown *Arizona Republic* reported that "when Norton came to town last winter, he paid his expenses for more than three months waiting for his nomination to clear, rather than taking a salaried 'consultancy' that was offered him."[23]

During the period of delay over his confirmation, Norton III had been working with Secretary of Agriculture John Block and his staff on the important task of crafting the Farm Bill of 1985—officially called the Food Security Act of 1985—which was the major piece of federal legislation emanating from the USDA. He conducted the USDA's internal budget hearings, the process in which division heads were required to justify their spending and program ideas. The so-called Farm Bill was crafted once every five years, and it formed the administration's most vivid public policy and political statement on American agriculture as a whole. As Norton III described his priorities in his unofficial and, after May 1985, his official capacities, "our desire in the Republican administration was to try to gradually extricate government out of agriculture." He argued that the agricultural support and subsidy programs that were begun as a result of the Depression years in the 1930s had outlived their usefulness, and that in some ways the previous Farm Bill (1981) had done more harm than good. The Deputy Secretary argued that he wanted to phase out subsidies gradually, with the intention of buffering the impact on individual commodity groups. The ultimate goal was to remove government from agriculture.[24]

Meanwhile, as Norton III completed the transition from Deputy Secretary-Designate to Deputy Secretary, Doris discovered that she could volunteer for a White House job. She applied and went through the customary FBI clearance. Several Arizona neighbors were interviewed, and she was confirmed for that job much more easily than her husband was for his. At first she helped address Christmas cards, sent thank you cards, and answered the phones. Within a few months, she was moved to Nancy Reagan's press office where she clipped articles about the first lady, her family, and her children. Mrs. Reagan wanted to keep apprised of her image in the press, and asked to see the daily clippings no later than 4:00 p.m. each day. During the Norton's comparatively short time in Washington, DC, Doris enjoyed her volunteer job in the White House as well as the social and cultural offerings that were part of life in the nation's capital.[25]

At times during his stint in Washington, Norton served as Acting Secretary of Agriculture when Secretary Block was absent from the capital, and sat in on cabinet meetings next to the Oval Office, with President Reagan two seats away. The cabinet meetings always started at 2:00 p.m., and all members of the cabinet, "George Schultz, Jim Baker, Caspar Weinberger, and all the different secretaries, were to be there at ten minutes before 2:00 p.m." The double doors to the Oval Office would always be closed when cabinet members entered, and all would stand around and talk. Then the doors would open and the President would walk in, "not at 1:59, not at 2:01. At 2:00!"[26] Norton enjoyed observing Reagan and appreciated his insistence on punctuality.

At his first cabinet meeting, Norton III stood with his back toward the double doors, talking with one of the cabinet secretaries. He failed to hear the Oval Office doors open. When he saw the other cabinet secretaries head toward to their chairs, he swung around and bumped into President Reagan. "I almost knocked him down!" Norton recalled. "Oh my God, Mr. President, I'm sorry," he said, dismayed. The president replied, "Well, nothing to worry

John and Doris with Vice President and Mrs. George H. W. Bush, 1985.

about. You didn't knock me down." That incident and several other interactions with the president convinced Norton that Reagan was really a genuine human being who didn't take himself too seriously. "He was a down-to-earth guy."[27]

Another dimension to the job involved social events surrounding professional gatherings convened in Washington. "When you are the number-two guy in the department," Norton III asserted later, "you're invited to all these trade association meetings in Washington—and there were a lot of them." The Meat Growers, Michigan Cherry Growers, Georgia Peach Growers, and Vidalia Onion Growers, among countless others, kept Norton III and Doris occupied on weeknights. "It got to be kind of a grind after a while," Norton complained, "I mean, you work all day and then you'd love to go home and put your feet up and watch the evening news, but you've got to go to some damn reception and stand there with a cocktail in your hand, and play nice guy to some goofball from Michigan; but that's the game of politics." The social duties associated with the job of Deputy Secretary of Agriculture, in some instances, rubbed Norton III the wrong way.[28]

Norton III and USDA professionals and staffers worked feverishly through June preparing the 1985 Farm Bill. He negotiated and helped craft the language that went into what he considered "a well-reasoned, comprehensive bill" that reflected President Reagan's conservative priorities.

The bill, however, would face formidable opposition in the person of Congressman Jamie Whitten (D-Mississippi), who had been elected and re-elected from Mississippi's second district since 1941. In 1950 he was named chairman of the Agriculture Subcommittee on Appropriations, and in 1979 he had risen to chairman of the entire Committee on Appropriations. As Norton III pointed out, Congressman Whitten "was very powerful. In fact, many people in Washington felt he was the second most powerful man in the government next to President Reagan, because being chairman of the House Subcommittee on Agriculture . . . *and* chairman of the House Appropriations Committee meant that few, if any nominees for positions of influence escaped his scrutiny." Whitten clashed with the Reagan administration on many policy matters, including agriculture. He voted against Reagan's economic plans, tax cuts, increased defense spending, balanced budget initiative, tort reform, welfare reform, missile defense system, abortion restrictions, and even the Persian Gulf War. In the debate surrounding the Farm Bill of 1985, Congressman Whitten kept his constituents in northern Mississippi foremost in mind, and they were largely poor southern farmers.[29]

To Norton III's enduring frustration, Congress never held a hearing on the Reagan administration's 1985 Farm Bill. Whitten and House majority Democrats reinstituted the previous program with even more generous subsidies. Norton III allowed that the bill, passed in December 1985, was "socialistic." Deputy Secretary Norton stayed in Washington, DC, for the pre-Christmas signing ceremony, then flew to Phoenix for the holiday break. In Phoenix he expressed his disillusionment with the political process, and while at home, assessed his businesses, which had a "few problems, but nothing too troublesome."[30]

He returned to Washington, DC, in January. Increasingly frustrated with the chasm between the Reagan administration's articulated philosophy of agriculture and its failure at implementation, he began to look at a return to his businesses in Arizona and California.[31] In addition to his palpable disenchantment with the political process concerning the Farm Bill of 1985, numerous changes at the USDA gave Norton even more misgivings about his government service. On January 7, 1986, John R. Block, Secretary of Agriculture and Norton III's immediate superior at the USDA, sent President Reagan a letter of resignation. "Dear Mr. President," Block's letter began, "For the past five years, it has been my great honor and privilege to serve as your Secretary of Agriculture. The proudest moment of my life was when you asked me to be a member of your Cabinet. It is, therefore, with great regret that I today herewith submit my resignation to be effective Friday, February 14, 1986." In his pro-forma response, President Reagan wrote, "It is with deep regret and heartfelt appreciation for your service to my Administration that I accept your resignation as Secretary of Agriculture, effective February 14, 1986." At the press conference announcing his resignation, Block indicated he was weighing "a number of options in the private sector but had no final decision on which one to select." One of President Reagan's four holdovers from his first term—Defense Secretary Caspar Weinberger, Commerce Secretary Malcolm Baldrige, and Housing and Urban Development Secretary Samuel Pierce—remained. Block gave no mention of his preference for a successor.[32]

As Block left Washington, DC, Norton III assumed the role of Acting Secretary of Agriculture on February 14, 1968. Yet he, too, had decided to leave the administration to return to private business. President Reagan had already signaled his intention of appointing fellow Californian Richard Lyng, who had preceded Norton III as Deputy Secretary of Agriculture, to the Secretary's post.[33] The usual protocol was for a new Secretary to appoint his own professional staff, and this, too, weighed into Norton's reasoning. Moreover, the notion of two residents from the Pacific Southwest serving in the USDA's top two posts would raise the hackles of Midwestern congressmen and their constituents. Regional balance and representation from the traditional "farm states" were considerations for any administration dealing with USDA political appointments. Ultimately, Norton III served as Acting Secretary for five days, and on February 19, 1986, resigned from the post that he had held for a little over nine months.[34]

Interestingly, after Norton had submitted his official letter of resignation, but prior to his return to his corporate offices in Phoenix, President Reagan, through his Chief of White House Personnel, Robert Tuttle, contacted Norton. "You know," Tuttle said, "we've been talking over at the White House here, and we'd like to do something nice for you, and we've got something here that you might be interested in. Would you like to become Ambassador to Argentina?" Initially, Norton III responded, "Let me think about it." He went home, discussed it with Doris, and after careful consideration, concluded that he really wanted to head home to Arizona. He told Tuttle, "You know, I don't suppose very many people would turn that one down, but I am going to respectfully decline. Thank you for a wonderful offer, but I'm going home to Arizona." Tuttle, half-expecting Norton's rejection of the offer, graciously told him, "Well, I understand but I thought if you wanted it, you could have it." In 1989, Norton III and Doris visited the US Embassy in Buenos Aires and attended a dinner at a table that was "as long as a football field with all these dignitaries." He turned to Doris and said, "You know, if I'd been the Ambassador to Argentina, we'd be doing this four or five nights a week." Doris responded, "Oh, thank you for not doing that!"[35]

Norton III came away from his USDA experience with greater insight into the inner workings of government. The penchant for the administration to talk about farm policy one way but act another was especially irksome. He suspected that the administration backed away from its conservative principles in order to retain Republican midwestern Senate seats in the fall election. "It's critical to patronize the Midwest in this political year [1986]; it's not wise politics [to have westerners in the two top department slots] in light of the critical senate races," he suggested to one California reporter.[36] Of the thirty-four Senate seats up for election, Republicans held twenty-two, and fourteen were "Farm Belt sensitive." Norton III, in the short time he worked in the federal bureaucracy, grew isolated within the administration. His oft-stated adherence to a market-oriented approach to farm policy conflicted with President Reagan's political pragmatism. When asked if there was a wide disparity between theory and practice in the administration concerning agricultural subsidies, Norton III answered, "As wide as the Pacific Ocean, and this gulf," he reiterated, "was directly related to the Midwest senatorial elections." [37]

Without question, Norton III remained forthright during his short stint as Deputy Secretary of Agriculture. His candid approach and inability to adopt "nothingspeak" was at stark variance with Washington, DC's political culture. One of the most successful farmers in the West and an outspoken critic of federal agricultural subsidy programs, he was not a favorite among midwestern politicians, and found himself at odds with powerful Senator Robert Dole (R-Kansas). Dole, who immediately responded to Lyng's nomination to succeed Block with statements championing the notion that one of the two top spots at USDA should hail from a farm-belt state, became, frankly, an irritation to the agribusiness leader from the West. Norton III also expressed regret and disappointment over his nine months in office after the failure of the administration's farm bill to gain passage in 1985. "My hopes of taking part in a program to gradually get the government out of agriculture," he told *Arizona Business Gazette* reporter, David Poppe, "were gradually eliminated by the bill, which provided for tremendous agricultural-subsidy increases."[38] These subsidies, Norton III concluded at the end of his tenure in government, had made for a generation of farmers dependent not on the market for their profits, but on the government. Even worse, subsidies encouraged farmers to overproduce, and they discouraged competition. Clearly, by the spring of 1986, life in Arizona appeared more attractive than ever, and he returned to his agribusiness enterprise with a fresh perspective and renewed commitment.

·11·

AN ACCURATE VISION

◇◇

"I did have a good and accurate vision of the extreme value and the very limited quantity of water that's available in the Southwest."

JOHN R. NORTON III, AT HIS OFFICE IN PHOENIX, ARIZONA, JULY 22, 2011.

"He revolutionized irrigated agriculture in the Southwest and has championed education, health care, and the arts. John Norton has added much to the quality of our lives."

HONORABLE DENNIS DECONCINI, US SENATOR RETIRED (D-ARIZONA),

PHOENIX, ARIZONA, APRIL 2, 2011.

I n the spring of 1986 John R. Norton III returned to Phoenix to reassert control over his vast agribusiness. Farming operations were going well, and the few problems he found were quickly addressed and corrected.[1] Norton III picked up where he left off and for the next year focused on maintaining the high quality of his company's produce, grains, and livestock, while at the same time improving his harvesting and delivery processes. In addition, he reestablished his role in community affairs, including his service on a variety of boards of directors and committees, while supporting his chosen cultural and political causes. Arizona, meanwhile, began to benefit from its legal and legislative victories: first *Arizona v. California* and the subsequent legislation related to it; then the Colorado River Basin Project Act of 1968—which, nearly two decades later, resulted in the completion of the Central Arizona Project (CAP) through the Salt River Valley and thence southward to Tucson.[2]

Indeed, when Norton III returned to Arizona, the first drops of Colorado River mainstream water had been diverted to delivery systems in the Valley—including some of the systems his grandfather had worked on in the nineteenth century. What had originally been intended as a massive recla-

John R. Norton III in the fields inspecting a lettuce crop.

mation project to sustain Arizona's agricultural economy now appeared to provide equal, if not greater, benefits for domestic and industrial water users. Norton III marveled that he and other central Arizona farmers had moved to Blythe three decades earlier because they could not conceive that CAP would ever materialize. And though agricultural uses would compete with urban and domestic uses for rights to this valuable new asset, he expressed gratitude to the legal and political leaders who had fought for CAP.

As Arizona's future took on a new trajectory with this fresh flow of water into the state's interior, Norton III's father, who had provided so much, passed away on May 17, 1987. Upon his passing, Norton III told family and friends that his father valued hard work and made sure that his son, during his teens, worked summers either in the packing sheds or on the ranches. Norton III realized that without his father's guidance and support he could never have reached such financial and professional heights. His father's passing, moreover, forced him to reevaluate his successes, goals, and priorities.

Operating the huge agricultural empire that was the J. R. Norton Company had become an enormous challenge, and Norton III began thinking about disentangling himself from this key dimension of his life's work.[3] Disposing of the operations and the materiel that comprised J. R. Norton Company posed daunting logistical obstacles. Nevertheless, as he told an interviewer in 2001, Norton had reached the decision that he "wanted to smell the roses."[4] He began phasing out of the year-round farming operation—primarily lettuce—in 1989. "Lettuce was by far the most demanding, because that's every day," Norton III asserted. "You have to harvest it when it's ready; you don't wait a week. It's a business that requires constant attention."[5] It took the better part of six years to extract himself from lettuce and other operations that required his daily and undivided attention, although in subsequent years he continued growing some of the less demanding crops.

In the fall of 1989 Norton ceased growing lettuce in Salinas and several other areas where the company produced seasonal rather than year-round crops. On July 25, 1989, the *Salinas Californian*, the region's leading newspaper, ran a front-page headline: "J. R. Norton Quits Salinas." When a reporter questioned him about the company pulling out of the area after twenty years, Vice President for Finance Owen Cotton said, "We're significantly phasing our operations down; the layoffs will be significant. . . . We're really looking at the whole way we've got our capital invested; too much is devoted to the fresh lettuce markets." Cotton's response revealed the company's intention to reallocate resources and invest in other vectors of the regional economy.[6]

Norton III retained property—two to three thousand acres—in the Palo Verde and Imperial Valleys and continued to farm there, though he no longer grew lettuce or row crops; only alfalfa, wheat, and citrus. Robert Micalizio, J. R. Norton Company's manager of the Palo Verde operations who had worked with Norton III in Blythe since 1960, continued to oversee the farms in Imperial and Palo Verde. The citrus harvest on those remaining properties was contracted to the Coachella Valley Citrus Company in Indio, who moved their crews to Norton III's land in the fall to harvest citrus—mostly lemons. The company then shipped the harvest to its packing houses in Coachella Valley. As Norton put it, this pared-down operation was "turnkey, easy, nothing to it," and after 1990 he described himself as a "gentleman farmer."[7]

According to close observers, over the past two decades Norton III has been as busy with his new business interests as he was prior to divesting himself of the bulk of his farming and ranching operations. These new projects include investment and real estate development, like his stakes in Sun Lakes and Westcor, Incorporated, among several others. His service on company boards of directors, both for-profit and nonprofit, as well as his increasing involvement in philanthropy have shaped his life over the past twenty years. Support for education, the arts, health care, and conservative political causes have also occupied his time since his disengagement from the daily grind of

John R. Norton III.

agribusiness in 1990. Significantly, many of these activities predated his service in the federal government and spanned the period 1960 to the present.[8]

From 1975 to 1981, for example, Norton III served as a member of the Advisory Board of the Smithsonian Institution. When Congress created the Smithsonian in 1846, it vested the administrative responsibility in a Board of Regents comprising the Chief Justice of the United States, the Vice President of the United States, three members of the US Senate, three members of the House of Representatives, and nine citizens. Over time, the citizen board increased its influence and became a real policy and decision-making force at the Smithsonian. Norton III was one of the fortunate citizens chosen to serve on the board. During his six-year term Norton supported the creation of a major outreach mechanism, the *Smithsonian* magazine. He recalled the spirited service of the eighth secretary of the Smithsonian, S. Dillon Ripley, who, upon the end of Norton III's sixth year of service, wanted him to recommend his successor. Like Norton, the candidate should hail from the Southwest. John provided the names of three distinguished Arizonans: Sandra Day O'Connor, then an Arizona Appeals Court Judge; James Simmons, Chairman and CEO of United Bank of Arizona; and Gay Firestone Wray, a prominent scion of the Firestone family. The Smithsonian selected O'Connor at their April 1981 meeting; she was to replace Norton on the first of July. In May, however, President Reagan appointed O'Connor to the Supreme Court, so she could not accept her Smithsonian responsibilities and had to defer. The impish Ripley, when he learned of Justice O'Connor's historic appointment, fired off a one-line handwritten note to Norton III: "Dear John, You sure know how to pick 'em!" Ultimately, the appointment went to Gay Firestone Wray, who served on the board for six years, eventually becoming president.[9]

In early 1975, Arizona Public Service (APS), the state's largest private utility, invited John Norton to join its board of directors. He made the motion for the feasibility study that eventually led to the construction of the Palo Verde Nuclear Generating Station, which brought electricity online in 1986. He had to resign when he went to Washington, but was welcomed back when he returned to Arizona. He served on that board until 1999, when he turned seventy years old. He also rejoined the board of Ramada Inns in 1986, after his stint at the USDA. He was on the board three years later when Ramada split its business, selling the hotel/restaurants to Hong Kong-based New World Development Company for $540 million. The casinos, including the Tropicana Atlantic City and the Tropicana Las Vegas, were reconfigured as a new, publicly traded company called Aztar Corporation. In 1995 Norton voted with the majority of the board when the new corporation expanded to include riverboat casinos in Caruthersville, Missouri and Evansville, Indiana. As he did with APS, he retired from Aztar's board in 1999 when he turned seventy.[10]

In 1978 the Phoenix 40, an organization comprising the region's top business and civic leaders who wrestled with major problems facing Arizona, asked Norton III to join the influential three-year-old group.[11] Norton described the Phoenix 40 as "a civic group of business leaders in the community who met for the purpose of discussing the problems of the community and tried to do something about them." It worked with the state legislature and city councils, and served as a kind of "value-added" dimension to public policy leadership and governance; according to Norton "they accomplished a lot of good things from the 1970s through the 1990s." John Norton became chairman in 1991 and enjoyed his year of service. The organization's attitude—

that civic leaders should not only be concerned with economic growth and development but also with ethics and with the needs of their communities—struck a chord with him. On January 8, 1993, as Norton III completed his membership in this group of Arizona leaders, it officially changed its name to "Greater Phoenix Leadership," and touted its steadfast commitment to collaborate with business and government in order "to promote initiatives, public policy, and projects that strengthened the future of Arizona." The organization clearly reflected not only John Norton III's values, but also his commitment to the region's overall growth and development.

In 1994, Norton III met the controversial, eccentric, yet brilliant leader of the for-profit education movement in the United States, Dr. John Sperling, at a seminar addressing decriminalization of marijuana. Norton's libertarian views on personal freedom and opposition to government intervention into the private lives of individuals prompted him to attend the meeting. Sperling, who amassed a fortune after founding the University of Phoenix, part of the publicly-traded Apollo Group, in 1976, admired Norton's comments at the confab, and soon asked him to serve on Apollo's board of directors. Norton knew that Sperling, who actively financed the decriminalization of marijuana, supported this then-unconventional cause because he had benefited from the drug's pain-reducing effect during his bout with cancer in the 1960s. While their views on decriminalization of marijuana comported to some extent, Sperling's outspoken views on almost every other public policy issue placed him squarely on the opposite side, politically, from John Norton III. Yet they forged an unlikely working relationship from the mid-1990s to the mid-2000s. During his decade-long service on the Apollo board, Norton III grew to appreciate the benefits and business model of the for-profit educational institution, which provided a much-needed service to adults who otherwise might have missed out on furthering their education.[12]

Two related organizations enabled John Norton to network and socialize with like-minded business leaders during the 1980s and 1990s: the Young Presidents Organization (YPO), which he joined in 1968 when he was thirty-nine years old, and the Chief Executive Organization, which YPO "graduates" aspired to after they turned fifty years of age. The former group, founded sixty years ago, not only introduced Norton to a peer network, it also enabled him to circulate in an environment in which he could freely discuss the subtleties and nuances of his business. He made lifelong friends in YPO, and his active participation and contributions to the group resulted in his invitation to the Chief Executives Organization in 1979.[13]

Norton III enjoyed the annual CEO conventions, called "universities" because of the mix of education and socializing that comprised these often elegant gatherings. Doris often joined him at various CEO events. In 1989, Norton was asked to run the annual CEO Convention that fall, which would be held in San Francisco at the elegant St. Francis Hotel. Norton III arranged the speakers, entertainment, and all the related activities for the five-day program. He crafted an invitation letter that reflected his sensibilities and values at the time: "It has been said, 'Every man should be allowed to love two cities, his own and San Francisco!'" He touted the Bay area's well-known attributes with a familiar refrain, "Those who lose their hearts to San Francisco, do so with good reason—an ideal location, picturesque topography, a spring-like climate, particularly in October, and a people unique for warmth, openness, and a diversity of culture."[14] Norton III announced that the forum would have a strong and diverse educational program, with particular emphasis on business and personal development and a subtheme of sharing and personal relationships. He promised participants that they would be challenged, entertained, stimulated, and renewed.

Sessions that attracted CEOs and spouses addressed the political and economic challenges of the day, and tours of San Francisco's unforgettable historic sites and cultural venues were forum highlights. The weeklong conference, which began with registration on Sunday, October 15, 1989, featured a keynote address by Vice President Dan Quayle. Sessions like "How to Be an Efficient Philanthropist" and "Do Something with Your Talents" generated overflow attendance. A glorious pre-conference reception on Sunday and a Monday evening moonlight cruise on San Francisco Bay following the first day's classes and seminars were huge successes.

On late Tuesday afternoon, at 5:04 p.m., October 17, 1989, as Norton III and Doris prepared for the evening's reception and entertainment in their thirty-second-story suite, John Norton heard what he thought was a thunderclap. Suddenly the building was shaking, and he was thrown across the bed. Doris and he rushed to the window and looked across Pershing Square at the I Magnin building. Sheets of glass rained on the street below, and the couple realized they were in the Great San Francisco Earthquake of 1989. "People on the street were running in every direction," Doris recalled, "and the top of the building felt like it would break off on the next swing; it was very scary." A 7.0 Loma Prieta earthquake, the worst since the great quake of 1906, rocked northern California.

The 1989 quake struck with tremendous force, cancelling Game 3 of the World Series between the Oakland A's and the San Francisco Giants, buckling Interstate 880 and the Bay Bridge, and causing hundreds of deaths and thousands of injuries. The World Series game, scheduled to start at 5:35 p.m. at Candlestick Park, slightly south and east of downtown San Francisco, enabled an audience of millions to witness the first major earthquake in the United States to be broadcast via live television. Thousands of people were in the stadium when the quake occurred. Experts credited the timing of the series as a fortunate coincidence that prevented massive loss of life in the Bay Area. The key was the fact that many people had left work early or stayed late to gather in after-work group viewings and parties, reducing traffic that otherwise would have been bumper to bumper on the collapsing freeways. Though initial estimates figured hundreds of people had died in the Interstate 880 bridge collapse, the final count was forty-two.

A few miles away, at the St. Francis, elevators jammed, the lights flickered and went dark, and water and plumbing shut down. The Nortons left their suite, descended thirty-two floors to the ballroom, and began to converse with a handful of dazed convention attendees who had gathered in the immediate post-quake confusion. Conference speaker Dr. Mortimer Feinberg, an organizational psychologist, observed a group of homeless men who had come into the hotel for shelter. These men, "used to living in Union Square Park, repeatedly calmed everyone with reassuring words of comfort. They were the ones at the bottom of the social ladder and here they were, typically bedraggled and forlorn, rising heroically to the occasion. . . . In addition, they refused all money offers." Similarly, the stout hotel staff brought food, drinks, and even recruited some street entertainers for the bewildered CEOs and their spouses.[15]

Incredibly, Norton III and his committee chairmen were able to secure buses the next morning, and the bulk of the convention left the destruction and went to Reno, Nevada, where one of the major hotels opened two hundred rooms for the displaced conventioneers. Three couples, including the Nortons, remained behind in San Francisco to take care of the details and logistics tied to nature's scuttling of the convention. Doris commented, "The hotel was so generous and concerned and willing to do anything to help us, and they would put out these feasts and champagne; whatever we wanted since we had to stay behind." For the next year, Norton III and Doris received

Nancy Reagan at the Goldwater Institute in Phoenix, Arizona, with
John R. Norton III and others.

letters from throughout the country, praising them for the cool, calm, and
efficient manner in which they handled the emergency situation at the CEO
University in San Francisco.[16]

The year prior to the earthquake in northern California, the Goldwater
Institute, with the blessing of the recently retired US Senator and conservative icon, Barry Goldwater, was established. The new think tank proclaimed
it was a government watchdog that was committed to expanding free enterprise and liberty. Norton III and a group of his politically active conservative
friends supported the institute, which has focused its advocacy—through the
publication of research, litigation, and media exposure—on economic freedom and personal liberty. In recent years, thanks in part to a significant gift
from Norton III, the Goldwater Institute, through its Scharf-Norton Institute for Constitutional Litigation, has initiated several legal actions that have
sought to promote constitutionally limited government. Norton III, who has
typically described himself as "libertarian," nevertheless became deeply in-

volved in the think tank in 1993, when he joined the executive committee. In 1994-1995 he served as chairman of the board, and was one of those responsible for hiring Jeff Flake as director. Flake has since served Arizona in the House of Representatives, and was elected to the US Senate in 2012. Norton III has remained involved both as a donor and board member, as the Goldwater Institute's influence on Arizona public policy has increased over the past decade.[17]

Clearly, Norton III synthesized a variety of careers in the past two decades: agribusiness visionary, education benefactor, patron of the arts, public servant, and acclaimed photographer. He has photographed subjects all over the world, and his images have been featured in numerous shows. His support of photography in Arizona has taken several forms, among them the John R. Norton III Vision Award, established in 2012, which is presented each year to a member of the Arizona photography community for lifetime achievement. Additionally, the John and Doris Photo Gallery at the Phoenix Art Museum, which opened in 2006, represents another of his noteworthy contributions to the community.

Besides his professional accomplishments and countless awards, another dimension of his portfolio has been his generosity to his fellow man. An incomplete list of John and Doris's charitable donations include St. Joseph's Hospital, where Norton III served on the board of directors for several years and donated $1 million for a healing garden; the Phoenix Art Museum, where he provided a gift of $1.4 million for a photo gallery; and the University of Arizona Art Gallery, where his $400,000 donation enabled that department to retain and display the NASA paintings of Paradise Valley artist Robert S. McCall. He also endowed a faculty chair in Prostate Cancer Research at the University of Arizona Medical School. A set of gifts to Norton III's alma mater, the University of Arizona, deserves special mention. In 2000 he and Doris created the John R. and Doris S. Norton Endowed Chair for Fathers, Parenting, and Families in the School of Family and Consumer Sciences, College of Agriculture and Life Sciences. And, in 2004, the Nortons donated $4 million to the program, and the school was named in their honor: the John and Doris Norton School for Family and Consumer Sciences. According to Norton School director Dr. Soyean Shim, a concern for society, a focus on hands-on experience, and cutting edge research unite the school's programs, whether one is interested in helping troubled families, teaching parenting skills, or directing international imports for a major retailer of consumer goods. Indeed

John R. Norton III photographing in Germany.

at a recent conference at the Norton School, top domestic companies and trade groups like Google, Microsoft, Walmart, the Arizona Food Marketing Alliance, Kohl's, Nestle, Purina, and others gathered to consider the newest product developments and marketing strategies.

The Nortons have made other substantial gifts to such organizations as Bourgade High School, the Boy Scouts of America, All Saints Episcopal Day School, Xavier School for Girls, and many other organizations. In short, over the past decade John Norton has given millions for educational, medical, and cultural programs that have benefitted Arizona and the Southwest.[18] These and other examples of Norton III's vision and generosity have contributed to a healthier and more prosperous region.[19]

John Norton III maintains that his achievements in agribusiness, civic stewardship, and philanthropy were all built on foundations provided by his father and grandfather. When asked about his proudest accomplishment, he issued what seems to be a curious answer, but it gets to the heart of understanding life in the arid regions of the Southwest. "I guess I'm proudest of the fact that my basic hunches about life and my intuitions were quite good, although my timing wasn't always perfect," he said after some thought. "I knew when I went to Blythe as a very young man, that water in the Colorado was going to be gold someday." And fifty-five years later he noted that Los Angeles and the Metropolitan Water District of Southern California (MWD) essentially completed the purchase of a good portion of the water in the Palo Verde Irrigation District in order to take it to Los Angeles." The water available in the PVID was "the reason I elected to go over there to farm in 1955." In the Salt River Project, he asserted, "20,000 acres out of the 250,000 under the project is all of the farming that is left." "When I was a kid," he recalled, "it was just the other way around. It was 20,000 in the cities, and 250,000 farming. But I did have a good and accurate vision of the extreme value and very limited quantity of water that's available in Arizona and southern California."[20]

More than any other issue, the effective distribution of water made Norton's agribusiness career possible. Unquestionably the fortunes of his Arizona pioneer family were based on water, or, more specifically, its diversion onto land. As a business leader who followed in the agricultural and ranching footsteps of his father and grandfather, he supported various reclamation projects and their expansion: the Salt River Project, the construction of Coolidge Dam on the Gila River, and the San Carlos Irrigation Project. He approved of local water-user associations, like PVID and his hometown SRP, in supporting and maintaining operation of federal reclamation projects. In the 1960s he became a student of *Arizona v California* (1963) and a proponent of Arizona's cause in the case. The case, Norton III observed, was one of the most complex and hotly contested in the history of the Supreme Court. In the course of eleven years and at a cost of $5 million, over 340 witnesses and 50 lawyers produced 25,000 pages of mind-numbing testimony. When a divided court announced its opinion on June 3, 1963, and its decree on March 9, 1964, a greatly modified legal framework governed the use of Colorado River water previously claimed by California. The ruling, Norton III knew, paved the way for what he had previously labeled a "pipe dream"—CAP. As he continued his sixteen-hour days in Blythe in the mid-1960s, he observed the five years of legislative wrangling that brought forth the Colorado River Basin Project Act of 1968, which sanctioned construction of CAP. The case also served as the capstone to legendary Senator Carl Hayden's unprecedented fifty-seven-year career in Congress (US House of Representatives, 1912-1927; US Senator, 1927-1969). Norton III and his fellow agriculturalists expressed their unanimous approval as hundreds of miles of canals, pumping stations, and water delivery systems began to wind their way through Arizona's Sonoran desert in the 1970s and 1980s." For three generations of Nortons in Arizona, the authorization of CAP marked the end of a sometimes antagonistic relationship with the US government, and the beginning of a governmental infrastructure that administers the use and distribution of Colorado River water within Arizona. Norton's assessment of the situation in 2011 was on point: "Instead of fighting for more of the resource [water] based on perceptions of fairness and past ill-treatment, the state moved aggressively to develop the resource it had won; 2.8 million acre-feet of mainstream water." Cooperation came at a price, however: in accordance with the 1968 legislation, Arizona's rights were made junior to California's and Nevada's, meaning shortages in times of drought hit Arizona first.[21]

On several occasions in recent years, John Norton III has lamented the decline of irrigated agriculture as a way of life, though he considered it inevitable, and even essential, in sustaining a tax-paying civilization in the Pacific Southwest. Cities and suburbs in the region, especially since the 1960s and 1970s, have increased their consumption at the expense of agricultural and rural users. He was not surprised when, with the passage of the Colorado River Basin Project Act of 1968, Arizona municipal leaders stepped to the forefront as supporters of CAP. They viewed it as the best source for augmenting dwindling water supplies. As Norton III put it: "What began as an agricultural rescue project became, by the 1960s, a lifeline to Arizona's growing cities and, by the way, a substantial gain to the state's tribal residents." He continued, "These debates, to me, prefigured a steady decline in agriculture over the next fifty years and beyond, and it was at this point that I started thinking about redirecting or reconfiguring my financial growth strategies."[22] He realized also that Arizona's tribes, cities, and farmers needed to play equal roles in the use and distribution of the benefits of CAP, and how Arizona stakeholders managed their hard-won good fortune depended upon sound financing, the efficient administration of CAP construction, and an open relationship with the federal government.

Shortly after passage of the 1968 legislation, but prior to Norton's return to Arizona in 1972, the thorny problem of the simultaneous need for Congress to appropriate monies for the US Bureau of Reclamation in the Department of the Interior as well as for related administrative issues pertaining to the construction of CAP remained to be addressed. Moreover, the Department of the Interior needed assurances that the government would be repaid for construction costs. In 1970 Congress appropriated the first $1.2 million to commence CAP construction, but the US Office of Management and Budget (OMB) nixed the release of funds until a repayment contract between the Secretary of the Interior and Arizona water users had been executed. Only then did Arizona water leaders react. After considering several alternatives, they took legislative action, and on June 16, 1971, created the multicounty Central Arizona Water Conservation District (CAWCD) as the agency that would oversee repayment to the federal government.[23]

In 1977 President Jimmy Carter, in a political miscalculation of mammoth proportion, initiated a campaign against nineteen western water projects, including CAP. Carter's mishandling of western water policy, according to Norton III and others then involved in irrigated agriculture, was one of a host of problems that beset the president and further cemented a growing Republican domination of Arizona and other portions of the Pacific Southwest. Ultimately Carter, whom Norton III considered "politically tone deaf to western water issues," backed down. Even after the project was reinstated, in October of 1979 Secretary of the Interior Cecil Andrus informed Governor Bruce Babbitt that Arizona must demonstrate a greater and more serious commitment to controlling groundwater in the state before he issued final recommendations for CAP allocations.

The groundwater issue, as Norton III stated time and again, was the reason he and other Arizona farmers moved to Blythe and its secure water rights. According to Norton III, who was a confidante of Governor Babbitt's on water and agricultural issues, the Democratic governor may have used the the threat of federal obstruction to unify Arizona water interests in getting groundwater use under control. For their part, the Carter administration, and especially Secretary Andrus, reasoned that because CAP had been promoted as a rescue project to alleviate the overdrafting of groundwater that had been going on for more than half a century, Arizona had to demonstrate that it maintained the political will to curtail the use of groundwater in exchange for surface-water deliveries from CAP.[24]

Babbitt and a cooperative Republican-dominated legislature, led by the pragmatic and effective Burton Barr in the Arizona House of Representatives, responded with the adoption of the Groundwater Management Act of 1980. John Norton III weighed in throughout the process and served as a confidential consultant to Babbitt, since the governor knew he was the leading irrigated-agriculture player in the region. The act established a specific timeline for reduction and elimination of groundwater pumping in environmentally compromised areas by creating Active Management Areas (AMAs) and the cumbersomely labeled Irrigation Nonexpansion Areas. Within the newly created AMAs, development was restricted to areas with an "assured water supply," which meant an area with sufficient water for the next hundred years of development. Thus any municipality with a CAP contract automatically had a hundred-year assured water supply designation.[25]

In 1986, then-Governor Babbitt speculated what Arizona would have been like had it lost the great Supreme Court case and there were no CAP. "It isn't so much what Arizona . . . would be like without CAP, but what sort of Arizona we'd be seeing in 2000," he told a writer for *Arizona Highways*. "The absence of the Central Arizona Project would be terribly evident. We'd see an Arizona where agriculture had been ruthlessly eliminated by law— agriculture would be out of production to prevent an overdraft of the water supply. Sometime in the early twenty-first century, certainly by 2025, other things would begin to change. We'd have unacceptable water quality, earth fissuring, drastic groundwater problems. Ultimately, we would have to limit, progressively, certain types of water usage and impose dramatic limitation on growth."[26]

In the 1980s and 1990s Norton III served on the board of the Central Arizona Project Association (CAPA), a private, nonprofit association composed of agricultural, business, and industrial professionals who understood that the CAP was fundamental to the economy of the entire state. Organized on July 1, 1946, it raised money, provided research and legal assistance, lobbied for and publicized the project.[27] As the CAP neared completion in the early 1990s, serious problems arose. First, the ultimate cost of the canal vastly exceeded original estimates. The repayment contract between the CAP and the federal government was exceedingly complicated, and was interpreted differently by Arizona and the United States. As a result, in 1994, when the notice of completion triggered Arizona's obligation to repay its share of the cost of the canal, the state and federal repayment calculations differed by nearly $1 billion. Norton III was not surprised by this disparity, and he informed Arizona leaders that they were in for a long and complex period of struggle with federal administrators.

In addition to the repayment issue, Norton and his agribusiness brethren were acutely aware of a problem within the original concept for charging CAP customers, which was that any water not used by Arizona cities would be paid for by Arizona farmers. The farmers, therefore, had entered into a series of contracts called "take or pay" obligations, which required them to pay for certain portions of the water supply. Norton III objected to a controversial aspect of this repayment mechanism. Under CAP, the debt to the federal government for agricultural water bore no interest, but the cost of repaying just the principal alone meant that the water was far more expensive than had been originally anticipated. Additional debt also accumulated within those irrigation districts that had constructed delivery systems to distribute CAP water to farms. At the same time, the CAP area's principal irrigated crop—cotton—had dramatically fallen in value on international markets. As a result, demand for CAP deliveries to agriculture dropped precipitously from 500,000 acre-feet to 50,000 acre-feet in 1993 and 1994, and agricultural customers were threatening to file bankruptcy.[28] Norton III expressed profound disappointment at this development, especially in the agribusiness sector that had been so good to him for forty years.

Even more disheartening for the rollout of this highly anticipated federal water, when CAP deliveries commenced to Tucson, the largest municipal customer, the pipes spewed water containing what many consumers described as "brown gunk." Some Tucsonans thought the gunk was a plot by Phoenix residents, located "upstream," to poison them. The problem was actually that Tucson had switched overnight from a groundwater-based system to an entirely surface-based system, and there were serious engineering and water quality impacts. In fact, Tucson consumers responded with emotion, and through the initiative process angrily shut off all direct delivery of CAP water.

The little-known CAWCD had to respond to all of the above-mentioned challenges simultaneously. It attempted to negotiate a compromise with the federal government over the repayment costs, but quickly realized that the federal government was less interested in the amount of repayment than in the question of how much CAP water could be allocated to Arizona's Indian tribes. The claims of Indian tribes to water in central Arizona stemmed from the US Supreme Court case *Winters v. United States* (1908). The water rights claims in Winters were based on a doctrine which held that when the federal government created the reservations, it implied that it had reserved enough water to irrigate all the practicably cultivated acreage within them. The claims in central Arizona, Norton III realized, were mainly claims on the Gila River and its tributaries, the Salt and the Verde. But since the Gila's waters were fully appropriated by SRP, no water was available for those claims. The federal government's plan, therefore, was to acquire some of Arizona's CAP appropriation and use it to satisfy Indian claims. In 1995 and 1996, after much discussion, negotiations to address repayment and Indian claims collapsed. In those negotiations, Norton III recalled with irony, the federal government was represented by his old friend Bruce Babbitt, the US Secretary of the Interior, and his Assistant Secretary, Betsy Rieke, the former director of the Arizona Department of Water Resources. The dispute over CAP repayment and Indian rights resulted in a lawsuit filed by CAWCD against the United States in 1995. Four years later, after hundreds of pages of deposition transcripts, motions, interrogatories, court hearings, and marathon settlement negotiations, a tentative agreement was announced. CAP's repayment obligation was fixed at $1.65 billion and a total of 653,000 acre-feet of water was reallocated to satisfy tribal claims, principally the Gila River Indian Community in central Arizona. In effect, Indians controlled 48 percent of CAP water.[29]

Concerning agricultural water deliveries, which drew Norton III's attention, then-Governor Fife Symington appointed a sixteen-member Governor's Task Force that met during the first six months of 1992. Its chief charge was to propose strategies for use of Arizona's CAP entitlement. This first Task Force disbanded in disarray; several members claimed they needed more information. In response to this initial misstep, the governor's office and the Arizona Department of Water Resources requested another study that resulted in "An Economic Assessment of Central Arizona Project Agriculture," crafted by Paul N. Wilson of the University of Arizona's Department of Agricultural and Resource Economics. Wilson suggested restructuring economic formulas to recognize a "target price." In effect, this meant that CAP would sell water to agriculture below even the marginal cost of delivery, with the difference made up by Arizona municipalities. CAP implemented this program in 1995 and in exchange for less costly water, agriculture was asked to give up its long term CAP allocations. This program proved to be a success and CAP agricultural deliveries bounced back to more than 500,000 acre-feet in 2001. Yet the seeds for agriculture's ultimate denouement had been planted ,and Norton III noted that this CAP deal for agricultural water institutionalized the end of irrigated agribusiness (and thereby his way of life in Arizona) fifty years hence.[30] This future, he lamented, was Arizona's "brave new water world," in which agriculture played little, if any, role.

Regarding the "brown gunk" issue in Tucson, the city slowly worked its way through its problems with CAP deliveries. With the pronouncement of "Start Your Pumps" on May 3, 2001, Mayor Robert Walkup announced the delivery of clean—and potable—water to his constituents. However, in part because the issues outlined above—Indian water, agricultural costs, and Tucson's reticence—Arizona, in the mid-1990s, was not using its full CAP allocation. California, as Norton III knew well, continued to vastly overuse its 4.4 million acre-feet because it had a right to take any water Arizona left in the river. This led Arizona to adopt additional innovative water management institutions to foster using as much Colorado River water possible.[31]

One of these institutions, "water banking," involved taking CAP water and recharging it underground in central Arizona so that the state would protect itself from its junior status in times of drought in the river. Ground-water banking was also a deliberate mechanism to keep Arizona's allocation out of the hands of predatory California. Norton III, who wholeheartedly approved of this innovation, witnessed SRP, CAWCD, and Arizona municipalities agree to purchase excess Colorado River water from CAP and put it underground in central Arizona. This could be accomplished through direct recharge—allowing water to seep into aquifers—or through a process called "indirect" recharge—which meant that the water could be sold to farmers at a heavily subsidized price and the groundwater that the farmers would otherwise have pumped would be counted as "recharged" surface water, which then could be used in the future by cities or whoever had paid the subsidy.

In 1996 Governor Symington signed legislation establishing the Arizona Water Bank Authority as a new and separate state agency. Its purpose: to acquire excess CAP water and bank it in central Arizona. Norton III allowed that this was a brilliant move in western water diplomacy because the Arizona Water Bank was also authorized to make deals with neighboring states to allow them to bank water in Arizona for their future water needs. This new mechanism proved especially effective with Nevada, enabling Arizona to secure a critical ally in the always-challenging relationships over Colorado River water.

Norton III, always sensitive to the groundwater issues that caused him to remove to Blythe, approved of another water banking mechanism created in 1993. The Central Arizona Groundwater Replenishment District (CA-GRD) was another testament to Arizona water leaders' ingenuity. It was designed to make it possible for developers to build housing subdivisions where they did not have direct access to CAP water or to a municipality with a CAP contract. The CAGRD expended its powers and spheres of influence—with legislative approval—in 1999. CAWCD administers CAGRD and the latter has the ability to acquire excess CAP water from the former and store it underground on behalf of developers seeking to build subdivisions without direct access to CAP water. The subdivisions join the CAGRD and each house is subject to an assessment to pay for the water put underground. And by joining CAGRD a subdivision is thus deemed to have the key criterion in establishing a hundred-year assured water supply. Though the CAGRD has been criticized as opening a loophole in the Groundwater Management Act's link between renewable water supplies and urban development, it has, since its implementation, proven to be more popular than originally anticipated. More than 165,000 lots had been enrolled in this groundwater management mechanism by 2005 and since that time, thousands more have joined.[32]

In a recent discussion, Norton III referenced his grandfather and father, wondering aloud how they were able to forge an existence in such an inhospitable environment that required so much physical labor to survive. Its defining characteristic was aridity, he avowed. "One thinks of Frank Herbert's 1965 science fiction classic, *Dune*," he explained, "because that fictional world was so dry that its culture was built around individual hoarding of water by nomads who wore special suits to conserve their fluids and they also recaptured water from their dead." This was an example of "aggressive self-reliance," and my grandfather and dad exemplified that quality." Actually, late twentieth and early twenty-first century Arizona and southern California—Norton III's professional arena—were not entirely devoid of water, and in fact, his life and career offered a contrast to the earlier worlds of his forbears. His agribusiness of regional extent, his government service, his community stewardship, and his philanthropy were testament to this contrast. Water, he realized, became a shared enterprise, especially during his grandfather's lifetime, and during his father's lifespan and his; Americans created irrigation and conservation districts and countless other institutions to obtain, protect, and manage water supplies. In fact, though Norton III championed limited government since he first voted in 1950, he witnessed the power of government to build vast public plumbing systems like CAP, and he and other leaders of the Pacific Southwest have crafted laws, contracts, decrees, and decisions to give shape and order to the sharing of our most precious resource. Norton III and others who owed their distinguished careers to irrigated agriculture celebrated those who focused their energies on this collective endeavor. For at least the period spanning the lives of John R. Norton, John R. Norton Jr., and John R. Norton III—over 150 years—water has provided the Pacific Southwest's defining consensus. As the three generations of Nortons doubtlessly realized in their respective times, they always needed more water, they would use all they could get, they would stretch its uses as far as they could, and they would fight anyone who tried to take it away.[33]

CONCLUSION

◇◇◇◇◇◇◇◇◇◇◇◇◇◇◇◇◇◇◇◇◇◇◇◇◇◇◇◇◇◇◇◇◇◇◇◇◇

For three generations, spanning a century and a half, the Norton family grew to value water and its central role in the growth and development of the American Southwest. From graders, scrapers, and mule power, to pickup trucks and packing sheds, to twenty-first century computers and software that can predict the future cost of produce, water provided a palpable thread of intergenerational continuity. So too did the Nortons' agricultural lifestyle, based upon irrigated agriculture, the innovation that enabled civilization to gain traction and flourish in an otherwise inhospitable environment. This continuity, however, involved technological innovation over time, and the bucolic rough-hewn world of John R. Norton differed markedly from the labor intensive produce-growing and cattle-ranching lifestyle of Norton Jr., and contrasted sharply with the corporate efficiency-minded and scientific world of crop diversity and year-round lettuce harvesting that shaped Norton III's remarkable career. Sheep, cattle, feedlots, produce, citrus, and grains, in a more palpable way, linked three generations of Nortons, where their activities were circumscribed by the vast watershed of the Colorado River system. Besides water, an indomitable work ethic that inhered in all three John R. Nortons—sixteen-hour days were the norm—linked the three generations.

In the broadest sense, the Nortons were unapologetic visionaries who pinned their hopes on "pipe dreams" like the Salt River Project of the early twentieth century or the Central Arizona Project at its end. The half-century struggle with California and the federal government, oftentimes led by wrong-headed state leaders who quibbled about states' rights and federal power, ultimately played out to Arizona's enduring benefit. Naturally, Norton and Norton Jr.'s generations were largely concerned with protecting their respective agricultural economies from the vagaries of the desert and the always-present specter of water shortages. But, when possible, they expanded their enterprises, and in the case of Norton III the expansion was regional in extent.

As it turned out an urban-suburban civilization has sprung from the use of the region's surface waters and groundwater supplies, and Gerald Nash, one of the leading interpreters of the modern West, has reminded us that one of the prime characteristics of the late twentieth and early twenty-first century

West has been its urban aspect. The leading historian on western water development, Norris Hundley Jr., concurs with Nash's analysis and ties the West's decidedly urban character with water: "It is clear … that the growth of the arid West's urban centers, with their universities, museums, art galleries, drive-ins, freeways, and other symbols of civilization has been intimately connected with success at obtaining water." Norton III, whether innately or through careful observation, realized the urban trend in water use and wisely made appropriate adjustments in his agribusiness operations over the past two decades. As Norton III stayed ahead of the curve in his economic endeavors, and removinghis holdings from agriculture and moving them to energy and real estate, he noted that modern Arizona's ability to transport water hundreds of miles from the mainstream of the Colorado to the urban expanses of Phoenix and Tucson has meant more than the survival of these distinctive cities and their respective populations. It has signaled the stabilization and, in some instances, the further development of selected agricultural and mining economies. Finally, federal reclamation, embraced by all three generations of Nortons, and in its recent iteration as CAP, has eliminated—psychologically at least—one of the barriers to future growth in Arizona: the fear of water shortages.

One of the important consequences of putting water to beneficial use in the region was its crucial byproduct, hydroelectric power. Public and private power generation, whether for groundwater pumping, or as a matter of public policy, attracted the attention of each generation, though Norton III, as a board member of Arizona Public Service, maintained the closest relationships in this realm. The commodity revolutionized the quality of life during the first third of the twentieth century, thus touching the lives of Norton and Norton Jr. Later, Norton III, whether musing over the debates concerning power generated at dams on the Colorado or observing conflicts between advocates of private power against proponents of "inexpensive" public power, was engaged in this important issue. Indeed local, state, and federal agencies, reflected in groups like the Palo Verde Irrigation District, Imperial Irrigation District, City of Los Angeles, State of Arizona, Salt River Project, or the US Departments of the Interior or Agriculture, among others, battled constantly over the production, administration, and use of electrical power.

Civic engagement, public stewardship, and political activism also connected the three generations. These activities took various forms: boards, committees, government service, elected office, and philanthropy. One of the distinct themes that illuminated much about the Nortons was how they reflected the region's evolution in politics from the late nineteenth century to the early twenty-first. Norton, the southern Democrat who won election to the Maricopa County Board of Supervisors and sought to rival Carl Hayden in turn-of-the-century Democratic caucuses, reflected mainstream Arizona political culture at the turn of the twentieth century. Norton Jr., not as politically active as his father or son, nevertheless expressed his "Pinto Democrat" leanings when he chaired the "Democrats for Wilkie" effort in the 1940 presidential election. He was well on his way to Republican registration. And Norton III, perhaps the most active and astute of the three in political affairs, rose to a prominent position in the presidential administration of Ronald Reagan—Deputy Secretary of Agriculture—and continued to champion conservative and libertarian causes upon his return to the private sector in 1986.

In several instances this intergenerational political evolution was underscored by a critical philosophical debate about the role of the federal government in the region. The evolving relationship between East Coast seats of power and the nascent institutions of the Southwest, as well as questions of the federal government's proper role in addressing the Southwest's most fundamental environmental issue—these large political and cultural processes lay at the very heart of the Norton family narrative. All three generations were directly involved in a great historical undertaking: the transformation of America's desert regions into viable places to live and work.

In 1883, youthful John R. Norton embarked on a task that held enormous symbolic and historical significance. The Arizona Canal, as Supreme Court Judge and Territorial Governor Richard Sloan attested, led to the development of the Valley and demonstrated that water resource development pioneers like Norton possessed "broad vision." Five years after completing that monumental accomplishment, Norton participated in another signal event, the Breakenridge Survey, which located the Tonto Reservoir site where Roosevelt Dam was built two decades later. Before he reached the age of fifty he had witnessed an evolution that moved from local water cooperatives to corporate water development to the implementation of federal reclamation. He watched former President Theodore Roosevelt speak at the dedication of that iconic structure, which reconfigured the environment and economy of the Salt River Valley. Directly and indirectly, his private and public engagement in water issues played a crucial role in triggering a century of growth and progress. John R. Norton forged a remarkable existence and cut a unique swath in the history of Arizona agriculture.

Norton Jr. created his own rugged path to agribusiness stability and success, though ultimately he viewed his accomplishments in this economic realm
as stepping-stones toward a career in ranching and livestock. His first wife,
Earlene, mother of Norton III, was a brilliant woman ill-suited to the cowboy
culture, and although their marriage ended, Earlene and John Jr. addressed
their situation in a manner that never detracted from the healthy growth and
development of their son. Norton Jr. built upon the foundation of goodwill
engendered by his father, and weathered the Depression years with hard work,
extraordinary resolve, and no small amount of luck. In partnership with the
Arena family in California, by the end of the ruinous 1930s Norton Jr. had,
incredibly, built a produce business that made him a leader in the industry.
Although he indulged his passion for livestock after World War II, he kept his
hand in the more predictable world of produce and shipping. He provided his
son with a solid, even auspicious, foundation for entry into the changing world
of agribusiness.

Norton III—thanks to his parents' thoughtfulness, access to higher education, work ethic, and values, both inherited and learned, made the most of
his situation. With a brilliant, Stanford-educated mother, he quickly embraced
the value of educational excellence and intellectual rigor. From his father he
learned the ethics and challenges of the produce business and livestock raising,
and developed a love and appreciation for Arizona's vast ranching enterprises
and grazing areas. Fortunately, he "was a kid who liked to work," which served
him well during his teen years. He unified the disparate worlds of intellectual
achievement and physical labor. His education at Stanford and at the University of Arizona and his extracurricular success as a Rodeo Team standout reflected a perfect synthesis of parental influences. His ROTC training, his brief
experience in Iwo Jima, where he resisted pressures from superiors in order
to perform his duty, and his stateside service all provided an appropriate bur-
nishing to what was arguably a well-rounded military and formal education.
His marriage to Doris Schaefer Norton has introduced a fourth generation of
Nortons to the region: his three children, Michael, John, and Melanie.

Knowing well the seemingly endless problems relating to groundwater
depletion on Arizona's side of the Colorado River, he focused his attention on
water rights, and in 1955 embarked on his seventeen-year run at John Norton
Farms. Introducing innovations from and within the technological, scientific,
and marketing dimensions of the business, he prospered and served on the
board of the Palo Verde Irrigation District. His Blythe enterprise quickly grew

east and west; he was both an Arizona and a California grower with an expansive enterprise that ultimately produced unprecedented financial rewards. He became a leader and champion of the industry and a crusader for legislative reform. This activism, in the industry and in political circles, led him to the Reagan administration and an eye-opening stint as deputy secretary of agriculture.

He returned to Arizona and refocused his energies and resources. Since stepping away from the grind of year-round lettuce production and restructuring his economic activities and investments, Norton III has served well not only his family and fellow man, but also the legacy of his father and grandfather. Community service and support for health, cultural, and educational institutions have occupied his time in recent years, along with certain private pursuits: photography and international travel with his children and two grandchildren.

Water resource management and agribusiness provided the three generations of Nortons with unity, continuity, and purpose. And though water dedicated to agriculture in this region will diminish over time—and Norton III may be the last of the agribusiness titans in the American Southwest—the Nortons played a significant and telling role in building a vibrant and dynamic civilization on the edge of the American desert.

Notes

Introduction

[1]Robert Kelly, *Battling the Inland Sea: American Political Culture, Public Policy, and the Sacramento Valley, 1850-1986* (Berkeley and Los Angeles: University of California Press, 1989).
[2]Walter Prescott Webb, "The American West: A Perpetual Mirage," *Harper's Magazine*, 2.
[3]Gerald D. Nash, *The American West in the Twentieth Century: A Short History of an Urban Oasis* (Albuquerque: University of New Mexico Press, 1985), 6; Edmund Wilson, *The American Jitters: A Year of the Slump* (New York: Charles Scribner's Sons, 1932). See especially Wilson's comments on Los Angeles, 224-244.

Chapter 1: On the Edge of a Desert Empire

[1]United States Senate, 51 Cong., 1 Sess, Report 928, Part 5, *Report of the Special Committee of the US Senate on the Irrigation and Reclamation of Arid Lands*, Vol. IV (Washington, DC: Government Printing Office, 1890); Jack L. August Jr., *Vision in the Desert: Carl Hayden and Hydropolitics in the American Southwest*, 1999), 18; Frederick H. Newell, *The History of Irrigation Movement: First Annual Report of the US Reclamation Service* (Washington, DC: Government Printing Office, 1903). The USGS, under the direction of the Secretary of the Interior, would survey the arid regions of the West to determine to what extent the lands could be reclaimed by irrigation and to select sites for reservoirs for water storage and flood prevention. Significantly, the law withdrew from public settlement both the reservoir sites and the lands susceptible to irrigation from the water stored in them.
[2]Jack L. August Jr., "Carl Hayden: Born a Politician," *Journal of Arizona History (JAH)* 26 (Summer 1985), 130-132.
[3]See Donald Worster, *Rivers of Empire: Water, Aridity and the Growth of the American West* (New York: Pantheon Books, 1985); Gerald Nash, *The American West in the Twentieth Century: A Short History of an Urban Oasis* (Albuquerque: University of New Mexico Press, 1985); John Opie, "Environmental History of the West," in Gerald Nash and Richard Etulain, *The Twentieth Century West: Historical Interpretations* (Albuquerque: University of New Mexico Press, 1992).
[4]The resolutions were passed on April 7, 1889.
[5]Newell, *History of the Irrigation Movement*, 4, 5; *Phoenix Daily Herald*, April 9, 16, 1889.
[6]The survey's expenses came to $599.67.
[7]McClintock was to be the scribe and write up the expedition.
[8]See Breakenridge's autobiographical account, William M. Breakenridge, *Helldorado: Bringing the Law to the Mesquite* (Boston: Houghton Mifflen, 1928). An edited version, by Richard Maxwell Brown appeared in 1992, Richard Maxwell Brown, ed., *Helldorado: Bringing the Law to the Mesquite* (Lincoln: University of Nebraska Press, 1992). Also see James H. McClintock, *Arizona: A History of the Youngest State 1540-1915* (Phoenix, 1916). Breakenridge had a colorful history in the military and law enforcement. He was born on December 26, 1846 in Watertown, Wisconsin, and traveled to the Pike's Peak mining area when he was just fifteen years old. Three years later, in 1864, he joined Company B of the Third Colorado Cavalry for service in the Civil War. He fought in the battle of Sand Creek and other encounters. In 1876 he arrived in Arizona Territory, ending up in Tombstone in 1880 where he was appointed a US Deputy Marshall under Sheriff Johnny Behan. In Cochise County he was courteous and prudent and resorted to gunplay only as a last resort. An excellent marksman, he was rarely challenged by outlaws. As a US Marshall he had unusual authority in the area. He was present

during the famous Gunfight at the OK Corral, was on friendly terms with the Clanton faction, and of course, working under Johnny Behan, he was perceived as opposing the Earps, writing in his 1928 memoir that Wyatt Earp was a desperate character. In 1888 he changed course and accepted the position as Surveyor of Maricopa County. He was soon tasked—along with Norton and McClintock—with searching for a suitable dam site above Phoenix.

[9]*Phoenix Daily Herald*, August 12, 1889. McClintock was born February 23, 1864 in Sacramento, California. After attending public schools in San Francisco and Berkeley, he moved to Arizona to help his brother, Charles, who was founder and co-owner of the first newspaper in the valley, the *Salt River Valley Herald*. This paper was later absorbed into the *Arizona Republican*. In 1881, after his brother died, McClintock moved to Globe, ultimately becoming editor of the Globe paper. Sometime between 1883 and 1886 he moved to Prescott—then the territorial capital. He found work in the office of the adjutant general at Fort Whipple and was there during the last military campaign against Geronimo. McClintock, then twenty-two years old, enrolled in the Territorial Normal School in Tempe and graduated with a teacher's certificate. At the same time he owned a farm a few miles southeast of town and found time to work on the local newspaper. In 1889, after a year teaching in remote Pleasant Valley in northeastern Arizona, he moved to Phoenix and threw his full weight and writing skill behind the crucial problem of the day: water. McClintock worked with Norton in several guises; he was a freelance writer for the Maricopa County Board of Supervisors and the Phoenix Chamber of Commerce, Arizona correspondent for the *Los Angeles Times* and other papers, and an active state representative in the National Irrigation Congresses of the 1890s and early 1900s.

[10]Ibid., August 14, 15, 1889. Later he and friend Bucky O'Neill joined the Rough Riders and McClintock suffered machine gun wounds to his leg. In June of 1902 he became a colonel in the Arizona National Guard. As a writer, however, McClintock made his greatest contributions, repeatedly praising the virtues of the Salt River Valley. In 1901 when his Rough Rider comrade, Theodore Roosevelt ascended to the presidency, another Rough Rider colleague, Colonel Alexander O. Brodie was appointed Governor of the Territory of Arizona. McClintock hoped that he, too, would receive a political appointment to high office but was somewhat disappointed that he received the honorary appointment as Phoenix postmaster general, an office he held from 1902 to 1914. It was during this period that McClintock established himself as Arizona's finest historian. With Woodrow Wilson's election to the presidency in 1912, McClintock eventually lost his postmaster job but plunged into preparation of a two-volume history of Arizona: *Arizona: The Youngest State*. It remains one of the best-written histories of the state. Perhaps his highest honor came in 1919 when Thomas E. Campbell, the state's first Republican governor, appointed him state historian. During the Coolidge administration McClintock was once again appointed postmaster and continued writing for newspapers and magazines. In 1930, at age sixty-six, he produced an outstanding series of radio shows sponsored by the Union Oil Company. The name of the series, "Forward Arizona," reflected the spirit of McClintock's life. He died at age seventy on May 10, 1934.

[11]August, "Born a Politician," *JAH*, 120; Howard R. Lamar, *The Far Southwest, 1846-1912: A Territorial History* (New Haven: Yale University Press, 1966), 415-417; Karen Smith, "The Campaign for Water in Central Arizona, 1890-1903," *Arizona and the West* 23, 2 (Summer 1981) 127-148; Stephen Shadegg, "The Miracle of Water in the Salt River Valley, Part 1," *Arizona Highways* 18, 7 (July 1942), 8-17; Hal Moore, "The Salt River Project: An Illustrious Chapter in US Reclamation," *Arizona Highways* 37, 4 (April 1961), 2-15.

[12]Robert Hunter Norton was of Scotch-Irish descent and Annie was English. Norton's parents

were descendants of farming families in Virginia and Kentucky. His father was a farmer, machinist, and inventor who developed new machinery for woolen mills.

[13]Earl Zarbin, *Two Sides of the River: Salt River Valley Canals, 1867-1902* (Phoenix: Salt River Project, 1997), 86.

[14]Merwin L. Murphy, "W. J. Murphy and the Arizona Canal Company," *Journal of Arizona History (JAH)*, 23 (Summer 1982), 140.

[15]Laura Fulwiler Murphy, Diary, Box 1, Folder 1, Folder 2, "Personal/Business Correspondence, 1871-1882," William J. and Laura Fulwiler Murphy Collection, 1871-1983, (Murphy Collection) Arizona Collection (AC), Arizona State University (ASU), Tempe, Arizona. William J. Murphy, contractor and land developer, was born August 23, 1839. He served in the Union army under General Sherman at the Battle of Atlanta and was honorably discharged on July 24, 1865. Murphy married Mary C. Bigelow of Nashville, Tennessee, and settled there after the war. In April 1871 Mary died leaving William with two children, George and Lucy. In 1874 William married Laura Jane Fulwiler and in 1880 the family moved to Flagstaff, Aubrey Valley, and Prescott, Arizona. During this period Murphy contracted with railroad companies, including the Atlantic & Pacific, for grading and roadwork services. In 1883 he won the contract for construction of the Arizona Canal and agreed to receive compensation in canal stock and land. As a result Murphy needed to raise capital from out-of-state sources in order to meet payroll and construction expenses. Murphy died on April 17, 1923 and Laura passed away on May 21, 1943.

[16]Laura Fulwiler Murphy, Murphy Collection, Box 1, Folder 2, Diary, December 26, 1881, AC, ASU.

[17]Zarbin, *Two Sides of the River*, 86, 87.

[18]Michael Logan, "Head Cuts and Check Dams: Changing Patterns of Environmental Manipulation by the Hohokam and Spanish in the Santa Cruz Valley, 200-1820," *Environmental History* 4 (July 1999), 405-430.

[19]Jack L. August Jr., and Grady Gammage Jr., "Shaped by Water: An Arizona Historical Perspective," in Bonnie G. Colby and Katharine L. Jacobs, *Arizona Water Policy: Management Innovations in an Urbanizing, Arid Region* (Resources for the Future Press: Washington, DC, 2007), 11; Douglas Kupel, *Fuel for Growth: Water and Arizona's Urban Environment* (Tucson: University of Arizona Press, 2003), 2-5.

[20]Lamar, *The Far Southwest*, 223.

[21]Michael C. Meyer, *Water in the Hispanic Southwest: A Social and Legal History, 1550-1850* (Tucson: University of Arizona Press), 23. See, also, Roger Dunbier, *The Sonora Desert: Its Geography, Economy and People* (Tucson: University of Arizona Press, 1970). For its importance to the natural and human history of the Southwest, the Salt and Gila River have inspired surprisingly few books. Two of the best known are Edwin Corle, *The Gila: River of the Southwest* (New York: Holt Rinehart, and Winston, 1951) and Ross Calvin, *River of the Sun* (Albuquerque: University of New Mexico Press, 1951). Corle's book is useful but dated, reflecting an ideology of conquering the wilderness. Other noteworthy accounts are M. S. Salmon, *Gila Descending* (Silver City, New Mexico, 1985); Edmunds Andrews et. al. *Colorado River Ecology and Dam Management* (Washington, DC: National Academy Press, 1991); Arizona Rivers Coalition, *Arizona Rivers: Lifeblood of the Desert* (Phoenix: Arizona Rivers Coalition, 1991); Richard Berkman and W. Kip Viscusi, *Damming the West* (New York: Grossman, 1973); Charles Bowden, *Killing the Hidden Waters* (Austin: University of Texas Press, 1977); Philip R. Fradkin, *A River No More: The Colorado River and the West* (New York: Alfred A. Knopf, 1981): Paul

Horgan, *The Great River: The Rio Grande in North American History* (New York: Rinehart and Company, 1954); H. B. N. Hyne, *The Ecology of Running Waters* (Toronto: University of Toronto Press, 1977); Ed Marston, *Water Made Simple* (Covelo, California: Island Press, 1987); Frank H. Olmstead, *Gila River Flood Control* (Washington, DC: Sen. Doc. No. 426, 65 Cong. 3 Sess., Government Printing Office, 1919); Rich Johnson, *The Central Arizona Project, 1918-1968* (Tucson: University of Arizona Press, 1977); Tim Palmer, *Endangered Rivers and the Conservation Movement* (Berkeley and Los Angeles: University of California Press, 1986); Jack L. August Jr., *Vision in the Desert: Carl Hayden and Hydropolitcs in the American Southwest* (Fort Worth: TCU Press, 1999); Jack L. August Jr., *Dividing Western Waters: Mark Wilmer and Arizona v. California* (Fort Worth: TCU Press, 2007); John Wesley Powell, *Lands of the Arid Region of the United States* (Washington, DC: Government Printing Office, 1879); Marc Reisner, *Cadillac Desert: The American West and its Disappearing Water* (New York: Viking Press, 1986); Salt River Project: *Taming the Salt* (Phoenix: Salt River Project, 1979); John Walton, *Western Times and Water Wars* (Los Angeles and Berkeley: University of California Press, 1992); Frank Welsh, *How to Create a Water Crisis* (Boulder: Johnson Books, 1985); Donald Worster, *Rivers of Empire* (New York: Pantheon, 1985).

[22]See Thomas D. Hall, *Social Change in the Southwest, 1350-1880* (Lawrence: University Press of Kansas, 1989); Donald W. Meinig, *Southwest: Three Peoples in Geographical Change, 1600-1970* (New York: Oxford University Press, 1971). Specialists in Southwest history, who are numerous, have yet to concur on cultural consequences of chronology, and the overall prehistory of North America. The field undergoes substantial revision every decade. The longtime dean of Southwestern archeology, Emil Haury, was one of the first scholars to produce large-scale studies of the region. See Emil Haury, *Hohokam: Desert Farmers and Craftsmen* (Tucson: University of Arizona Press, 1976); Emil Haury, *Prehistory of the American Southwest* (Tucson: University of Arizona Press, 1986); Emil Haury, *The Archeology and Stratigraphy of Ventana Cave, Arizona* (Tucson: University of Arizona Press, 1966). Also, one should consult Suzanne K. Fish, et. al. eds., *The Marana Community in the Hohokam World* (Tucson: Anthropological Papers of the University of Arizona No. 56, 1993).

[23]The literature is extensive concerning Spanish exploration in the region. Without question Herbert Eugene Bolton's work during the first half of the twentieth century set the standard. See, for example, Herbert Eugene Bolton, *Anza's California Expeditions* 5 vols. (Berkeley: University of California Press, 1930); "The Early Expeditions of Father Garces on the Pacific Slope," *The Pacific Ocean History*, ed. Morris Stevens (MacMillan: New York, 1917); *Guide to the Materials for the History of the United States in the Principal Archives in Mexico* (Washington, DC: Carnegie Institution, 1913); "The Mission as a Frontier Institution in the Spanish-American Colonies," *American Historical Review* 23 (1917), 42-61; and *Rim of Christiandom: A Biography of Eusebio Francisco Kino, Pacific Coast Pioneer* (New York: MacMillan, 1936).

[24]See Emil Haury, Harold S. Gladwin, E. B. Sayles, and Winifred Gladwin, *Excavations of Snaketown, Material Culture* (Globe, Arizona: Medallion Papers No. 25, 1937); John L. Kessel, *Friars, Soldiers, and Reformers: Hispanic Arizona and the Sonora Mission Frontier, 1767-1856* (Tucson: University of Arizona Press, 1976).

[25]Kessel, *Friars, Soldiers, and Reformers*, 90-115; Sidney B. Brinckerhoff and Odie B. Faulk, *Lancers for the King: A Study of the Military System of Northern New Spain, with a translation of the Royal Regulations of 1772* (Tempe: Arizona Historical Foundation, 1965). The expedition, comprised of roughly two hundred people, traveled from Horcasitas, Sonora, to San Francisco.

[26]A solid source for this transitional period is James Officer, *Hispanic Arizona, 1536-1856*

(Tucson: University of Arizona Press, 1996). Also a good general source is Thomas Sheridan, *Arizona: A History* (Tucson: University of Arizona Press, 1996).

[27]The four companies of cavalry and one company of infantry stationed there fed themselves by hacking two hundred acres out of the dense bottomland along the river, but their farm failed. Thereafter, civilians, particularly those who settled in the Salt River Valley, supplied the four hundred and seventy soldiers and their horses. See Sheridan, *Arizona: A History*, 199.

[28]Douglas Kupel, *Fuel for Growth: Water and Arizona's Urban Environment* (Tucson: University of Arizona Press, 2002), 34; *Phoenix Herald*, May 4, 1878. Swilling recorded his filing in Prescott because it was the county seat for that portion of the Salt River Valley. Maricopa County was created in 1871 from portions of Yavapai County. A miner's inch is a measurement of flowing water and in Arizona equals 0.025 cubic feet per second (cfs). Importantly, this earlier unit of measurement was not fixed and definite, and varied from place to place. The flow was measured by passing water under six inches of head through a one-inch square opening. By 1871 the ditch was carrying two hundred cubic feet of water per second and could irrigate four thousand acres. It later became the Salt River Valley Canal, which ran straight through the center of Phoenix.

[29]*Arizona Miner* (Prescott), September 3, 1870; March 2, 1872; April 27, 1872; November 23, 1872.

[30]*Arizona Miner* (Prescott) August 27, 1870; August 31, 1872.

[31]*Phoenix Herald*, January 26, 1878; September 28, 1878; October 17, 1879; October 27, 1879.

[32]Kupel, *Fuel for Growth*, 34-35.

[33]See Wallace Stegner, *The Gathering of Zion: The Mormon Trail* (Lincoln: University of Nebraska Press, 1992); Howard Lamar, *The Far Southwest, 1848-1912: A Territorial History* rev. ed. (Albuquerque, University of New Mexico Press, 2000); Hubert Howe Bancroft, *History of Utah, 1540-1886* rev. ed. (New York: Bookcraft, 1964).

[34]The original Desert Land Act of 1876 called for adequate irrigation but did not specify how officials of the land office could determine what was adequate. The omission created vast opportunities for fraud. Speculators paid people to make claims and plow a few furrows and claimed that the furrows were ditches.

[35]See Jack August, "The Buckeye Irrigation Canal: A Legal and Institutional History" (Buckeye, Arizona: Buckeye Water Conservation and Drainage District, 2008), 7-12; Richard White, *It's Your Misfortune and None of My Own: A New History of the American West* (Norman: University of Oklahoma Press, 1996), 138-139; Jon A. Souder and Sally K. Fairfax, *State Trust Lands: History, Management, and Sustainable Use* (Lawrence: University Press of Kansas, 1996), especially chapters 1 and 2.

[36]*Arizona Gazette* (Phoenix) June 16, 1881; December 15, 1881; *Phoenix Herald*, May 20, 1880; August 17, 1881; June 19, 1884.

[37]"It Was Arizona's Night: The Governor of that Territory Banqueted at Bath Beach," *New York Times*, July 29, 1891. In the article, Logan, described as the ex-president of the New York State Bar Association, hosted an event for Arizona Territorial Governor John Irwin to discuss the "size, the growth, and the resources of Arizona . . . from various standpoints . . . over coffee and cigars that followed a complimentary dinner given by Walter S. Logan to the Honorable John S. Irwin and H. H. Logan of Phoenix, Arizona."

[38]Geoffrey P. Mawn, "Phoenix, Arizona: Central City of the Southwest, 1870-1920," (PhD dissertation, Arizona State University, 1979), 80-81, 108; *Arizona Gazette*, May 10, 1883; Zarbin, *Two Sides of the River*, 86, 87. Regarding those involved in the early formation of the company, the *Arizona Gazette* ran the following story it picked up from the Tucson *Arizona*

Daily Citizen: "Among the parties interested in the Arizona Canal, Phoenix, are Messrs. W. A. Hancock and Kales & Lewis of Phoenix; the Bank of California and the Anglo-California Bank of San Francisco, together with a powerful Eastern syndicate. Hon. Clark Churchill is President; Kales & Lewis, Phoenix, Treasurer; and R. F. Trott, Superintendent."

[39]In creating Arizona Territory in 1863, Congress had reserved from each township sections 16 and 36 to benefit public schools. Then, in 1871, Congress gave the Texas & Pacific Railroad Company the odd numbered sections for forty miles on each side of its proposed route across central Arizona. This meant that only sixteen sections (10,240 acres) of the thirty-six sections (23,040 acres) in each township along the projected line of the Arizona Canal would be open for entry. If the canal were constructed as outlined above, settlers could conceivably file entries only on approximately 90,000 acres.

[40]John Dickinson, "Birth of the Salt River Project," unpublished manuscript, Arizona Historical Society, Library, (Tucson, Arizona, May 17, 1972) 40-41.

[41]Murphy, "W. J. Murphy and the Arizona Canal Company," *JAH*, 40-43.

[42]*Weekly Phoenix Herald*, June 4, 1885; *Arizona Gazette*, November 20, 1883; *A Historical and Biographical Record of the Territory of Arizona* (Chicago: McFarland and Poole, 1896), 230; Larry Schweikert, "'You Count on It': The Birth of Banking in Arizona," *JAH*, 22 (Autumn, 1981), 349-368.

[43]*Arizona Republican*, April 18, 1923; *Arizona Gazette*, November 20, 1883; *Weekly Arizona Herald*, June 4, 1885; W. J. Murphy to Laura, August 21, 1884, AC, ASU.

[44]Entries are in the Gila and Salt River Meridian Land Base Line, Arizona, General Land Office Records, Record Group 49, National Archives and Records Service. *The Statutes at Large of the United States of America* (Washington, DC: Government Printing Office, 1877), Vol. 19, 377, 392, 393; *US Statutes at Large*, 36 Cong., 2 Sess., Chap. 338, 333; William O. Hendricks, *M. H. Sherman: Pioneer Developer of the Pacific Southwest* (Corona Del Mar Foundation: Corona Del Mar, California, 1971), 6-7; Paul W. Gates, "The Homestead Law in an Incongruous Land System," in Vernon Carstensen, ed., *The Public Lands: Studies in the History of the Public Domain* (Madison: University of Wisconsin Press, 1982), 316, 318. Early land laws were often misused. By using the Preemption Act, the Homestead Act, and the Timber Culture Act, for example, a person could acquire a total of 480 acres of public land. The Desert Land Act boosted this to 1,120 acres.

[45]The same pattern of dummy entrants tied to Murphy emerged after Congress repealed the Texas & Pacific Railroad subsidy in February 1885 due to the failure of the company to build. Thus the odd sections of land were thrown open to settlement and Laura Murphy wrote her husband, who was peddling Arizona Canal Company bonds in New York, "The odd sections I suppose will be thrown open soon. . . . Are there any you want entered on Desert [Land] Act and what names can you use?" She continued on with her admonition and mentioned three adjoining sections in what is now Phoenix, beginning on Thirty-Second Street on the west and continuing to Fifty-Sixth Street on the east and between Thomas Road on the south and Indian School Road on the north. See Laura Murphy to W. J. Murphy February 23, 1885, W. J. and Laura F. Murphy Collection, Special Collections, Hayden Library, Arizona State University, Tempe, Arizona.

[46]Another solid account on the early history of Phoenix and the Salt River Valley is Karen Smith, "From Town to City: A History of Phoenix, Arizona, 1870-1912," (MA Thesis: University of California at Santa Barbara, 1978); John W. Caughey, "The Insignificance of the Frontier in American History," *Western Historical Quarterly*, V (1974), 13-14; Smith, *The*

Chapter 2: Corporate Water

[1] An outstanding illustration of the culture of violence on the American Southwestern frontier can be found in Robert Utley, *Billy the Kid: A Short and Violent Life* (New York: Bison Books, 1989). Also it appears that Norton became an incorporator and president of Phoenix Hay and Grain Company on March 16, 1882, thus predating his permanent move to the Salt River Valley. The business was located at the corner of Jefferson and First Streets in Phoenix. Even though he worked in northern Arizona at the time he carried on in the general business of selling hay, grain, and farm products with fellow incorporators Henry E. Kemp, vice president, W. D. Hamman, secretary treasurer and manager, John O. Spicer, and T. J. Trask, who worked with him on the canal project.

[2] Richard E. Sloan, *Memories of an Arizona Judge* (Stanford: Stanford University Press, 1932), 19.

[3] Murphy, "W. J. Murphy and the Arizona Canal Company," *JAH*, 147-148.

[4] Laura F. Murphy to W. J., August 23, 1844, W. J. Murphy Papers, ASU.

[5] The section of land, owned by Murphy, became the Ingleside Tract. Murphy planted trees and citrus orchards and it became a kind of local, unofficial park on the county road eight miles northeast of Phoenix. The "unofficial park" status was short-lived, though Murphy had big plans for the area. He built what was the Valley's first "resort," the Ingleside Club, later renamed the Ingleside Inn. Since there was no air conditioning or cooling the Inn was open only in the winter. It operated until the early 1940s. Then, two women turned the facility into a private school for girls. The school operated until 1957 and in 1959 it was sold to a developer who tore it down and put up apartments that have been converted into condominiums. Regarding the site that Norton and his crew were unable to displace, in 1902 a small hydroelectric plant was built over the falls and in 1910 construction began on a larger 850-kilowatt plant at the falls. This plant was in continuous use until 1950. Then, in 2003, the Salt River Project rejuvenated the area and called it the "Arizona Falls Hydroelectric Project," which included an improved park (G. R. Herberger Park), dance floor, waterfalls, and a 750-kilowatt hydroelectric generator designed as an educational facility for local schools.

[6] Christy was a banker from Des Moines, Iowa and knew the Chief Engineer of the Atlantic & Pacific, who introduced him to Murphy. The two got along well and Christy accompanied Murphy to Phoenix to look over the project. Since his project was not being supported by the existing bank Murphy proposed that Christy move to Phoenix to establish a new bank. Christy remained in Phoenix for the rest of his life and later, in 1891, was appointed Territorial Treasurer by Governor John Irwin. Christy arguably was the first person to gain a degree of serious financial recognition of Arizona by Eastern money markets, selling more than $2 million of bonds in New York to fund the debts of towns, counties, and the Territory, debts that had been mounting due to pressing water issues.

[7] *Arizona Gazette*, September 25, 1883; W. J. Murphy to Laura, no date (This letter was apparently written in August-September 1883), W. J. Murphy Collection, AC, ASU.

[8] Christy was born in Trumbull County, Ohio, February 14, 1841. When he was thirteen the family settled on a farm in Osceola, Iowa, where at the age of seventeen, he began teaching for three years at a nearby country school. In July 1861 he enlisted in the Union Army and rose rapidly in the ranks, fighting at Shiloh and Lookout Mountain. At the Battle of Jonesboro, July 29, 1864, Christy was wounded four times while leading a saber charge. In spite of wounds

in both shoulders and through his left hand and arm, he led still another charge against the Confederates. The following day he was captured and sent to a hospital in Newman, Georgia. Six months later he was paroled and exchanged. His wounds were so serious that he carried his left arm in a sling for three years. Shortly after being mustered out of the military in 1865 Christy bought a business college in Des Moines, Iowa, and met and married Carrie Bennitt. He became active in Republican politics and in 1872, at the tender age of thirty-one, he was elected State Treasurer of Iowa and was reelected to a second term in 1874. Then he became cashier and a director of the Capital City Bank of Des Moines and continued in that capacity until 1881. Then he organized the Merchants' National Bank of Des Moines, and served as cashier there. One evening, on the way home from Republican party meetings he caught cold and contracted pneumonia. This was followed by violent asthma attacks. On his doctors order he and his family moved to Arizona in August of 1882 and purchased a ranch forty-five miles north of Prescott. Fortunately, his brother-in-law, E. J. Bennitt had lived in Yavapai County for several years, working as a surveyor, civil engineer, and also for the Goldwater mercantile firm. After eighteen months, Christy had regained his health and was successful in raising cattle and cultivating crops on his ranch.

[9]Papers of W. J. Murphy, cancelled notes to J. W. Dodge, AC, ASU.

[10]Zarbin, *Two Sides of the River*, 85-88.

[11]Evidence of the haste in the overnight reorganization was that the new stock certificates had "Valley Bank" penned in and "First National Bank" crossed out. See stock certificates, numbers 11, 14, 15, 21, First National Bank of Phoenix, AC, ASU; *Weekly Phoenix Herald*, April 17, 1884; *Arizona Gazette*, April 14, 1884.

[12]*Arizona Republican*, April 18, 1923. It remains unclear how many subcontractors continued to work after the attempted attachment but Simms stayed on for some time, though he quit late in the summer 1884. In April 1884 he headed to San Francisco to procure necessary irons and timbers for putting in the dam and headgates at the head of the Arizona Canal. But it appears that in the summer of 1884, Simms quit. On August 19, 1884, Murphy wrote, "Shall have my hands quite full now. Simms has abandoned his contract. He had been trying to get me to pay him the final 10 percent of what work he has already done. He is not entitled to them—till all his excavation is completed and his anxiety to get them made me suspicious and now failing to get the bonds he admits he is not going on with his contract." See W. J. Murphy to Laura, August 19, 1884, AC, ASU.

[13]Murphy, "W. J. Murphy and the Arizona Canal Company," *JAH*, 156. It appears also that Murphy was indebted, financially at least, to J. W. Dodge of San Francisco. He wrote much later that "In relation to Mrs. Dodge I will only say that to her dead husband more than to any one man I am indebted for success in completing the canal. When others in San Francisco believed the lying stories afloat, he said I have faith in the canal and in Murphy and he not only worked with others to get money for me but he borrowed $10,000 to loan to me. Mrs. Dodge was in full sympathy with her husband in his efforts to assist me and she has been very lenient and a considerate creditor. When she needed money to settle up the estate she could have used the securities she held of mine but would not for the reason that [the] party who offered to take them would have sold them out at a great loss to me." In fact, two cancelled notes indicate that on January 20, 1884, W. J. Murphy executed a note for $25,000 to J. W. Dodge, putting up for collateral five $1,000 bonds. Interest was set at 2 percent per month but reduced to 1.5 percent on May 10, 1884. A second note for $10,000 to J. W. Dodge was secured by twenty $1,000 bonds. The notes were finally paid off on February 11, 1888.

[14]Tritle, who previously had sought office in Nevada, had experience in arid land economics and gained the confidence of President Chester A. Arthur, who appointed him Governor of Arizona Territory. Tritle served from 1882-1885 when President Grover Cleveland, a Democrat, assumed the presidency.

[15]Frederick A. Tritle, "Appendix 27: Extracts from Report of Hon. F. A. Tritle, Governor of Arizona, for the Year 1884," in Joseph Nimmo Jr., Chief, Bureau of Statistics, Navigation and Commerce, *Report on the Internal Commerce of the United States* (Washington: Government Printing Office, 1885) ,244-247

[16]Ibid.

[17]Murphy, "W. J. Murphy and the Arizona Canal Company," *JAH*, 162. Murphy spent the month of January 1885 in New Haven, where he hoped to make "arrangement for some temporary help." He confided to Laura that "The success of building the canal seems assured already, and now comes the other part, to make a success of disposing of the securities. That part ought to be the easier part, but I am not sure that it will prove to be."

[18]*Arizona Gazette*, June 1, 1885; Murphy, "W. J. Murphy and the Arizona Canal," *JAH*, 164.

[19]*Weekly Phoenix Herald*, June 4, 1885. The reporter overstated his case here. While Murphy's spirit was willing, his bonds were weak. Murphy was not part of the triumphal inspection; instead he was in New York City hounding investment bankers and potential investors, trying desperately to sell bonds or borrow money on them to head off the creditors who were hounding him on all fronts.

[20]*Arizona Gazette*, May 28, 1885.

[21]For information on canals in the Salt River Valley before 1900 see E. F. Young, compiler for the United States Bureau of Reclamation, *Early History of the Salt River Project*, unpublished typescript, 1917, Salt River Project Archives, Tempe, Arizona; George Strebel, "Irrigation as a Factor in Western History" (PhD dissertation: University of California at Berkeley, 1965).

[22]Murphy's Peoria land sales demonstrated how his relinquishment system worked. Arizona Canal Company engineer Andrew Barry filed an entry on April 25, 1883, and it was filed by Churchill and Murphy. Barry signed a relinquishment, and the General Land Office canceled the entry on March 6, 1886. Six days later, Samuel Bartlett, a Peoria grain dealer, filed an entry on the same section. The land entered by Barry and Bartlett was Sec. 4-T2N-R2E, Forty-Third to Fifty-First Avenues and Glendale to Northern Avenues. Bartlett did not own water rights when proof of irrigation was made. After Bartlett's death the land was transferred to Murphy's Glendale Land Company for $1.00 on October 27, 1903.

[23]Significantly, at this time Murphy owned no land in the Salt River Valley nor was he an agent of the federal government nor authorized to sell the public domain. He was not actually "selling" land. After his dummies filed entries on the public domain they signed and gave to Murphy a document called a "relinquishment," which restored the land entry back to the federal government. Upon receiving a relinquishment the US General Land Office reopened the claim for entry by any other party. The challenge lay in the timing; when Murphy forwarded the relinquishment to the General Land Office he had to ensure that the person to whom he had "sold" the land would be able to make the new entry. The finder's fee may have ranged from $1,000-$1,500 per section. Also, the goal of selling one thousand water rights was never reached. See William Christy to W. J. Murphy, August 25, 1885, W. J. Murphy Papers, AC, ASU. Deed Books in the Maricopa County Recorder's Office (MRCO) reveal 744.8125 acre-feet of water rights issued by the Arizona Canal Company between January 24, 1886 and March 15, 1907. For an overview see Earl Zarbin, "Desert Land Schemes: W. J. Murphy and

the Arizona Canal Company," *JAH*, 42 (Summer 2001), 163-164.

[24]All was not perfect on the Arizona Canal for the first decade of operation. William Bartlett, for example, explained how the distribution system worked to the benefit of Murphy and his partners to the detriment of the water users under the project. "As land owners," Bartlett fumed, "we are entitled to a certain amount of water which we have not received and we have not received it for two reasons . . . because Murphy as manager has never cleaned out the canal and it is not in shape to carry the amount of water necessary to satisfy the water rights the company has sold and . . . because a large percentage of the water that starts in the canals is diverted before it reaches our land for the use of Murphy and his friends." The latter complaint plagued virtually every canal venture in the Salt River Valley during the last quarter of the nineteenth century. See Zarbin, "Desert Land Schemes," *JAH*, 165.

[25]*Arizona Republican*, January 4, 1899. The first order of business, once the Arizona Canal Company controlled the other three canal companies, was to remove them as plaintiffs in a February 1887 lawsuit against the Arizona Company for diverting water from the Salt River. Then, Norton worked to integrate the infrastructure and distribution systems regulating water use and distribution.

[26]*Phoenix Gazette*, May 19, 1887.

[27]See August, *Vision in the Desert*, chapters 1 and 2.

[28]The standard work on Newlands is William D. Rowley, *Reclaiming the Arid West: The Career of Francis G. Newlands* (Bloomington: Indiana University Press, 1996). Newlands was the driving force behind the so-called Newlands Reclamation Act of 1902. The measure established the nation's reclamation policy and had a profound effect on the development of the western economy in the twentieth century. It empowered the Department of the Interior to build reservoirs and reserve public lands for farmers in the arid areas of the West, and it gave the Secretary of the Interior unprecedented administrative authority. It was considered a radical measure designed to both conserve and utilize resources within western states.

[29]Michael Duchemin, "Introducing the Urban Form: The Arizona Improvement Company in Phoenix, 1887-1890," (MA thesis: Arizona State University, 1992), 37-49. Murphy's and Christy's business practices aroused suspicions from the outside investors in AIC. Newlands was especially critical of Murphy's business practices and all the outside investors by 1889-1890 ridded themselves of their shares. Meanwhile, Murphy continued his questionable bait and switch practices utilizing the imperfect federal land laws designed to encourage settlement on the public lands. Though the General Land Office attempted to investigate alleged wrongdoing under the Desert Land Act, they were woefully understaffed and their special agent proved impolitic at best. Historian Harold H. Dunham wrote: "The special agents could not effectually stop the rampant fraud on the public domain because of inadequate numbers, vastness of territory covered, lack of reform in the land system itself, temptations of bribery, and removal for thwarting powerful interests." See Harold H. Dunham, "Some Crucial Years in the General Land Office, 1875-1890," in Vernon Carsensen, ed., *Public Lands: Studies in the History of the Public Domain* (Madison: University of Wisconsin Press, 1968), 181, 195.

[30]According to his grandson, Murphy crafted an apologia for his famous progenitor in 1982. "It would be nice to be able to say that this worthy gentleman [Murphy] soon raised all the money he needed and then returned to Phoenix with all his worries over, but it wasn't to be. W. J. would wrestle with these problems for years. This is not to imply that W. J. did not find his way out of the financial mire; indeed, he remained in Arizona, registered a long list of accomplishments that contributed materially to the Salt River Valley, and prospered."

Murphy, "W. J. Murphy and the Arizona Canal Company," *JAH*, 166. Also, a good overview for the social and economic impact of the Depression of the 1890s is H. W. Brands, *The Reckless Decade: America in the 1890s* (Chicago: University of Chicago Press, 2002).

[31]In 1896 also, W. J. Murphy laid out the subdivision of Orangewood and planted citrus there, as had done with Ingleside in 1891. He planted several hundred acres of citrus and farm crops in the Glendale area, and is credited with revolutionizing the Valley's agriculture industry through introduction of commercial citrus growing. He also organized a citrus growers association and collaborated to build a citrus packing house. Because of his interests in the canals and agriculture industry, Murphy laid out the town of Glendale, and then persuaded friends from Peoria, Illinois to found the town that bears that name. Recongizing the need for transportation to these outlying areas he developed Grand Avenue from Five Points to Peoria. He also formed a company to operate streetcars and was awarded the franchise in Phoenix. He then laid tracks from the Southern Pacific Depot, on Seventh Street and Washington Street west to Seventh Avenue, north to Five Points, and out Grand Avenue to Six Points.

[32]Ibid., 541.

[33]William H. Goetzmann, *Exploration and Empire: The Explorer and Scientist in the Winning of the American West* (New York: W. W. Norton, 1967), 530.

[34]Some of Powell's relevant works that pertain to this topic: John Wesley Powell, *Report of Explorations in 1873 of the Colorado River of the West and its Tributaries Under the Direction of the Smithsonian Intuition* (Washington, DC: Government Printing Office, 1879); *Report on the Lands of the Arid Regions of the United* States, second edition (Washington, DC: Government Printing Office, 1879).

[35]See John Wesley Powell, "Report on the Lands of the Arid Region of the United States," *House Executive Document 73*, 45 Cong. 2 Sess. (1878); John Wesley Powell, "Reservoirs in Arid Regions of the United States," *Senate Book of Arizona* (San Francisco: Payot, Upham and Company, 1878), 66; Norris Hundley Jr., "The Politics of Reclamation: California, the Federal Government, and the Origins of the Boulder Canyon Act," *California Historical Quarterly* 52 (Winter 1973), 294; Smith, *The Magnificent Experiment*, 6-7; August, *Vision in the Desert*, 71-72;

[36]See Jack L. August Jr., "The Future of Western History: The Third Wave," *Journal of Arizona History* 26 (Spring 1986).

[37]One of the best accounts of the Carey Act in operation can be found in Mark Fiege, *Irrigated Eden: The Making of the Agricultural Landscape in the American West* (Seattle: University of Washington Press, 2000). In fact, Arizona was not eligible for Carey Act funds until 1909 when Congress extended its provisions to the territories. By that time, however, its provisions were not practicable for Arizona. Joseph Maull Carey, Wyoming governor and senator was born on January 19, 1845 in Milton, Delaware. He was educated at Union College in New York and the University of Pennsylvania. In 1867 he was admitted to the bar and practiced law in Philadelphia before heading to Wyoming Territory in 1869 where he was appointed the first US Attorney of the newly-organized territory. President Ulysses S. Grant then appointed him as associate justice of the Wyoming territorial Supreme Court, a position he held from 1872-1876. He then went into Wyoming politics; mayor of Cheyenne, 1881-1885 and territorial delegate to Congress, 1885-1891. In 1890 he introduced the act establishing Wyoming statehood. He authored the Carey Act, which was embraced largely in Idaho and Wyoming, but not in Arizona. He was then elected United States Senator and served in that office until 1895. In1910, he switched parties from Republican to Democrat and was elected governor of Wyoming in 1910 and served a four-year term. He died on January 17, 1937.

[38]*Arizona Republican*, January 16, 1891. Etta had a sister, Mattie and two brothers, John and Robert. John Wright worked as a grader for Maricopa County. Robert became a Methodist cowboy preacher, and traveled over northern Arizona, sermonizing from ranch to ranch.

[39]August, "Born a Politician," *JAH*, 128.

[40]Sims Ely, *The Lost Dutchman Mine* (New York: William Morrow and Company, 1933). Sims Ely's son, Northcutt "Mike" Ely, graduated from Stanford Law School and became California's lead attorney in *Arizona v. California*. See August, *Dividing Western Waters*, 55.

[41]McClintock, *History of Arizona*, 432; US Congress, House Committee on Indian Affairs, Hearings before the House Committee on Indian Affairs, *Compilation of Laws Relating to Indian Irrigation Projects*, 2 vols. (Washington, DC: Government Printing Office, 1919), 303; Arthur Powell Davis, *Irrigation Near Phoenix, Arizona*, Water Supply and Irrigation Paper of the US Geological Survey, No. 2, House of Representatives Doc. 342, 54 Cong., 2 Sess., 1897.

[42]See *Arizona Republican*, April 20, 1863; December 20, 1864; January 13, 1895; *Arizona Gazette*, December 21, 1898; Mawn, "Phoenix, Arizona," 223-224.

[43]Carl Hayden to Charles Trumbull Hayden, March 29, 1897; Hayden Family Letters Collection, Carl Hayden Papers Collection; Carl Hayden, "Remarks at the Tenth Annual Banquet of the National Reclamation Association," Hayden Biographical File, Carl Hayden Papers, Collection, Special Collections, Hayden Library, Arizona State University, Tempe, Arizona.

[44]See Donald Pisani, *To Reclaim a Divided West: Water, Law, and Public Policy, 1848-1902* (Albuquerque: University of New Mexico Press, 1992), 56, 55.

[45]*Arizona Republican*, July 4, 1896.

[46]Frederick Newell, "The History of the Irrigation Movement," from *First Annual Report of the United States Reclamation Service*, June 17, 1902, to December 1, 1902, House of Representatives Doc. No. 79, 57 Cong.. 2 Sess.,Washington, DC, 7.

[47]*Arizona Gazette*, December 29, 1898; Samuel P. Hays, *Conservation and the Gospel of Efficiency: The Progressive Conservation Movement, 1890-1920* (New York: Athenaeum, 1975). This classic work discusses the rivalry between the various government water agencies. This concern with bureaucratic turf also impacted the development of the Salt River Valley as the Department of Agriculture's Division of Irrigation Investigations challenged the authority of the US Reclamation Service.

Chapter 3: Westward Tilt

[1]Norton received 1808 votes, while the Republican received 1063 and the Populist, 553. See *Arizona Republican*, November 17, 1896

[2]*Arizona Republican*, March 14, 26, May 12, 1891.

[3]*Arizona Republican*, September 14, 1895. The other commissioners were E. M. Boggs and George Woodford. Louis C. Hughes arrived in Arizona for health reasons in 1871. In short order he achieved professional and public distinction in varied ways. For example, Hughes was admitted to practice law before the Arizona Territorial Supreme Court on January 17, 1873. He was a city councilman for Tucson in June of 1872 and later a county attorney on April 15, 1873. He was appointed as Attorney General of the Arizona territory, but resigned after only a year. In March of 1877 he started the *Daily Bulletin*, which later became the *Triweekly Arizona Star*, and finally the *Arizona Star* on June 26, 1879. In the 1880s he was a United States Court Commissioner and served on the board of managers for the World's Fair in Chicago in 1891. He also preached on street corners in the 1890s since he served in the Salvation Army

in Tucson, always promoting the liberal causes of temperance, women's suffrage, Arizona statehood, and other progressive causes. With the election of Democrat Grover Cleveland for a second term in 1892, Hughes was appointed to the governorship on April 14, 1893. He could not address the Arizona legislature for almost two years because his appointment fell on the tail end of the last session of the year. Hughes and other early progressives, like Norton, wanted to clean up the elections process because voters were being bought with alcohol. He also endorsed women's suffrage, the secret ballot, and expressed dismay that most of the laws in Arizona were being ignored. He lamented the fact that most laws on houses of prostitution, furnishing liquor to minors, punishing adultery, and the following of the Sabbath law were ignored. He wanted the establishment of a board of control for government agencies like the Territorial Prison and the Insane Asylum. Hughes worked, unsuccessfully, for the establishment of a Board of Immigration to encourage business to come to Arizona. Irrigation and the booming lumber industry in Flagstaff was a priority and that is where Norton fit into his plans. In a strange occurrence Tohono O'odahm Indians had captured nine wild camels, domesticated them, and bred them to over one hundred. They were found to be from the original 1850s stock and Hughes had plans to grow the herd. However, in 1896, Hughes was removed from office due to partisan rancor for his progressive views and public policies. Nevertheless, Norton hooked his political star to one of the most active governors in territorial history. During his three-year tenure, Hughes reduced the territorial debt and created a nonpartisan board of control. He was on the Board of Regents in 1897 and eventually returned to Tucson and ran his newspaper, the *Arizona Star*. He favored joint statehood with New Mexico in 1904 and 1905, when that issue swept the two adjoining territories. He left the *Arizona Star* in 1907 after a libel suit was filed against him. He and Norton were friends of George W. P. Hunt and his family and he was invited by Theodore Roosevelt to the christening of the ship the USS *Arizona*, but refused to attend because champagne was used in the ceremony. He died on November 24, 1915 in Tucson and was buried in Evergreen Cemetery there.

[4] Ibid., January 7, 1898. During this period Norton remained as superintendant of the Arizona Canal until he retired in early January 1899. See *Arizona Republican*, January 4, 1899, "A New Superintendant." The announcement read: "Mr. Dan McDermott, who has been employed by the Arizona Canal Company for many years as zanjero, has been appointed superintendant of the canal to succeed John R. Norton. Mr. McDermott assumed the duties of his new position on Monday. He is a capable man and will make a good superintendant."

[5] *Arizona Republican*, March, 28; June 6; December 3, 1900; January 5, 1901. In fact, Norton was among the organizers of the Maricopa County Cattleman's Association in 1897, when fifty ranchers met at the Hotel Adams on Central Avenue in Phoenix. He was appointed to the bylaws committee and to a second committee to collaborate with cattlemen in all parts of the valley so that they worked for common goals. It was an all-embracing group; every cattleman in the valley was urged to attend the meeting, "whether his herd consists of five cattle or five thousand." At the organizational meeting, Norton was elected to the executive committee, and to serve as a delegate to the National Livestock Show in Denver on December 25-25, 1897, at which the burning issue of the day—grazing on public lands—was to be debated.

[6] *Arizona Republican*, November 11, 1899. The noted cattleman and later Arizona historian, Will C. Barnes, chaired many of the Arizona Livestock Association meetings at this time, including the November 10, 1899 meeting noted above. Will Croft Barnes (1858-1937) first came to Arizona as a cavalryman and went on to become a rancher, state legislator, and conservationist. From 1905 to 1935, his travels throughout the state, largely on horseback, enabled him to

gather the anecdotes and geographical information that came to constitute *Arizona Place Names* which was reissued in revised format in 1988. See Will C. Barnes, *Arizona Place Names* (Tucson: University of Arizona Press, 1988). For this first toponymic encyclopedia of Arizona, Barnes compiled information from published histories, federal and state government documents, and reminiscences of "old timers, Indians, Mexicans, cowboys, sheep-herders, historians, any and everybody who had a story to tell as to the origin and meaning of Arizona names." The result is a book of oddments, humor, and now-forgotten lore.

[7] *Arizona Republican*, December 18, 1898; January 5, 1899.

[8] I. H. Parkman, *History of Buckeye Canal* (Buckeye, Arizona: Buckeye Irrigation Company, 1957), 1; *Arizona Capitol Times*, May 5, 1987. The three Buckeye pioneers possessed neither writing utensils nor paper in the last of their excursions and therefore posted notice by taking an axe and hewing a smooth flat place on a willow tree and in charcoal wrote their temporary location notice. Subsequently they headed back to Phoenix, built themselves a triangle, returned to the willow tree and posted legal notice on May 28, 1885. Over the next month Jackson, Spain, and Mitchell conducted a feasibility study using the only technology available to them; a triangle. A triangle was made out of one-by-four lumber in the form of a letter A with the cross near the bottom. The two legs were generally sixteen to twenty feet apart at the bottom. From the apex of the A a cord with a weight called a plumb bob attached was suspended. Then, they placed the feet of the triangle on stakes drive down in a pond of water so the top of the stakes were at the exact water level. The string and the plumb bob hung at the center of the cross bar and by raising one leg an inch at a time the plumb bob swing was marked on the cross bar at each rise. After this, the process was reversed and the other side marked, thus providing the operator a complete scale to determine the rise and fall of the land at each set. As the machine had to be calibrated every twenty feet and the proper calculations made at each set, it was obviously a slow and exacting operation. They ran three miles of land and were satisfied their project was feasible.

[9] A miner's inch did not represent a fixed or definite quantity of water being measured generally by arbitrary standard of various ditch companies. For this reason a miner's inch is 0.020 cfs (1/50th of a cfs) in Idaho, Kansas, Nebraska, South Dakota, North Dakota, New Mexico, Utah, Washington, and southern California, while it equals 0.025 cfs (1/40th of a cfs) in northern California, Arizona, Nevada, Oregon, and Montana. Colorado maintains that a miner's inch equals 0.026 cfs (1/38th of a cfs) and British Columbia asserts that a miner's inch equals 0.028 cfs (1/36th of a cfs). The miner's inch was primarily intended for measuring small quantities of flow and it eventually gave way to the second foot or cubic foot per second (cfs) unit that is commonly used in the US to measure flow. To determine the amount of water, early water users used an inch box, a special type of free flowing orifice. A miner's inch is the quantity of water which discharges through a square inch opening under a prescribed head. The number of miner's inches is equal to the area of the opening in square inches. This remains a mystery. In the preceding sentences, the openings is one square inch, so "the number of miner's inches is equal to the area of the opening in square inches" puts us back at one square inch. In effect, in Arizona this amounts to about 1.5 cubic feet per minute or 11.25 gallons per minute.

[10] E. H. Spain, who had previous experience on the Arizona Canal construction project served as head linesman for the survey. Meanwhile, the Buckeye Canal Company, on July 24, 1886, filed notice of location of 38,000 inches of more water to be added to the original 12,000, totaling 50,000 miner's inches of water. This filing was recorded in Phoenix on October 8,

1886, at about the same time of the original canal survey, which extended from the dam at the Agua Fria to the lower end of Arlington near the Gillespie Dam site, was completed. Engineers envisioned water being carried across the Hassayampa River by means of a four-foot-high sand dam built across the river. According to plans, the water ran in above the dam until it was full, then it ran out on the other side. For nearly twenty years this was the means of carrying the water across the river, though the dam washed out every time when unusually large runoffs occurred.

[11]These 6,000 inches were represented by seventy-five shares of stock that had already been issued. Significantly, on October 13, 1888, two weeks prior to entering into the contract, O'Neill formed the company and it was under his direction that a good substantial dam at the head of the canal across the Gila commenced construction. The diversion would be sufficient to keep the 6,000 inches or seventy-five shares of water in the canal for delivery at all times "for a period of twenty-one years and thereafter at a price not to exceed $1 per inch per year. The new Buckeye Irrigation Company further agreed to keep the water in the canal for at least twenty days out of every month except for the years 1889 and 1890 where they might turn out all the water to enlarge and expand the canal. On March 1, 1889, the Buckeye Canal Company received the final and official transfer of O'Neill-owned property.

[12]Major Edward T. Wilton to William O'Neill, January 8, 1889, Box 1, File 3, Buckeye Water and Conservation District Archives (BWCDD), Buckeye, Arizona. Wilson went on to say that a canal forty-eight miles long could be constructed at a cost of $225,907.00 and would cover 29,120 acres east of the Hassayampa, 34,580 west of the surveyed land and 48,000 more of unsurveyed land; a total of 111,700 acres.

[13]Charles N. Fowler to William Barnes, January 23,1889, Box 1, File 3, BWCDD; Parkman, *History of Buckeye Canal*, 4.

[14]Churchill also made other observations: "Arizona courts recognize the first appropriation in a river of all the water appropriated up to extent of his appropriation." Thus, he allowed, first in use, first in right, or the doctrine of prior appropriation. Further, he wrote that "an inch of water in Arizona is considered enough to irrigate two acres of land—after land has been irrigated a few years an inch will irrigate three or four acres; to the ability of the Buckeye Irrigation Co. to continuously and perpetually supply the amount of water its appropriation entitles it to.... The annual rain fall around Salt River Valley is 7.5 inches while in the vicinity of Prescott it is 15.18 inches."

[15]See Act of October 2, 1888, and Amendatory Acts of March 2, 1889, August 30, 1890, and Section 17 of the Act of March 3, 1891, which takes a corrective approach to the seemingly restrictive elements in the 1888 law. Chamberlin wrote that the 1888 law would apply to the Buckeye District, the homesteads or entry men could not gain title to their land, or file on more so it would be unwise for his company to invest directly or indirectly in the Buckeye Valley.

[16]See Jack L. August Jr., "Carl Hayden: Born A Politician," *JAH* (26, 2), 117-144.

[17]Parkman, *History of Buckeye Canal*, 5.

[18]Ibid., 8.

[19]On September 27, 1900, the Maricopa County Sheriff issued a deed against the old company on a judgment obtained by Moultrie on February 24, 1900, for the sum of $11,292.23 and assigned to the new company and thus the old Buckeye Canal Company passed into oblivion.

[20]Donald Pisani, *Water and the American Government: The Reclamation Bureau, National Water Policy, and the West, 1902-1935* (Berkeley: University of California Press, 2002), xiii.

[21]In fact in 1894 alone 1,400 strikes involving more than 5,000 workers visited the country.

[22]See, for example, Robert Higgs, *Crisis and Leviathan: Critical Episodes in the Growth of American Government* (New York: Oxford University Press, 1987). On the impact of the 1890s depression review, David Thelen, "Social Tensions and the Origins of Progressivism," *JAH* 56 (September 1969), 323-341; David Thelen, *The New Citizenship: Origins of Progressivism in Wisconsin, 1885-1900* (Columbia: University of Missouri Press, 1972); David Wrobel, *The End of American Expansionism: Frontier Anxiety from the Old West to the New Deal* (Lawrence: University Press of Kansas, 1993).

[23]Frederick H. Newell, "Autobiography," unpublished manuscript, no date (circa 1927), American Heritage Center, University of Wyoming, 23, 34-38.

[24]Davis, *Irrigation Near Phoenix*, 66; Newell, *History of Irrigation Movement*, 4. Arthur Powell Davis was born February 9, 1861 in Decatur, Illinois, attended public schools and graduated in 1882 from Kansas State Normal School, then secured a degree at Columbian University (now George Washington University) in 1888. The year he graduated from Kansas Normal he was employed as assistant topographer of USGS by his uncle, John Wesley Powell. This appointment began a career of government service that spanned four decades and earned him an international reputation as an hydraulic engineer. Before he retired Davis personally planned and supervised the construction of more than one hundred dams, and fifteen thousand miles of irrigation and drainage canals in sixteen states benefiting from the Reclamation Act. These dams and waters brought forth crops on more than three million acres.

[25]*Arizona Gazette*, April 5, 1899.

[26]Ibid., June 1, 4, 7, 1899.

[27]The Phoenix and Maricopa County Board of Trade was the successor to the Phoenix Chamber of Commerce. McCowan presented the report on behalf of the Water Storage Committee before the board of trade on the evening of May 1, 1900. In addition to discussing various methods to conserve water and increase yields, the report examined four ways to raise moneys for reservoirs: federal appropriations, federal government cession of lands to states and territories, private enterprise, and voting bonds to be sold by Maricopa County. They recommended the latter course. S. M. McGowan, "Report to the Phoenix and Maricopa County Board of Trade," April 10, 1900. See also *Arizona Republican*, May 2, 1900; *Arizona Gazette*, May 2, 1900.

[28]*Arizona Republican*, January 9, 1901. At the January 8, 1901 meeting where Norton weighed in on the Hudson Company dam site issue, the price of $200,000 was discussed. McGowan said the company was entitled to something, while others argued that the site could be purchased for far less. Norton remained steadfast in his claim that Hudson should receive nothing.

[29]*Arizona Republican*, April 10, 12, 1900. Norton added, "I do not see how the taxpayers of the county can lose anything by bonding. If capitalists are willing to risk from two to two-and-a-half millions it seems to me that we can afford to risk a half million. I believe it would be a successful enterprise and men who would put up a couple of million dollars must feel pretty sure about it. If it should be a success the taxable value of property in this county would jump at once from $10 million to no less than $30 million. It wouldn't take long for the tax on this increased valuation to amount to as much as the bond issue and the county would have its preferred stock in the enterprise left. This is the public or county view of it. The benefit to the farmer, the businessman, and everyone else interested in the valley would be beyond calculation. As I said, something will have to be done, anyhow."

[30]McGowan, "Report to the Phoenix and Maricopa County Board of Trade"; Smith, *Magnificent Experiment*, 15.

[31]Several newspapers began calling Norton, "Captain" Norton because of his well-known fondness for deep sea fishing in California. One story, published on July 17, 1901 in the *Arizona Republican* read: "Captain John R. Norton had fished every day till the people of Long Beach began complaining of the smell of fish that lay around because there so many of them that couldn't be eaten. A committee waited on him and asked if he was just getting his way to an unsportsmanlike cruelty and wantonness, or if he was thinking of starting a fertilizer plant. . . . Captain Norton had eighty-one yellow tail and barracuda floundering around the boat. Captain Norton called Mr. C. J. Hall's attention to the marine reports published in the Los Angeles papers within the last week showing that the tidewater had never been so low. The Captain intimated that this phenomenon had been caused by the number of fish that had been taken out of the sea during his stay."

[32]*Arizona Gazette*, August 12, 15, 16, 28, 1900; *Arizona Republican*, August 12, 23, 1900.

[33]See G. Wesley Johnson Jr., "Dwight Heard in Phoenix: The Early Years," *JAH* 18 (Autumn, 1977).

[34]Fowler later became the first president of the Salt River Valley Water Users Association. He was born in Stoneham, Massachusetts on December 14, 1843, and attended public schools before going to Andover and Yale. He was a member of the Fiftieth Regiment, Massachusetts Volunteers during the Civil War and detailed in the US Signal Corps under General Nathaniel Prentiss Banks. By the time he discharged he had made up his mind to teach and did so in Danvers, Massachusetts. He later studied law in Boston but quit after one year to go into the publishing business. During the next twenty-seven years he published books in Chicago, Boston, and New York. Then, when he arrived in Arizona, he saw the arid desert, and its potential and for the next seventeen years he thought of little else other than water.

[35]Fowler's arguments that the British had reclaimed more than 35 million acres in India fell on deaf ears. Even those congressmen who supported the concept could visualize a successful project in which a dam would be located so far from its service area.

[36]Smith, *Magnificent Experiment*, 18-19; B. A. Fowler to Ethan Hitchcock, November 20, 1900, and Charles Walcott to Ethan Hitchcock, January 14, 1901, Records of the Bureau of Land Management, Record Group 49, Old Canal and Reservoir files, National Archives, Washington, DC.

[37]*Arizona Republican*, March 20, 1901.

[38] See William D. Rowley, *Reclaiming the Arid West: The Career of Francis G. Newlands* (Bloomington: Indiana University Press, 1996).

[39]Portions of the letter were reprinted in *the Arizona Republican*, January 25, 1901.

[40]Samuel P. Hays, *Conservation and the Gospel of Efficiency: The Progressive Conservation Movement, 189-1920* (New York: Athenaeum, 1975); William Lilley and Lewis Gould, "The Western Irrigation Movement, 1878-1902: A Reappraisal," in Gene Gressley, ed. *The American West: A Reorientation* (Laramie: University of Wyoming Press, 1968); Samuel P. Hays, *The Response to Industrialism, 1885-1914* (Chicago: University of Chicago Press, 1957); Otis Graham Jr., *The Great Campaign: Reform and War in America, 1900-1928* (Englewood Cliffs, New Jersey: Prentice-Hall, Inc., 1971): Leahmae Brown, "The Development of National Policy With Respect to Water Resources," (PhD dissertation: University of Illinois, 1937); Jack L. August Jr., *Vision in the Desert: Carl Hayden and Hydropolitics in the American Southwest* (Fort Worth: TCU Press, 1999).

[41]August, *Vision in the Desert*, 26-27. Roosevelt was on a hiking trip with his family and was summoned from Mount Thawahus in the Adirondacks to Buffalo. At 3:15 p.m. that day, at the Ansley Wilcox Mansion, 641 Delaware Avenue, in Buffalo, forty-three-year-old Theodore

Roosevelt became the twenty-sixth President of the United States and the youngest man to hold the office up to that time. See also D. Jerome Treton, "Theodore Roosevelt and the Arid Lands," *North Dakota Quarterly* 36 (Spring 1968), 21-28; Pisani, *To Reclaim a Divided West*, 312-315.

[42]Pisani, *Water and the American Government*, 17-35; August, *Vision in the Desert*, 27; Smith, *Magnificent Experiment*, 23.

[43]Carl Hayden represented south side water interests and later became Maricopa County Sheriff, Member of the US House of Representatives, (1911-1927), and US Senator from Arizona, (1927-1969). Carl Hayden to Sallie Davis Hayden, February 8, 1903, "Speech Delivered Before Salt River Valley Water Storage Committee," Holograph notes, Hayden Family Letters Collection (HFLC), ASU; Rowley, *Reclaiming the Arid West*, 104; Statutes at Large of the United States of America, 1901-1902, vol. XXXII, part 1 (Washington, DC: Government Printing Office, 1903), 390; William Cronon, "Landscapes of Abundance and Scarcity," *Oxford History of the American West*, ed. Clyde A. Milner III, Carol A. O'Connor, and Martha A. Sandweiss (New York: Oxford University Press, 1994), 618.

[44]Salt River Project, *Taming of the Salt*, 68.

[45]Peggy Heim, "Financing the Reclamation Program, 1902-1919: The Development of a Repayment Policy," (PhD dissertation: Columbia University, 1953); *Arizona Republican*, June 3, July 4, 1902; *Arizona Gazette*, January 12, 14, July 2, 1902.

[46]*Arizona Republican*, April 24, 1901.

[47]Jack L. August Jr., "History of the Buckeye Water Conservation and Drainage District," (Buckeye, Arizona: Buckeye Water Conservation and Drainage District, 2007), 55-90.

[48]*Arizona Republican*, June 16, 1901.

[49]Ibid., July 10, 1907.

[50]Joseph H. Kibbey, "Brief of Articles of Incorporation of the Salt River Valley Water Users Association, May 25, 1903," Salt River Project Archives, Salt River Project; Smith, *Magnificent Experiment*, 35. Specifically Kibbey took on the thorny problem of water rights. By using beneficial use—in this case meaning the amount of water required for proper irrigation—as the measure and limit of the water right, the articles attempted to limit the possibility that a landowner with prior rights to the natural flow would waste water by taking that water and project developed water. For example, a farmer with prior rights to ten acre-feet of water would not be able to take an extra acre-foot of project water to irrigate his alfalfa crop if the proper amount of water to irrigate that crop was ten acre-feet. Similarly, priority of right lost its primary importance in determining water rights, as the articles, craftily, attempted to harmonize water use among old and new settlers. But for landowners with old water rights, particularly under the Tempe and Mesa canals, the Articles of Incorporation radically changed local practices. Prior appropriation, though still extant, lost much of its power. So too, the notion that the individual should determine his own beneficial use of water was diminished. Where so-called floating water rights existed in the past, SRVWUA prohibited its members from transferring shares in the association unless the corresponding land was also transferred. The suggested cost per acre, $12.50, payable over a ten-year period, did not include the initial subscription charge to the water users' association. Therefore, the cost per acre charge was substantially higher than the average $1.50 per acre charge most landowners with vested rights were used to paying for delivery of natural flow.

[51]F. H. Newell to Secretary of the Interior Hitchcock, February 20, 1904, Records of the Bureau of Reclamation, Record Group 115, Salt River 1902-1919, series 261, National Archives, Washington, DC.

[52]Charles D. Walcott to E. A. Hitchcock, June 6, 1903, Records of the Secretary of the Interior, Record Group 48, Lands and Railroads, Reclamation, National Archives, Washington, DC; *Minutes*, Board of Governors of the Salt River Water Users' Association, April 6, 1903, Corporate Secretary's Office, Salt River Project.

[53]Fowler and Kibbey knew that selection of the Salt River as a project depended upon government acceptance of its organization and the executive committee carefully considered the content of the Articles of Incorporation. In essence, the articles sought to reconcile reclamation law with territorial vested rights. See *Arizona Republic*, December 25, 1982.

[54]*Arizona Republican*, January 4, 1904. Also, Norton worked with the City of Phoenix on a committee to finance a municipal water works bond. See *Arizona Republican*, January 29, 1903.

[55]*Arizona Republican*, May 24, 1903.

[56]Smith, *Magnificant Experiment*, 25.

[57] *Arizona Republican*, June 8, 1903.

[58] Charles D. Walcott to Dwight B. Heard, June 19, 1903, Charles D. Walcott Papers, Smithsonian Institution Archives, Record Unit 7004/Box 1, Smithsonian Institution, Washington, DC; An example of Heard's amendments to the articles can be read in *Arizona Republican*, April 19, June 21 1903; *Phoenix Daily Enterprise*, June 20, 1903. Heard's opposition was tenacious and riled Reclamation Service personnel. In essence, Heard traveled to Washington to meet with the Secretary to try to convince him to modify the SRVWUA Articles of Incorporation. Hitchcock's refusal to modify the articles effectively stopped Heard from pursuing the path of changing the fundamental direction of the Water User's Association. At the same time, Heard considered himself a friend of the water storage movement. Still, his earlier opposition excluded him from the early decision making of the new water organization.

[59]*Arizona Republican*, March 5, 1904. Fowler and Judge Kibbey addressed enthusiastic crowds at the celebratory event.

[60]*Arizona Republican*, May 2, 3, August 20, 1904; August, *Vision in the Desert*, 30. Norton continued to speak out on the issues of the day—he addressed various audiences about the dangers inherent for Arizona if it entered into joint statehood with New Mexico— and supported various bond issues and other infrastructure developments that enhanced reclamation in the Salt River Valley

[61]*Arizona Republican*, January 24, 1903.

[62]Nash, *The American West in the Twentieth Century*, 18.

[63] On this and other phases of western economic development see, C. R. Niklason, *Commercial Survey of the Pacific Southwest* (Washington, DC: Government Printing Office, 1930). For the decades prior to this discussion see Oscar Osborn Winther, *The Transportation Frontier in the Trans-Mississippi West, 1865-1890* (New York: Holt, Rinehart, 1974).

[64]*Arizona Republican*, August 5, 1905, February 8, 1906. In middle age Norton suffered from some kidney ailments and in September 1910 survived an acute appendicitis attack.

[65]Sheridan, *Arizona: A History*, 209-210.

[66]Ibid, 210.

[67]See Norris Hundley Jr., *Water and the West: The Colorado River Compact and the Politics of Water in the American West* (Berkeley: University of California Press, 1975); August, *Vision in the Desert*, 55.

[68]*Arizona Democrat*, March 20, 1911.

[69]*Arizona Republican*, March 19, 1911; *Arizona Democrat*, March 20, 1911. Roosevelt added some relevant reflections in his unedited remarks. He said "As soon as it was done [the National

Reclamation Act signed into law] I called Mr. Newell and I said, 'Now I want this work divided fairly. There will be great pressure by different senators and congressmen who will honestly think that their state has first claim, that they have a meritorious project, and as Arizona and New Mexico have not many senators or congressmen, and as I raised three-fourths of my regiment in New Mexico and Arizona I will take their place and now I want to see that they get a fair deal.' Mr. Davis and Mr. Newell answered at once that they were perfectly easy in Arizona. Mr. Newell said that there were two projects that they regarded as two of the most important, if not the most important, projects of the Reclamation Service. Mr. Newell and Mr. Davis took the keenest personal interest in everything connected with starting this work just as if they had been citizens of Arizona who were directly to be benefited by the proposed work and they couldn't have been more devoted to it or towards it or have served it or more conscientiously worked to see this policy adopted in a form that would make it of the widest and most far reaching benefit to the people of the Salt River Valley."

[70]By this time, Kibbey had also served as governor of Arizona Territory (1905-1910).

[71]August and Gammage, "Shaped by Water: An Arizona Historical Perspective," *Arizona Water Policy: Management Innovations in an Urbanizing Arid Region*, 13-15.

[72]John R. Norton III, oral history interview with Jack L. August Jr., December 4, 2009, Phoenix, Arizona, author's files; *Arizona Republican*, February 6, 1923. Among some of his other noteworthy activities Norton was the executive officer of the Taxpayers Economy League, an organization formed to work for the general reduction of taxes. Many tributes noted that Norton served three terms on the Maricopa County Board of Supervisors and for many years was the head of the volunteer firefighting organization formed here before modern equipment was available for the city.

[73]Last rights for John R. Norton were conducted by the local lodge of the Woodmen of the World of which Norton was a member. His wife, Etta Wright Norton, his two sons, Fred and John, and his daughter, then married to William Corpstein of Phoenix, attended the funeral at McLellan Chapel on north Central Avenue.

Chapter 4: John R. Norton Jr. and the Urban Oasis

[1]Geoffrey Mawn, "Phoenix, Arizona: Central City of the Southwest," (PhD dissertation, Arizona State University, 1979), 288-91; *Arizona Republican*, June 16, 1901; Harry Welch, "The Wonderful Roosevelt Dam of the Salt River Valley of Arizona," *National Irrigation Journal*, 2 (August 1910), 10-11; C. J. Blanchard, "The Call of the West: Homes Are Being Made for Millions of People in the Arid West," *National Geographic*, 20 (May 1909), 425-27; Bradford Luckingham, *Phoenix: The History of a Southwestern Metropolis* (Tucson: University of Arizona Press, 1989), 48-49. Similarly, outside investors promoted Phoenix and the Salt River Valley. As Marshall Field, a wealthy Chicago-based investor characterized it: "I am delighted with the country, and with Phoenix; both are far above expectations. I have just been writing to a friend in Chicago that this is the best country I have so far struck in the West." Field invested in Phoenix in 1896 when he gave $25,000 to a former Chicago businessman John C. Adams to help him construct a hotel.

[2] When the Roosevelt Neighborhood was developed, the use of brick, wood, stone, prefabricated components, and pressed and cast metal was commonplace in building construction in Phoenix. The completion of the transcontinental railroad in Arizona in 1883-84 made the use of imported materials common and relatively inexpensive. Local brick had been available since

1878 and the use of wood shingles, milled woodwork, and leaded and stained glass throughout the neighborhood stands as testimony to the availability of these materials. In fact, the majority of the buildings in Norton Jr.'s neighborhood were built of local, soft brick and almost 70 percent had a stucco finish—like the residence at 351 West Portland Street. See, for example, Bernard Michael Boyle, *Materials in Architecture in Arizona, 1870-1920* (Tempe: Arizona State University, College of Architecture, 1976).

[3]The Indian School line, for example, which was completed in 1900, ran through north side subdivisions owned by Phoenix land developers, Lloyd Christy, Dwight Heard, and William Thomas. Sherman also helped expand the physical dimensions of the city while at the same time padding his bank account by extending it to local attractions like the territorial capital building, the Indian School, the fairgrounds, and Phoenix Park—later known as Eastlake Park. See, for example, Richard Lynch, "The Relationship of Streetcars to Real Estate Promotion in Phoenix" (unpublished manuscript, 1973), AHF, Hayden Library, ASU; Lawrence Fleming, *Ride a Mile and Smile the While: A History of the Phoenix Street Railway* (Phoenix: Swain Publications, 1977).

[4]After Henry Ford introduced the mass-produced Model T in 1908, more Phoenix residents could afford to purchase a car. By 1913, when John Jr. was twelve years old, the city contained ten automobile dealerships. More cars meant more dispersion of population and commercial decentralization. Thomas Peterson, "The Automobile Comes to Arizona," *JAH* 15 (Autumn 1974), 249-68; Daniel Davis, "Phoenix, Arizona, 1907-1913 (unpublished manuscript, 1976), Department of Archives and Manuscripts, Hayden Library, ASU; Karen L. Smith, "From Town to City: A History of Phoenix, 1870-1912," (MA thesis: University of California, Santa Barbara, 1978), 122-138, 154-55; Sylvia Laughlin, "Iron Springs: Timeless Summer Resort," *Journal of Arizona History*, 22 (Summer 1981), 235-54.

[5]Luckingham, *Phoenix*, 52-53.

[6]Ibid., 53.

[7]Health seekers with wealth could check in to the Desert Inn Sanatorium located on "one hundred beautiful acres," about six miles west of Phoenix. The Desert Inn Sanatorium claimed to be good for incipient bronchial and pulmonary cases and convalescents for la frappe, pneumonia, and other debilitating cases. Whitelaw Reid, owner and editor of the *New York Times*, suffered from respiratory problems and he took several winters in Phoenix. Reid invited many of his well-to-do friends from throughout the country to visit Phoenix and the Salt River Valley.

[8]John R. Norton III oral history interview with Jack L. August Jr., August 11, 2009, Phoenix, Arizona, author's files.

[9]Local leaders struggled to gain advantages in urban development. The greatest challenge confronting late nineteenth and early twentieth century boosters and developers in the Salt River Valley was a stable and regulated supply of water. Following severe droughts and disastrous floods in the 1890s and 1900s Phoenix leaders and Salt River Valley agricultural leaders decided that water storage was the only way to address the "water problem." They formed the Salt River Valley Water Users' Association (SRVWUA) and this organization was able to take advantage of the National Reclamation Act of 1902 which resulted in the construction of Roosevelt Dam, which was completed in 1911. This and related efforts brought stability, water supply, and irrigation control thus enabling the area to assure agricultural growth into the future. And as the Salt River Valley prospered so did its urban manifestation, Phoenix. Indeed, the area became the leading agricultural producer in the Southwest. John R. Norton, as noted

earlier in the manuscript, recognized that the nation's first federal reclamation project also demonstrated the necessary contribution of the federal government to the growth of Phoenix.
[10] *Arizona Republican*, August 19, 1917; Sheridan, *Arizona: A History*, 212.

[11] Bradford Luckingham, *Phoenix: The History of a Southwestern Metropolis* (Tucson: University of Arizona Press, 1989), 74. *Arizona Republican*, February 20, 1919, January 1, 1920; A. George Daws, *The Commercial History of Maricopa County* (Phoenix: Daws Publishing, 1919); Geoffrey Mawn, "Phoenix: Central City of the Southwest" (PhD dissertation, Arizona State University, 1979). By 1918 cotton production replaced alfalfa growing as the leading industry in the Salt River Valley. In that year 72,000 acres were planted in cotton; up from 7,400 acres in 1916. By 1920 that number had increased to 190,000 acres or three-quarters of the irrigated land in the valley. Land prices rose astronomically. According to one observer, "In January 1916, eighty acres of land near Peoria sold for $60 an acre. In January 1919 it sold similar tracts for $500 an acre. That was an increase of 833 percent."

[12] *Arizona State Magazine* (March, 1919); Sheridan, *Arizona: A History*, 212-213. With irrigated land increasing in price by as much as 800 percent, dairying and alfalfa production became unprofitable for established farmers and astronomically expensive for those looking to purchase land. The patchwork of grain fields, date orchards, and citrus groves disappeared as monoculture spread across the Salt River Valley.

[13] Richard Wells, "Cotton Boom," *The Country Gentleman*, (Philadelphia: Curtis Publishing Company, April 17, 1920), 1. In 1955 the *Country Gentleman* was merged into the *Farm Journal*.

[14] In 1920, 90,560 farmers were in Arizona and in 1925 the number had shrunk to 71,954. Norton interview with the author, August 11, 2009.

[15] By 1929 more than five hundred carloads of oranges and grapefruit were being shipped from Phoenix to cities throughout the country.

[16] A variety of support services for agriculture—packing sheds and processing plants—sprung up along the railroad tracks in south Phoenix. These helped make the area a distribution point and a variety of organizations, led by the Arizona Cotton Growers' Association, made the city the agricultural and marketing center for the state. The SRVWUA also contributed substantially during this period to the redevelopment and expansion of agriculture in the area. Of special note was that early in the 1920s SRVWUA began a process of building hydroelectric power plants to generate more power and prevent a host of problems The association also constructed more infrastructure like storage dams below Roosevelt Dam. Mormon Flat Dam completed in 1925, created ten-mile-long Canyon Lake; Horse Mesa Dam, completed in 1927, created seventeen-mile Apache Lake; and Stewart Mountain Dam, completed in 1930, created twelve-mile-long Saguaro Lake.

[17] See, for example, David Myrick, *Railroads of Arizona*, Vol. 2 (San Diego: Howell-North Books, 1981); William S. Greever, "Railway Development in the Southwest," *New Mexico Historical Review* (NMHR) 32 (April 1957), 185-86; *Arizona Gazette*, June 16, October 1, 1923; *Arizona Republican*, October 15, 16, 1926.

[18] In fact, Phoenix had been slow on the uptake to enter the world of air transportation. In late 1925, after two years of promotional efforts, by the chamber of commerce and other booster organizations, the first iteration—the Phoenix Municipal Airport—opened on 160 acres of land six miles west of the city. It started with a whimper. Several private air strips were located throughout the Salt River Valley; the best was the Phoenix Commercial Airfield on Henshaw Road on South Central Avenue. There, in November 1927, the Aero Corporation of California operated Standard Airlines and promoted regular flights to Tucson and Los

Angeles. Meanwhile, unable to compete with the private air fields at this time, the City of Phoenix sold the Phoenix Municipal Airfield to an alfalfa farmer. See, for example, *Arizona Republican*, November 11, 1928, September 23, 1929; Ruth M. Reinhold, *Sky Pioneering: Arizona in Aviation History* (Tucson: University of Arizona Press, 1982), 86-91, 109-112; Ken Western, *Arizona Republic*, October 4, 2010; Jack August, *Arizona Republic*, October 7, 2010.

[19]Luckingham, *Phoenix*, 82, 86-87. The Arizona Biltmore cost a staggering $2 million to build and joined the nearby and already established Ingleside Inn (1910) as new destination locations. The Biltmore, however, set a new standard and was located six miles northeast of Phoenix between Squaw Peak and Camelback Mountain. It was designed by Albert McArthur, a Frank Lloyd Wright protégé. The hotel was an instant success and in July 1929 William Wrigley Jr., chewing gum magnate, owner of the Chicago Cubs, and a part-time winter resident of Phoenix, invested $1.7 million in the existing hotel company, which included architect McArthur's brothers and the Bowman-Biltmore Hotel Chain, and assured the Biltmore's future as one of the premiere hotel resorts in the nation. Furthermore, Wrigley built a mansion adjacent to the hotel and wealthy investors followed his lead: Louis Swift, Cornelius Vanderbilt Jr., and George Bartlett, among many others, represented the type of person that viewed Phoenix as a winter time destination.

[20]Michael J. Kotlanger, "Phoenix, Arizona, 1920-1940," (PhD dissertation, Arizona State University, 1983), 271-84; *Phoenix City and Salt River Valley Directory, 1930*, 10-14; Snell Biography (unpublished manuscript), AHF.

[21] See Bradford Luckingham, "Urban Development in Arizona: The Rise of Phoenix," *JAH* 22 (Summer 1981); Earl Pomeroy, *The Pacific Slope: A History of California, Oregon, Washington, Idaho, Utah, and Nevada* (New York: Knopf, 1966); Raymond H. Mohl, *The New City: Urban America in the Industrial Age, 1960-1920* (Arlington Heights, Ill: Harlan Davidson, 1985).

Chapter 5: Establishing Prior Rights

[1]Arizona's congressman Carl Hayden introduced the earliest bill calling for Colorado River flood control. See HR 9421, *Congressional Record*, 66 Cong. 1 Sess. (1919), 22, 24, 309; *Arizona Daily Star*, December 24, 1927.

[2]Committee on Water of the National Research Council, *Water and Choice in the Colorado River Basin* (Washington, DC: National Academy of Science Publications No. 1689, 1968), 1-47; Norris Hundley Jr., *Water and the West: The Colorado River Compact and the Politics of Water in the American West* (Berkeley: University of California Press, 1975), xiv-xvi; Norris Hundley Jr., "The Colorado River Waters Dispute, *Foreign Affairs* 42 (April 1964), 495; Beverly Moeller, *Phil Swing and Boulder Dam* (Berkeley: University of California Press, 1971), x-xi; Philip Fradkin, *A River No More: The Colorado River and the West* (New York: Alfred A. Knopf, 1981), 15.

[3]The best account of the Colorado's silt-carrying capacity appears in Norris Hundley Jr., "The Politics of Reclamation: California, the Federal Government, and the Origins of the Boulder Canyon Act," *California Historical Quarterly* 3 (January 1972), 229. For a more extensive analysis of silt, see US Department of Agriculture, "Silt in the Colorado and its Relationship to Irrigation," *Technical Bulletin No. 67* (Washington, DC: 1928).

[4]*Mohave County Miner* (Kingman) December 1, 1894, January 12, 19, and February 12, 1895; Hundley, *Water and the West*, 14.

[5]John Wesley Powell, "Report on the Lands of the Arid Region of the United States," *House Executive Document 73*, 45 Cong., 2 Sess. (1878); John Wesley Powell, "Reservoirs in Arid Regions of the United States," *Senate Book of Arizona* (San Francisco: Payot, Upham and

Company, 1878), 66; Hiram Chittenden, "Preliminary Examination of Reservoir Sites in Wyoming and Colorado," *House Document 141*, 55 Cong. 2 Sess. (1897), 58; Hundley, "Politics of Reclamation," *California Historical Quarterly*, 294.

[6]Hundley, "Politics of Reclamation," *California Historical Quarterly*, 299-300; Hundley, *Water and the West*, 53; Barbara Ann Metcalf, "Oliver Wozencraft in California" (MA thesis: University of Southern California, 1963), 81-96; *Weekly Arizona Miner* (Prescott) June 13, 1879; *Arizona Citizen* (Tucson) June 13, 1879.

[7]Jack L. August Jr. and Grady Gammage Jr., "Shaped by Water: An Arizona Historical Perspective," in Bonnie G. Colby and Katharine L. Jacobs, *Arizona Water Policy: Management Innovations in an Urbanizing, Arid Region* (Washington, DC: Resources for the Future Press, 2007), 12. As noted earlier, the most significant Spanish influence in Arizona and the Southwest was in the law. The single most important Spanish water legacy was prior appropriation: first in time, first in right.

[8]See Hundley, *Water and the West*, chapters 1 and 2; Hundley, "The Politics of Reclamation," *California Historical Quarterly*, 301-06.

[9]August, *Vision in the Desert*, 74.

[10]Hundley, "The Politics of Reclamation," *California Historical Quarterly*, 297-301; Arthur Powell Davis to J. B. Lippencott, October 10, 1902, Bureau of Reclamation papers, File 187, Colorado River Project, 1902-1919, Record Group 115, National Archives and Records Service (NARS); US Geological Survey, *First Annual Report of the Reclamation Service 1902* (Washington, DC, 1903), 106-7, 109; US Department of the Interior, *Fourteenth Annual Report of the Reclamation Service, 1914-1915* (Washington, DC, 1915), 323; House Committee on Irrigation of Arid Lands, "Hearings on the All-American Canal in Imperial Valley, California, HR 6044," 66 Cong. 1 Sess. (1919), 98-99. For the best short essay on Arthur Powell Davis see Gene Gressley, "Arthur Powell Davis, Reclamation, and the West," *Agricultural History* 42 (July 1968), 241-257.

[11]The seven Colorado River Basin states are Wyoming, Colorado, Utah, Nevada, New Mexico, Arizona, and California.

[12]Hundley, *Water and the West*, 53-82; *San Diego Union*, January 20, 1918; Rufus Von KleinSmid, "The League of the Southwest: What it is and Why," *Arizona: The State Magazine* 11 (May 1920). Additional support for reclamation on the Colorado came from a group meeting in Salt Lake City, Utah. The Soldiers, Sailors, and Marines Land Settlement Conference added another dimension of support when it endorsed an interior department proposal to reclaim several million acres in the West for veterans of World War I.

[13]For the best account of Phil Swing's activities before his election to Congress from California's Eleventh District see Moeller, *Phil Swing and Boulder Dam*, 3-19. See also, Hundley, *Water and the West*, 38-39.

[14]August, *Vision in the Desert*, 76, 79. For the early bills submitted by Hayden, Kettner, and Randall see *Congressional Record*, 65 Cong. 3 Sess. (1919), 2647, 2934, 3738; 66 Cong. 1 Sess. (1919), 22, 24, 1258. See also House Committee on Irrigation of Arid Lands, "Hearings on the All-American Canal in Imperial Valley, California, HR 6044 (1919), 7-8, 48-51, 94, 142, 285-287. Ultimately, Hayden had little trouble convincing chairman of the House Irrigation Committee, Moses Kincaid, to table the Kettner bill. Citing testimony of Reclamation Director Davis, objections from the Department of State and Treasury, and statements from noted irrigation specialist Elwood Mead, he sealed the bill's fate. Secretary of State Robert Lansing objected to the measure because, he ventured, an All-American Canal should be built

only after a treaty had been negotiated with Mexico. "Equity and comity" entitled Mexico to some of the river's flow, Lansing suggested, citing a treaty drafted in 1906 that awarded Mexico sixty thousand acre-feet annually from the Rio Grande River. Treasury Secretary Carter Glass criticized the bill in harsher terms, finding the financial features of the bill "untenable." He objected vigorously to the feature that compelled the federal government to underwrite the project by guaranteeing IID bonds. "If the project is meritous," he inveighed, "it should be funded with a direct appropriation of a specific amount for the purpose." Mead, Chairman of the California Land Settlement Board, added his objections. Veterans, he avowed, should receive preferential treatment and landholders should be limited to 160-acre single-family farms.

[15]*Congressional Record*, 69 Cong. 2 Sess. (1926), 5822; Dwight Mayo, "Arizona and the Colorado River Compact" (MA thesis: Arizona State University, 1964), 99-100; Hundley, *Water and the West*, 266; Moeller, *Phil Swing and Boulder Dam*, 75; *Arizona Daily Star*, March 3, 6, 1925.

[16]*Tombstone Epitaph*, June 8, 1922; August, *Vision in the Desert*, 80.

[17]In May 1920 Congress tabled Kettner's bill and approved, instead, the Kincaid Act, a compromise measure named after the House Committee chairman, which authorized the Secretary of the Interior to complete ongoing surveys with special attention paid to the needs of the Imperial Valley.

[18]August, *Vision in the Desert*, 80. Quotes included in Hundley, *Water and the West*, 96-98, 104.

[19]For an excellent biography of Carpenter see Ron Tyler, *Silver Fox of the Rockies: Delpheus E. Carpenter and Western Water Compacts* (Norman: University of Oklahoma Press, 2003).

[20]Jack L. August Jr., *Dividing Western Waters: Mark Wilmer and Arizona v. California* (Fort Worth: TCU Press, 2007), 33.

[21]Carpenter believed in the preservation of states rights in order to preserve the constitutionally mandated balance between state and federal authority. See also *Wyoming v. Colorado* 259 US 419 (1922).

[22]Hundley, *Water and the West*, 76, 78, 105, 106. Carpenter's role in subsequent Colorado River affairs was especially noteworthy. Although he called for federal cooperation in river development, he shared the long-cherished, jealously guarded position many westerners held concerning the supremacy of states rights. The Coloradan vigorously resisted the claim, argued in *Wyoming v. Colorado*, that the US government owned all the unappropriated waters in nonnavigable streams. "There should be no super-government imposition established," he wrote western congressmen, and he saw the federal government as an arbiter, not a dictator, in Colorado River affairs.

[23]Moeller, *Phil Swing and Boulder Dam*, 29; Hundley, *Water and the West*, 105, 110, 111. According to Phil Swing's biographer, Beverly Moeller, Hoover's appointment came after Phil Swing protested the rumored appointment of a "Denver man," presumably Delph Carpenter. Secretary of the Interior Albert Fall, according to Moeller, persuaded Harding to appoint Hoover.

[24]See Senate Document 142, appendix, "Problems of Imperial Valley and Vicinity," 67 Cong. 2 Sess. (1922), 238, 239; Hundley, *Water and the West*, 119-134; Moeller, *Phil Swing and Boulder Dam*, 21, 24, 25.

[25]Hydroelectric power generation, previously an incidental byproduct of reclamation endeavors, suddenly emerged as one of the most emotional, bitterly contested, and complicated issues of the entire effort. Commissioner Davis, in preliminary version of the Fall-Davis Report, left open the possibility that a private, municipal, or state agency might build the proposed dam

and sell the power generated at the site. Almost immediately, private power companies, like Southern Sierras Power and California Edison, saw the potential for enormous profits, and quickly filed with the recently created Federal Power Commission (FPC). Likewise, the City of Los Angeles, seizing upon public sentiment that private power companies were already charging unfair rates, and eager to secure affordable power for a burgeoning population, also applied to build a power generating dam at the site. The subsequent conflicts over the merits of private versus municipally operated power raged for months, polarizing contending interests, and as a consequence, hastened the demise of the League of the Southwest. Secretary Fall, however, closed this chapter of the power debate and ended preliminary speculation over who would operate the dam when, at public hearings in San Diego, he announced that the US government had ultimate authority to deal with that issue. Arizona's needs, while not as immediate as California's, nevertheless required hydroelectric power. It would be needed for groundwater pumping, Arizona's world-renowned copper industry, and increasing domestic demand. Moreover, power was needed to pump water from the Colorado the central portions of the state; an illusory dream of many Arizonans at the time.

[26]Jack L. August Jr., "Carl Hayden, Arizona, and the Politics of Water Development in the Southwest, 1923-1928," *Pacific Historical Review* 58 (May 1989), 197; Hundley, *Water and the West*, 142, 149, 200-203; William S. Norviel to Carl Hayden, February 26, 1926, Carl Hayden Papers Collection, Arizona State University, Tempe, Arizona.

[27]August, *Vision in the Desert*, 84; Carl Hayden to G. E. P. Smith, February 17, 1923, Box 598, Folder 10; Hayden to F. T. Pomeroy, December 22, 1924, Box 600, Folder 9, CHPC, ASU; House Committee on Arid Lands, "Hearings on Protection and Development of Lower Colorado River Basin, HR 11449," 67 Cong. 2 Sess. (1922), 18, 19; *Arizona Republican*, November 25, 1922; Hundley, "Colorado River Waters Dispute," *Foreign Affairs*, 195-200.

[28]Other basin leaders expressed optimism similar to Hamale. Delph Carpenter, for example, placed the compact signing in broader perspective when he wrote, "This is the exemplification of interstate diplomacy on so large a scale." Fellow Colorado water law expert, F. Ward Bannister called the pact, "a great document . . . a product of two years of labor by the best minds the states and nation had to give." In Arizona, newspaper editor and businessman Dwight Heard hailed the compact as a broad and necessary foundation for the erection of machinery that would ultimately determine the rights of the basin states and expedite development of the river.

[29]August, "Hayden and the Politics of Water," *Pacific Historical Review*, 197; E. C. LaRue to Hayden, February 20, 1923, Box 598, Folder 4; Hayden to Kean St. Charles, January 11, February 7, 1923, Box 598, Folder 10; Ottamar Hamale to Hayden, January 27, 1923, Box 598, Folder 9; Hayden to LaRue, February 12, 1923, Box 598, Folder 3; Hayden to Albert Leyhe, February 20, 1923, Box 598, Folder 3; Hayden to W. C. Wyatt, May 21, 1923, Box 598, Folder 2, CHPC, ASU; Congressional Record, 67 Cong. 4 Sess. (1923), 2710-2717. In February 1923 Arizona Congressman Hayden received an unsolicited opinion from Reclamation Service engineer E. C. LaRue. The seven-page missive delineated critical problems with the pact, yet LaRue, in the end, gave his lukewarm endorsement. "I have made this statement," he wrote to Hayden, "inasmuch as the seven states have agreed to something I suppose the Compact should be ratified. In making that statement, it was necessary for me to follow a big lump. If you want to know my honest opinion of the Compact I can give it to you in one word—rotten. The terms can never be enforced in actual practice."

[30]Dwight Heard, a staunch supporter of ex-Governor Thomas Campbell, ran numerous articles and editorials urging the Arizona state legislature to ratify the compact as quickly as

possible. "The federal government," he accurately predicted, "is going to establish flood control on the Colorado. That is going to be done . . . Compact or no Compact. And without the Compact," he continued, "we have reason to believe that it will be done with scant regard to the interests of Arizona in the waters of the Colorado, either for irrigation or power." See *Arizona Republican*, September 28, 1922, January 30, 1923; Malcolm Parsons, "Origins of the Colorado River Controversy in Arizona Politics, 1922-1923," *Arizona and the West* 4 (Spring 1962), 28-31, 37.

[31]Parsons, "Origins of the Colorado River Controversy," *Arizona and the West*, 30, 37; *Arizona Republican*, September 28, 1922; Hundley, *Water and the West*, 136, 137; N. D. Houghton, "Problems of the Colorado as Reflected in Arizona Politics," *Western Political Quarterly* 4 (December 1951), 638-639.

[32]See Marjorie Haines Wilson, "The Gubernatorial Career of G. W. P. Hunt of Arizona (PhD dissertation: Arizona State University, 1973); John S. Goff, *George W. P. Hunt and His Arizona* (Pasadena, California: Socio Technical Publications, 1973); Malcolm Parsons, "Party and Pressure Politics in Arizona's Opposition to Colorado River Development," *Pacific Historical Review* 19 (February 1950), 50-51; G. W. P. Hunt, "Why I Oppose Approval of the Colorado River Compact," Box 599, Folder 10, CHPC, ASU. One student of Hunt and the controversy concluded that the Arizona governor, like many other state and local politicians in Arizona in the 1920s, "ran on water for a long time." See Walter Rusinek, "In Response to the Compact: The Critical Opposition of George Hunt," unpublished manuscript, Small Collections File, Department of Archives and Manuscripts, ASU.

[33]*Arizona Republican*, January 7, 8, 9, 1923; *Arizona Daily Star*, January 8, 9, 1923.

[34]*Congressional Record*, 70 Cong., 2 Sess. (1928), 459, 466-472; Hundley, *Water and the West*, 268-270; Moeller, *Phil Swing and Boulder Dam*, 118-119, August, "Carl Hayden and the Politics of Water," *Pacific Historical Review*, 207.

[35]For an analysis of Arizona's debates over ratification of the Colorado River Compact see August, *Vision in the Desert*, chapter 5. Also, see *Congressional Record*, 69 Cong., 2 Sess. (1926), 5822; Dwight Mayo, "Arizona and the Colorado River Compact," 30-34.

[36]*Arizona Republican*, January 22, 1927; *New York Times*, January 22, 1927; House Committee on Rules and Administration, *Hearings on HR 9826*, 60 Cong., 2 Sess. (1927), 61-62.

[37]House Committee on Rules, *Hearings on HR 9826*, 64-70; *Arizona Republican*, January 22, 1927; *Imperial Valley Farmer* (El Centro), January 13, 1927.

[38]*Los Angeles Times*, May 29, 1928; *Los Angeles Examiner*, May 30, 1928; *Arizona Republican*, May 9, 1928; Moeller, *Phil Swing and Boulder Dam*, 115, 120; Hundley, *Water and the West*, 275-276; *Congressional Record*, 70 Cong., 2 Sess. (1928), 823, 836. The Boulder Canyon Project Act was signed into law on December 21, 1928.

[39]Hundley, "Politics of Reclamation," *California Historical Quarterly*, 292.

[40]This portion of the bill, in subsequent years, became most controversial. Hundley has argued that "thirty-five years later the US Supreme Court would misconstrue this action and decide that the Boulder Canyon Act provided for a statutory apportionment of the waters of the lower Colorado." According to Hundley, Congress was "merely suggesting a way in which lower states might settle their problems among themselves." Hundley, *Water and the West*, 270; Hundley, "Clio Nods: *Arizona v. California* and the Boulder Canyon Act—A Reassessment," *Western Historical Quarterly* 3, No. 1 (1972), 17-51.

[41]Judge Simon Rifkind oral history interview with Jack L. August Jr., October 9, 1986, author's files; *Arizona v. California* et. al., 372 US 546 (1963); *Congressional Record*, 70 Cong. 2 Sess.

(1928), 459, 466-472; Hundley, *Water and the West*, 170, 268-270; Moeller, *Phil Swing and Boulder Dam*, 118-120.

[42]See "Statement of Mr. Hayden, May 22, 1926," Box 598, Folder 13, CHPC, ASU.

[43]Ralph Murphy, "Arizona's Side of the Question," *Sunset Magazine* 56, 1 (1926), 34-37; Richard Newhall, "Arizona and the Colorado: How Not to Win a River by Trying Very Hard," *Phoenix Point West Magazine* 6 (1965), 31.

Chapter 6: Depression to Empire

[1]David A. Shannon, *Twentieth Century America: The Twenties and Thirties* II, 4th Edition (Rand McNally College Publishing: Chicago, 1977), 148-149; Luckingham, *Phoenix*, 101. When President Herbert Hoover took office in March 1929 the market had already acted in curious fashion; it skidded in 1926 but recovered by the end of the year. In 1927 the bull market began in earnest and soon grew even more exaggerated. In 1928, for example, Radio Corporation of America rose from $85 to $420. The year 1929 saw even wilder speculation take hold and buying stocks on margin further stimulated speculation.

[2]The literature on Herbert Hoover and the Great Depression is vast. See, for example, William Leuchtenburg, *Herbert Hoover* (New York: Times Books, 2009); David M. Kennedy, *Freedom from Fear: The American People in Depression and War* (New York: Oxford History of the United States, 2001); Amity Shaes, *The Forgotten Man: A New History of the Great Depression* (New York: Harper Perennial Books, 2008); John Kenneth Galbraith, *The Great Crash of 1929* (New York: Mariner Books, 2009).

[3]Shannon, *Twentieth Century America*, 149.

[4]See *Arizona Republican*, October 25, 26, 27, 1929; *Tucson Daily Citizen*, October 25, 26, 27, 1929.

[5]Unemployment in mining towns was severe; total population in the state dropped from 435, 573 in 1930 to 380,000 due in large part to the abandonment of mining towns. Many of these displaced miners moved to Phoenix or Tucson to look for employment in the urban centers. See Jay Edward Niebur, "The Social and Economic Effect of the Great Depression in Phoenix, Arizona, 1929-1934" (MA thesis, Arizona State University, 1967), 2-18.

[6]John R. Norton III oral history interview with Jack L. August Jr., August 11, 2009, Phoenix, Arizona, author's files.

[7]Norton interview with August, August 11, 2009.

[8]William H. Jervey Jr., "When the Banks Closed: Arizona's Bank Holiday of 1933," *Arizona and the West* 10 (Summer 1968), 127-1152.

[9]John R. Norton III interview with Jack L. August Jr., September 28, 2010, Phoenix, Arizona, author's files.

[10]Though the partnership began in 1932, the *Arizona Republic*, May 2, 1935, reported, "Arena-Norton Company, Los Angeles, applied to the Corporation Commission for a license yesterday to do business in Arizona. The company is incorporated in California." According to John R. Norton III, "The Arena Company wanted to have a branch in Arizona to grow produce because the season was a little different in California . . . it would broaden their season. . . . So they came and interviewed my dad who was destitute and offered him the job to run the Arena-Norton Company and he got 40 percent of the company. He must have been bargaining from a little more strength than the story yields but at least he had enough bargaining strength that they made him a 40 percent partner and then sent the youngest Arena brother, Sylvester, over here and A. T. Arena stayed in Los Angeles, where they had a big office. . . . This is my dad's

big break because he had been wiped out but he must have had a good enough reputation as a farmer that they sought him out and created a partnership."

[11]Norton interview with August, September 28, 2010.

[12]Norton oral history interview with August, September 28, 2010; Norton oral history interview with August, August 11, 2009. When young five-year-old Norton III went to enroll for kindergarten at Franklin School he tested out and entered first grade which resulted in his oftentimes being younger than his fellow students in grade school, high school, and college.

[13]*Arizona Republic*, July 13, 1939.

[14]Norton interview with August, September 28, 2010. Also for a colorful portrayal of the flood and the region see Jeanette Walls, *Half Broke Horses* (New York: Scribner, 2009).

[15]The Arena-Norton packing shed at Six Points was used until Norton sold his interests in the company in 1947. See Norton oral history with August, August 11, 2009.

[16]The *Arizona Independent Republic* of October 6, 1939 announced: "A certificate of incorporation was issued yesterday by the Arizona Corporation Commission to the Norton-Recker Cattle Company of Phoenix. Incorporators are John R. Norton Jr., Charles A. Recker of Phoenix, and Earl C. Recker of Mesa. Capital stock is 1,500 shares with par value of $100 each."

[17]Charles Recker farmed in Mesa and focused his activities on honeydew melons.

[18]Norton oral history interview with August, August 11, 2009; *New Times*, February 27, 1997. Philip Tovrea arrived in Arizona in 1883 and created a meat processing empire. According to Norton III, his father decided in 1941 or 1942 to go after one of his great loves and he talked the Arenas family into buying the Double O Ranch in Seligman from Bill Tovrea, who owned the Tovrea Packing Company. Tovrea wanted to sell, but the Arenas cared little for the cattle business. Somehow Norton Jr. persuaded his benefactors to purchase the ranch and Norton III asserted, "My dad thought he had died and gone to heaven." When Norton Jr. worked for the 3V, which ran from Seligman to the Grand Canyon and through to the Hualapai Reservation, it was a one million-acre-plus ranch, the biggest in Arizona. When he made a little money he couldn't wait to buy the Double O Ranch. As Norton III remembered, "If you drew a line from Seligman south to Prescott, the road ran on the east side of the ranch. It went west from there and it bordered the ranch that Ray Cowden had, just to the south of us. The Campbell brothers of Ash Fork had the ranch immediately east of us that went from our property all the way to Highway 89 that linked Ash Fork and Prescott. Everybody knew everybody then."

[19]Norton oral history interview with August, September 28, 2010.

[20]*Arizona Republic*, November 26, 2010; John Norton III oral history interview with Jack L. August Jr., November 28, 2010, Phoenix, Arizona, author's files. The swimming pool, more tennis courts, and a softball field were added in 1951 and in 1956 two lagoons were added.

[21]Norton oral history interview with August, August 11, 2009. Norton was eleven years old at the time of Wilkie's visit to the family home. Also see Steve Neal, *Dark Horse: A Biography of Wendell Wilkie* (Lawrence: University Press of Kansas, 1989); Wendell Wilkie, *One World* (New York: Simon and Schuster, 1943); Charles Peters, *Five Days in Philadelphia: The Amazing "We Want Wilkie!" Convention of 1940 and How it Freed FDR to Save the Western World* (New York: Public Affairs, 2005). Wendell Willkie never held a public office, yet he nearly became president of the United States. He was a Democrat until the fall of 1939, then switched parties and nearly captured the Republican party's nomination less than a year later. It was a meteoric rise and Willkie defeated establishment candidates like Thomas Dewey, Robert Taft, and Arthur Vandenberg. These Republican isolationists insisted that the war in Europe should not

concern the United States since two oceans protected the United States from the aggressors. Conversely Willkie had warned of the danger of a Europe controlled by fascists. Shortly before the GOP convened in Philadelphia, Hitler's armies swallowed Denmark, Norway, the Netherlands, and France. Willkie transformed the Republicans from isolationists into internationalists that stood together against Axis aggression. Roosevelt, although recognizing Willkie as a formidable political opponent, called his nomination a "godsend" because it brought national unity. Roosevelt's election to a third term—and Willkie's defeat—turned out to be the closest presidential race in a generation, and Willkie received more votes than any previous Republican candidate, setting a record that stood until Eisenhower's landslide election in 1952. And despite his defeat, Willkie grew in stature becoming Roosevelt's special envoy during World War II, first to London during the Blitz and later to the Middle East, to Russia, and to China. On the home front Willkie became the spokesman of the One World philosophy that influenced US foreign policy for a generation and the conscience of American politics, speaking out against isolationism, imperialism, and the persecution of minorities.

[22]Shannon, *Twentieth Century America*, III, 35.

[23]Norton oral history interview with August, September 28, 2010. Earlene was a smart woman, John III asserted. "She put her money into stocks and things. She was very close to one of the founders of Arizona Public Service Company when it was known as the Central Arizona Light and Power Company . . . she absolutely would never sell a share of that stock. It paid a nice dividend." Earlene was an intellectual. Her son, John III, remembers her as a "fountain of wisdom." Unfortunately, the social mindset at the time was that a woman whose marriage failed was herself a failure. Earlene bought into this and she had a very difficult time dealing with it. She never remarried.

[24]L. W. Pete Dewhirst oral history interview with John R. Norton III, Arizona Historical Society Oral History Programs, August 7, 2000, Phoenix, Arizona; August interview with Norton, August 11, 2009.

[25]*Arizona Daily Star*, May 8, 1955; *Western Livestock Journal* (December 1956), 8, 9. Swaps, who was later named the "Horse of the Year" in 1956, won the "richest ever" Kentucky Derby purse, $108,400. The total purse was $152,000. Ellsworth and Tenney were partners in the racing stable where Swaps was bred. They had two ranches in California, one in Colorado and one in Arizona. Swaps was their first entry in the Derby.

[26]At the time of his purchase of the Chino Valley property, Norton Jr. diversified his agribusiness portfolio and acquired the JV Ranch on the west side of Phoenix and developed it into a citrus farm where ultimately he employed between five hundred and two thousand employees.

[27]Norton's two-year-old Grand Champion, 00 Miss Domino 83, went on to Denver to win her class and she had previously won at the New Mexico State Fair, the Tri-State at Amarillo, Texas, the Utah State Fair, and the Polorama Show at Sacramento. See Dick Shaus, "After One Hundred Years of Fame: Del Rio Ranch is Now Producing Champion Polled Herefords," *Arizona Cattleog*, (May, 1960).

[28]Dewhirst interview with Norton III, August 7, 2000.

[29]For a short, comprehensive account of the water situation in Arizona during the 1940s see August, *Vision in the Desert*, 139-159.

[30]They endured what was described as the worst marketing year in the history of desert-grown lettuce in 1967 when overproduction depressed prices.

[31]Between 1909 and 1985, Buenos Aires Ranch changed ownership several times. It became one

of the most prominent and successful livestock operations in Arizona. From 1926 to 1959, the Gill family raised prize-winning racing quarter horses. During the 1970s and '80s, the Victorio Land and Cattle Company specialized in purebred Brangus cattle, which are well suited to hot, dry climates. John R. Norton III to Jack L. August Jr., March 29, 2011 (author's files).

[32]Norton Jr.'s other activities and associations included the Arizona Cattlemen's Association, Arizona Livestock Show, Maricopa County Sheriff's Posse, Pioneer Historical Stockmen's Association, President's Club of the University of Arizona, and Paradise Valley Country Club. He was also a member of the Phoenix Country Club, Moon Valley Country Club, and La Jolla Country Club.

Chapter 7: To the Other Side of the River

[1]John R. Norton III oral history interview with Mona McCroskey, Phoenix, Arizona, November 21, 2003, Sharlot Hall Museum Oral History Project, Prescott, Arizona. As Norton III recalled that summer of 1943: "The day after school was out my dad took me up there and put me out in camp with a cowboy. In that time of year, in early June, we were bringing the cattle from the desert down there around Hackberry desert country into the high country for the summer."

[2]Dewhirst oral history interview with Norton, August 7, 2001; August oral history interview with Norton, August 8, 2009.

[3]John R. Norton III to Jack L. August Jr., August 4, 2010, author's files.

[4]Dewhirst oral history interview with Norton III, August, 7, 2001.

[5]Norton oral history interview with August, September 28, 2010.

[6]The University of New Mexico finished second. See *Arizona Republic*, May 9, 1948; *Arizona Daily Sun* (Flagstaff), May 9, 1949.

[7]Members of the winning UA Rodeo team in 1949 were Norton III, Jack Thompson, his roping teammate, George Masek, Vernon Mounce, Dutch Shuman, and Don Martin.

[8]See *Arizona Daily Star*, May 17, 1949.

[9]Other recipients were Melba Shelton, Robert Stubbs, Lowell Rothschild, Bernard Phelan, and Ralph Deal.

[10]McCroskey oral history interview with Norton III, December 28, 2003.

[11]*Arizona Daily Star*, May 31, 1950. Garvey served as governor of Arizona from 1948-1951, assuming the office upon the death of Governor Sidney P. Osborn.

[12]Charlie Recker had come to Arizona from Rocky Ford, Colorado, in the southeastern part of the state, an area where many latter-day Arizona farmers began their produce careers. Many, like Recker, had grown melons and produce and these farmers came to Maricopa County because they believed they could grow and harvest melons earlier in the year (June-July instead of August-September). Recker moved to Mesa and began producing acres of honeydew melons. He partnered with Norton Jr. in what they renamed the X Bar One Cattle Company, located about fifteen miles south of Peach Springs, Arizona. Norton Jr. and Recker actually purchased the old Aubrey Valley Investment Company Ranch (AIC) which was on the plateau at 5,000 feet elevation. Then, the partnership purchased land west of the X Bar One that connected it to the Hualapai Valley on both sides of Hackberry, and west and northwest of that small northwestern Arizona community, on the north side of Route 66. What ultimately comprised the 150,000-acre ranch was called the X Bar One Cattle Company.

[13]Norton interview with August, August 8, 2009.

[14]Norton oral history interview with August, September 28, 2010.

[15]Doris Schaefer Norton oral history interview with Jack L. August Jr., June 6, 2011, Phoenix, Arizona, author's files; LeAnne Stevensen, "Doris's Story: A Lifetime of Memories, Yesterday, Today, and Tomorrow" (unpublished manuscript, September 1998), Norton Family Papers, Phoenix, Arizona. Doris Schaefer Norton was born September 9, 1927 and raised in Sioux Falls, South Dakota. Born into a strict German Lutheran family her mother was the organist at the local Lutheran church and that formed the framework of Doris's early life; church, choir practice, and child catechism on Saturday morning. She recalled ice skating on the pond not far from the family home with her older brother and sister, as well as her younger brother. The children shared one bicycle and a pair of roller skates. Doris and her sister were popular in high school and loved to dance, much to the consternation of the Lutheran minister, who often seemed to be directing his sermon at the Schaefer sisters. Her father was a liquor salesman. In one life-changing instance, he decided to accompany a fellow salesman to California and en route, arrived in Phoenix in February. According to Doris, "He had never seen orange trees and he literally wired my mother and said he had found Paradise and we are going to move and he stayed here [Phoenix] and started a little delicatessen." The rest of the family arrived after selling the Sioux Falls house and arrived in Arizona in 1946. The first thing Doris wanted was a horse; she had never had one. The family obliged and they kept the horse tethered to the back yard shed at 1530 East Maryland, where the Schaefer family lived. Ultimately, the horse ran off and Doris found a job at the Central Arizona Light and Power Company (CALAPCO). She entered UA in 1949, after meeting John at the dance in February 1948, and was a twenty-one-year-old freshman.

[16]Doris Norton oral history interview with August, June 6, 2011.

[17]Norton III recalled that one night he and Doris were sleeping and all of a sudden there was an odor wafting from the floor. A skunk had made its way under the house and John III shot the skunk but they had to abandon the home for some time. Also every dog at the ranch ran into a skunk or porcupine and that was always a one time occurrence.

[18]See McCroskey oral history interview with Norton, December 28, 2003.

[19]John R. Norton to Jack L. August Jr., February 7, 2011, author's files.

[20]Norton III oral history interview with August, August 8, 2009.

[21]Norton III oral history interview with McCroskey, December 28, 2003.

[22]He had hoped to head to Germany or Europe, because many of his fellow soldiers were heading there.

[23]Doris Norton interview with August, June 6, 2011.

[24]See Norton III oral history interview with August, September 28, 2010.

[25]John R. Norton to Jack L. August Jr., April 16, 2011, author's files.

[26]Norton III oral history interview with McCroskey, December 28, 2003; Norton III oral history interview with August, September 28, 2010.

[27]Norton III oral history interview with August, September 28, 2010.

[28]John R. Norton III to Jack L. August Jr., April 7, 2010, June 14, 2011; Norton III oral history interview with McCroskey, December 28, 2003.

[29]See Norton III oral history interview with McCroskey, December 28, 2003.

[30]See www.pvid.org. In recent years, the annual value of crops produced within the district has ranged from $60 million to nearly $160 million, excluding livestock. Additionally, in recent years, up to 35,000 head of sheep have wintered annually in the valley.

[31]The federal census for Blythe in 1950 revealed 4,089 in population while the 1960 census counted 6,022. In both censuses, over 50 percent of Blythe's permanent residents were

Hispanic.

[32]Briefly, from 1852 to 1877, steamboats moved north along the Colorado from Port Ysabel at the mouth of the river to Hardyville [Bullhead City] near the current site of Davis Dam. Mining supported this form of commerce and transportation and a freight line crossed the river at Ehrenberg, Arizona across the river from Blythe, delivered necessary supplies to the mining areas of Quartzite, Wickenburg, and Prescott, Arizona. One other reason for river traffic during this period was to supply the military posts in Arizona. When the Southern Pacific Railroad reached the port of Yuma in 1877, steamboat traffic on the Colorado virtually halted. Further, with the advent of the railroads within Arizona to the east, the wagon freight line diminished and, with the construction of Laguna Dam as part of the newly instituted federal reclamation program in 1909, river traffic ceased altogether. See Richard E. Lingenfelter, *Steamboats Along the Colorado, 1852-1916* (Tucson: University of Arizona Press, 1978).

[33]See Grady Setzler, *Another Wilderness Conquered: An Informal History of Blythe and the Palo Verde Valley, California* (Blythe: Palo Verde Historical Museum, 1968); Palo Verde Historical Museum, *Blythe and the Palo Verde Valley* (Charleston, South Carolina: Arcadia Publishing, 2005).

[34]Palo Verde Irrigation District Act of 1923 (Stats., 1923, ch. 452, p. 1069).

Chapter 8: An Expatriate's Dilemma: *Arizona v. California*

[1]John R. Norton III oral history interview with Mona McCroskey, January 27, 2004, Phoenix, Arizona, Sharlot Hall Museum Oral History Project, Prescott, Arizona.

[2]*Arizona v. California*, 283 US 423 (1931). The bill was dismissed without prejudice to a future action for relief in the vent that the dam was so operated as to interfere with Arizona's rights.

[3]August, *Vision in the Desert*, 140. Black Canyon was often publicly misidentified as Boulder Canyon. Also, Simon Rifkind, oral history interview with Jack L. August Jr., October 9, 1986, author's files; *Arizona v. California et. al.* 373 US 546 (1963); *Congressional Record*, 70 Congress, 2 Session (1928), 459, 466-472.

[4]August, *Vision in the Desert*, 140-141. Arizona's Colorado River Commission provides a case in point. A. H. Favour, a veteran of several commissions wrote Arizona Senator Carl Hayden about that agency's lack of effectiveness: "We haven't formulated any definite water policy pertaining to the Colorado." In 1935 the editor of Tucson's *Arizona Daily Star* noted that for over a dozen years, "factional disputes within the state have cost it a pretty price." And, in 1939, Favour, briefing newly elected Governor Robert T. Jones, again commented that "Arizona has gone through various administrations since 1922 with no very definite water policy." Thus, the Arizona Colorado River Commission, often inefficient, usually unprofessional, and with little political power, further reflected the state's problems within the context of evolving federal reclamation policy.

[5]In his report on December 5, 1960, Rifkind wrote, "On February 14, 1934, Arizona moved for speaking through Mr. Justice Brandeis, denied the application. One of the alternate grounds for decision was the incompetence of the evidence sought to be perpetuated. It was held that oral statements of negotiators of a treaty or compact not communicated to the ratifying body were not admissible to establish meaning." Rifkind, *Special Master's Report*, 7. See also, *Arizona v. California* 292 US 341 (1934).

[6]Craig Lowell Whetten, "The March on Parker Dam: Why Governor B. B. Moeur Declared Martial Law, the Consequences Thereof, and its Place in Arizona Water Rights History," unpublished manuscript, 2005; Benjamin B. Moeur to Ray Lyman Wilbur, February 16, 1934;

Ray Lyman Wilbur to Carl Hayden, February 28, 1934, Box 601, Folder 19, CHPC, ASU; Hundley, *Water and the West*, 285-288. For further accounts see *Tucson Daily Citizen*, July 1, 1931; *Arizona Daily Star*, May 29, 1931.

[7] 49 Stat. 1039

[8] *Arizona v. California*, 298 US 558 (1936)

[9] Rifkind, *Special Master's Report*, 8.

[10] Parker Dam was completed in 1938 at a cost of approximately $8.8 million. The original bid by the Six Companies, who constructed Hoover Dam, was $4.2 million. As it turned out, Parker Dam was one of the deepest in the world; the crest reached 320 feet and the excavation required digging to bedrock 235 feet below the river bottom.

[11] August, *Vision in the Desert*, 147-148.

[12] Hundley, Water and the West, 295-297; Carl Hayden to Jack Gavin, n.d., Box 614, Folder 7, CHPC, ASU. The implications of the Mexican Treaty were abundantly clear. Unless Arizona took steps to put mainstream water to beneficial use within the state, California and Mexico could claim a prior right to Arizona's claimed share. Moreover, Bureau of Reclamation engineers, during hearings on the Mexican treaty, underscored the point when they reported that supply figures were significantly less than previously believed.

[13] Senate Committee on Foreign Relations, *Hearings on Water Treaty with Mexico*, 79 Cong. 1 Sess. (1945) 1760; "Water Supply Below Boulder Dam," *Sen. Doc. 39*, 79 Cong. 1 Sess. (1945) part 1, 5-8; Hundley, *Water and the West*, 297; Carl Hayden to Sidney P. Osborn, February 15, 1943, Box 616, Folder 2, CHPC, ASU.

[14] The short-term historical context of these developments is significant and shed light on the Arizona senator's relationship with the Bureau of Reclamation. As a result of the water shortage emergency in Arizona during the late 1930s and early 1940s, Secretary of the Interior Harold Ickes supported Hayden's request to have Bureau of Reclamation engineers survey various routes to direct water from Parker Dam to central Arizona. In fact, on October 4, 1940, Hayden met with President Roosevelt to discuss the desperate need for new sources of water. Although the wet year of 1941 eliminated the need for emergency surveys, these steps anticipated later developments. In connection with these preparations, and the resumption of the drought a year later, the Arizona legislature, in early 1943, approved $200,000 for use by the Bureau to conduct investigations on various routes for an aqueduct from the Colorado to central Arizona. Also, according to Marc Reisner, a critic of the Bureau of Reclamation activities in the American West, Hayden supposedly exercised "near-despotic rule" over the Bureau's authorizing committees by World War II. During the war, moreover, Hayden was the ranking member of the Senate Committee on Post-War Economic and Policy Planning. Arizonans viewed their senator's position on this committee as crucial in their quest for water from the Colorado River. See Marc Reisner, *Cadillac Desert: The American West and its Disappearing Water*, rev. ed., (Vancouver: Douglas and McIntyre, 1993), 118; Gerald Nash, *The American West Transformed: The Impact of World War II* (Bloomington: Indiana University Press, 1985).

[15] US Senate, "Authorizing with Respect to Present and Future Need for Development of Projects for Irrigation and Hydroelectric Power," (1943), passim.

[16] August, *Vision in the Desert*, 161.

[17] Hearings, *Sen. Res. 304*, 12-15; *Arizona Republic*, August 22, 1944.

[18] This portion of the hearings were held in Florence, Arizona. Hearings, *Sen. Res. 304*, 50-54; *Arizona Daily Star*, August 4, 1944.

19US Bureau of Reclamation, *Comparison of Diversion Routes, Central Arizona Project Planning Report*, Project Planning Report No. 3-8b 4-0 (Washington, DC, 1945), 4-6; *Arizona Republic*, August 22, 1944.

20Bureau of Reclamation, *Report on the Central Arizona Project*, 13; *Arizona Republic*, February 6, 1948; Arizona Daily Star, February 6, 1948; Jack L. August Jr., "Carl Hayden's 'Indian Card': Environmental Politics and the San Carlos Reclamation Project," *JAH* 33 (Winter 1992), 397-422.

21Bureau of Reclamation, Report on the Central Arizona Project, 43; Walter Rusinek, "Bristor v. Cheatham: Conflict Over Groundwater Law in Arizona," *Arizona and the West* 24 (Summer 1985), 143-162. Indeed much remained to be resolved, including the groundwater issue which resulted in a long series of State Supreme Court decisions that had the effect of limiting the right of private ownership of groundwater while increasing the power of the state government in controlling withdrawals.

22McFarland, *Mac*, 207; Carl Hayden to Forrest Donnell, June 20 1947, Box 662, Folder 4, CHPC, ASU. In the House of Representatives, John Murdock (D-Arizona) submitted a companion bill to accompany S 433. HR 1598 was offered on February 3, 1947.

23August, *Vision in the Desert*, 165.

24*Newsweek Magazine*, March 6, 1950, described this aspect of the political process in vivid terms for its readers: "Carl Hayden, who almost never speaks to newsmen but is articulate enough in the cloakrooms, had a couple of helpful talking points. As chairman of the Rules Committee he is in charge of Capitol patronage in the Senate wing. This strategic spot enables him to block funds for any special committee investigations voted by the Senate. In addition, Hayden is the ranking member and de facto chairman of the Appropriations Committee, which passes on specific projects, sometime legitimate, sometimes porkbarrel, upon which the political life of many senators depends." *Arizona Republic*, February 22, 1950; US Senate, *Congressional Record*, February 21, 1950, 81 Cong., 2 Sess., 2101-102; Walter Bimson to Carl Hayden, February 22, 1950, Box 19, Folder 10, Carl Hayden Congressional Papers Collection (CHPC), Arizona State University, Tempe, Arizona

25*Phoenix Gazette*, October 25, 1950; Johnson, *Central Arizona Project*, 60, 67.

26Sims Ely, *The Lost Dutchman Mine* (New York: William Morrow and Company, 1953); Dennis McBride, "Sims Ely (1862-1954): The Boulder City Dictator," Stephens Media Group, 1999, 2.

27Northcutt Ely, "Doctor Ray Lyman Wilbur: Third President of Stanford and Secretary of the Interior," The Fortnightly Club, Redlands, California, December 16, 1994.

28"Overseer Northcutt 'Mike' Ely Dies at age 93," *Hoover Institution Newsletter* (Summer 1997). As executive assistant to Wilbur during the Hoover administration, Ely chaired a technical advisory commission to the Federal Oil Conservation Board in Washington from 1931-1933. Other significant legal benchmarks in his career: he was counsel to the governor of Oklahoma when he was negotiating the Interstate Oil Compact of 1935-35 and counsel to the governments of Saudi Arabia, Turkey, The People's Republic of China, Algeria, Malagasy Republic, Ethiopia, Grenada, and Thailand on mining and petroleum issues. He was also a member to the US delegation to the United Nations Conference on the Application of Science and Technology for the Benefit of Less Developed Areas in 1963 and the United Nations Conference on mineral legislation held in Manila in 1969 and Bangkok in 1973.

29Roger Ernst interview with Jack L. August Jr., March 23, 1983, Oral History Collection (OHC), Department of Archives and Manuscripts, Hayden Library, ASU; *Arizona Republic*,

278

September 7, 1950; Johnson, *Central Arizona Project*, 59-71.

[30]Representatives Murdock and Patten introduced HR 1500 and HR 1501, to support construction of CAP in the first session of the Eighty-Second Congress. See also, August, *Vision in the Desert*, 170.

[31]Howard Pyle interview with Jack L. August Jr., November 9, 1982, Tempe, Arizona, OHC, ASU; House of Representatives, Interior and Insular Affairs Committee, *Hearings on HR 1500, 1501*, 82 Cong., 1 Sess., 741-63; *Arizona Republic*, April 19, 1951, *Arizona Daily Star*, April 19, 1951, *Los Angeles Times*, April 19, 1951.

[32]*Arizona Daily Star*, May 31, 1951; US Senate, *Congressional Record*, 82 Cong., 1 Sess., 5974-6210. This version of S 75 not only authorized a Supreme Court test of water rights but also provided that construction could not begin while any suit was pending before the Court.

[33]For an analysis of the tempestuous legislative history of the CAP bills from 1947-1952 see August, *Vision in the Desert*, 159-174.

[34]Perry Ling and Burr Sutter, who each served as special counsel to the Arizona Interstate Stream Commission, were associates at Snell & Wilmer.

[35]Moeur was the son of former Arizona Governor B. B. Moeur. *Arizona Daily Star*, August 14, 1952; *Arizona Republic*, August 14, 1952; Wayne Akin oral history interview with Jack L. August Jr., September 18, 1982, Oral History Collection, Department of Archives and Manuscripts, Hayden Library, ASU; Hundley, "The West Against Itself," in Weatherford and Browns, eds., *New Courses for the Colorado River*, 30; Donald Worster, *Rivers of Empire: Water, Aridity, and the Growth of the American West* (New York: Pantheon, 1985). At the start of the suit, Governor Pyle signed the motion and complaint in which the Arizona Interstate Stream Commission asked the Supreme Court to take jurisdiction in the battle over the Colorado River. With Pyle at the signing were Moeur, Secretary of State Wesley Bolin, AISC Executive Secretary Ray Killian, Arizona Attorney General Fred O. Wilson, Chief Assistant Attorney General Alexander B. Baker, and Perry M. Ling, of Snell & Wilmer who served as Special Counsel to AISC.

[36]*Arizona Republic*, August 14, 1952; *Los Angeles Times*, August 14, 1952.

[37]Frank Snell and Mark Wilmer met Senator McFarland in the late 1930s, when he was serving as Pinal County Superior Court Judge, admired his legal and political skills, and campaigned vigorously for him in the Democratic primary of 1940, when he upset sitting Democratic Senator Henry Fountain Ashurst. McFarland went on to win the general election and entered the Senate in 1941.

[38]*Los Angeles Times*, August 3, 1952.

[39]*Arizona Republic*, August 14, 1952; J. H. Moeur, Special Counsel, Arizona Interstate Stream Commission, "The Colorado River Controversy," memorandum, February 1, 1951, Arizona Interstate Stream Commission (AISC), Arizona State Library, Archives and Public Records, Phoenix, Arizona. Nixon had been recently tapped as the Republicans' vice presidential candidate with General Dwight D. Eisenhower.

[40]While case transcripts document this phase, Governor Ernest McFarland's unique role in the final phases of the preliminary hearings is described in the recent biography of James E. McMillan Jr., *Ernest W. McFarland* (Prescott, Arizona: Sharlot Hall Museum Press, 2004), 310-313.

[41]Ibid., 310-311.

[42]*Arizona Republic*, June 9, 1984. Avery reflected on this instance and others upon McFarland's death.

[43]US Supreme Court, *Arizona v. California*, 350 US114 (1955). Court records indicate that John P. Frank and Ernest W. McFarland argued the cause for the State of Arizona, complainant. On the brief were Robert Morrison, Attorney General of John H. "Hub" Moeur, John Geoffrey Will, Burr Sutter, Perry Ling, and Theodore Kiendl.

[44]McMillan, *Ernest W. McFarland*, 311-313. *Arizona v. California*, 350 US114 (1955).

[45]See Paul Kleppner, "Politics Without Parties: The Western States, 1900-1984" in Gerald D. Nash and Richard Etulain eds., *The Twentierth Century West: Historical Interpretations* (Albuquerque: University of New Mexico Press, 1989), 295.

Chapter 9: We Must Not be Indecisive Lest We Be Ineffective

[1]John R. Norton III to Jack L. August Jr., June 22, 2011, Phoenix, Arizona, author's files.

[2]"United's chairman uses company plane for 'close management,'" *Outlook: The Management Magazine of the Produce Industry* 9, 6 (November-December 1982), 33.

[3]Norton oral history interview with McCroskey, January 27, 2004.

[4]Ibid.

[5]Norton III oral history interview with August, September 28, 2010.

[6]Doris Norton interview with August, June 6, 2011; Melanie Norton oral history interview with August, June 26, 2011, author's files.

[7]See Deborah Cohen, *Braceros: Migrant Citizens and Transnational Subjects in the Postwar United States and Mexico* (Chapel Hill: University of North Carolina Press, 2011). The Bracero Program was initially prompted by a demand for manual labor during World War II and began with the US government bringing in a few hundred experienced Mexican farm laborers to harvest sugar beets near Stockton, California. To place this in historical context, in the 1930s, during the Great Depression over 500,000 Mexican Americans were deported or pressured to leave, during what was called the Mexican Repatriation. Thus there were fewer Mexican Americans available when labor demand returned with World War II. Upon the initial foray in Stockton, the program soon spread to cover most of the United States with the notable exception being Texas, which initially opted out of the program in preference to an "open border" policy, and was denied braceros by the Mexican government until 1947 due to perceived mistreatment of Mexican laborers. There was also a railroad bracero program, independently negotiated to supply US railroads initially with unskilled workers for track maintenance but eventually to cover other unskilled and skilled labor. By 1945, the quota for the agricultural program was more than 75,000 braceros working in the US railroad system and 50,000 braceros working in US agriculture. The railroad program ended with the conclusion of World War II but at the behest of US growers, who claimed ongoing labor shortages, the program was extended under a number of acts of congress until 1948. Between 1948 and 1951, the importation of Mexican agricultural laborers continued under negotiated administrative agreements between growers and the Mexican government. On July 13, 1951, President Harry Truman signed Public Law 78, the provision that Norton III operated under, which was a two-year program that embodied formalized protections for Mexican laborers. The program was renewed every two years until 1963 when, under heavy criticism, it was extended for a single year with the understanding it would not be renewed. After the formal end of the agricultural program in 1964, there were agreements covering a much smaller number of contracts until 1967, after which no more braceros were granted.

[8]"Norton Blythe Cantaloupes Advertisement," File 1, Box 2, John Norton Farms Collection, Phoenix, Arizona.

9 "Norton Advertisements for Phoenix Market," File 2, Box 1, John Norton Farms Collection, Phoenix, Arizona.

10 *Palo Verde Valley Times* (Blythe) January 7, 1960. The article, reflecting the tight-knit community's intimate relationships, included the comment, "The valley's outstanding young agriculturalist and his wife, Doris, are the parents of three children—Michael, five years old, Johnny Jr., and daughter Melanie, ten months." On the selection committee was Fred Denewiler, manager of the Blythe branch of the Farmers and Merchants Bank. Denewiler, in 1958, floated a loan without question to Norton III after a particularly rough harvest and had he not provided a loan based on a handshake, Norton allowed that he could well have not planted a crop that season.

11 "Champion Polled Hereford," File 6, Box 2, John Norton Farms Collection, Phoenix, Arizona.

12 Norton III oral history interview with August, August 11, 2009.

13 Norton III knew the pilot in high school when both attended North High.

14 For an examination of the broader movement to sustain species, as in the case of the Buenos Aires Ranch, see Richard Knight, Wendell Gilget, and Ed Marston eds., *Ranching West of the 100th Meridian: Culture, Ecology, and Economics* (Washington, DC: Island Press, 2002); Dan Daggett and Jay Dusard, *Beyond the Rangeland Conflict: Toward a West that Works* (Layton, Utah: Gibbs Smith Publishers, 2000).

15 Norton III oral history interview with Dewhirst, August 7, 2001. When asked, "What was the magnitude of your operations?" Norton III replied, "Well at the peak we were farming about 25,000 acres of irrigated land. The Blythe area was about 10,000 or 11,000 acres. We had several thousand here in Arizona. We had the San Tan Ranch in southeast Chandler and we had some produce land out in the Luke area. We also grew lettuce in Marana and Willcox. We had an operation in Aguila for a number of years in the 1960s and 1970s. We grew lettuce in the different areas—each of which is a different climatic zone. Then Willcox is 4,000 feet above sea level, so we were able to produce lettuce in different periods in the desert. In mid-summer, from late May to the first of October, you have to be in the Salina, California area, which is a cool coastal valley next to the ocean."

16 Roy L. Elson, Administrative Aide to Senator Carl Hayden and Candidate for the United States Senate, 1955-1969, Oral History Interviews, July 27 to August 21, 1990, Senate Historical Office, Washington, DC, 103.

17 *Arizona Republic*, June 4, 1963.

18 John Rhodes to Carl Hayden, January 25, 1963, Box 3, Folder 8, CHPC; Wayne Aspinall to Stewart Udall, November 27, 1962, Box 3, Folder 10, CHPC, ASU; *Washington Post*, January 22, 1962; *Arizona Republic*, January 22, 1962; *New York Times*, January 22, 1962; United States Secretary of the Interior, "News Release: Secretary Udall Announces Study for Regional Solution of Water and Power Problems of the Pacific Southwest," January 22, 1963. See also, Helen Ingram, *Water Politics: Continuity and Change* (Albuquerque: University of New Mexico Press, 1990), 46-48.

19 In 1977 SDG&E submitted an application to the United States Nuclear Regulatory Commission, one year before the Three Mile Island accident. The State of California refused to allow the utility to commence construction in the absence of federally demonstrated and approved technology for permanent disposal of radioactive wastes. The project was cancelled that year symbolizing a troublesome decade for nuclear power advocates. Indeed the 1970s were a pivotal period in the California's history with nuclear power, pitting environmentalists and

nuclear advocates. California voters initially refused to pass a 1972 proposal placing a five-year moratorium on nuclear plant construction but conservation and environmental groups worked throughout the decade to halt construction of several proposed nuclear plants, especially near fault lines. In 1975, Jerry Brown replaced Ronald Reagan as California's governor, and the California Committee for Nuclear Safeguards qualified Proposition 15 for the state ballot. This stringent proposition, which ultimately failed at the polls, would have prohibited licenses for any new power plants until public 'proof' of an effective radioactive waste disposal system was discovered. In 1976, just before this measure failed at the polls, Governor Brown passed three nuclear safeguard laws; one of which included the provision that the Resources Conservation and Development Commission of California and the Legislature determine at lease one method of disposing of radioactive waste material safely. Most utilities supported the laws in an attempt to ward off Proposition 15. By the end of the decade, SDG&E had abandoned any hope of constructing the plant. For a thorough review of this effort see San Diego Gas & Electric, Sundesert Nuclear Plant Collection, 1974-1980, San Diego State University, Department of Special Collections, San Diego, California. See also *Los Angeles Times*, October 17, 2001.

[20]John R. Norton III oral history interview with Jack L. August Jr., July 22, 2011, Phoenix, Arizona, author's files.

[21]*Los Angeles Times*, October 10, 2001.

[22]Norton III oral history interview with McCroskey, January 27, 2004.

[23]Karl Eller Center for the Study of the Private Market Economy, College of Business and Public Administration, November 30, 1984, Tucson, Arizona.

[24]Ed Robson with Tom Horton, *Outrageous Good Fortune: The Autobiography of Ed Robson*, (TH&D, Hamilton, Glenwood New Jersey: 2006), 142-143. Early on in the partnership with Smith, Norton III bought him out and when Sun Lakes sold all of its land holdings, Norton III and Robson garnered significant profits.

[25]Norton III oral history interview with Dewhirst, August 7, 2001.

[26]United Fresh represents the interests of member companies throughout the global, fresh produce chain, including family-owned, private and publicly-traded businesses as well as regional, nation, and international companies and their partners. See www.wga.com and www.unitedfresh.org.

[27]As Norton III described this important point in the process, "So they argued violently over the fact that it's one thing to say that the cattleman hadn't been paid when the brand is on the animal and the animal is still out there in the corral. They haven't knocked it over the head yet. But once you've pulled the hide off the animal, and the brand is gone, there is no way to tell whose animal is whose." Norton III oral history interview with McCroskey, January 27, 2004.

[28]Wilson's attorney general, A. Mitchell Palmer, excoriated the meat packing industry and threatened an antitrust suit in February 1920. Palmer succeeded in having the "Big Five" packers—Armour, Cudahy, Morris, Swift, and Wilson—to agree to a consent decree under the Sherman Antitrust Act of 1890 which had the effect of driving packers out of all nonmeat production, including stockyards, warehouses, wholesale, and retail meat. That action notwithstanding, agitation for legislation to regulate the packers persisted into the administration of President Warren G. Harding, and the US Congress passed the Packers and Stockyards Act on August 15, 1921 as HR 6230 and the law went into effect in September 1921. In ensuing years, the Act's scope expanded to regulate the activity of livestock dealers, market agencies, live poultry dealers and swine contractors as well as meatpackers. The Packers

and Stockyards Act of 1921 (7 USC. §§ 181-229b; P&S Act).

[29]John R. Norton III to Jack L. August Jr., November 14, 2010, author's files. In contrast to an earlier era, Norton III allowed, "Today half of your produce goes to Kroger, Albertson's, Wal-Mart, and the like but twenty years earlier a lot of our produce was going to terminal markets in New York, Chicago, Boston, and Philadelphia; then the produce was disbursed to merchants. And the big chains did not dominate everything like they do now."

[30]By this time Norton III balanced the challenges of his chairmanship with the operation of J. R. Norton Company with access to aircraft. He credited the company's new Gulfstream Commander Jetprop 980 with enabling him to navigate quickly and efficiently between Phoenix and other parts of Arizona, New Mexico, central, and southern California. By the late 1970s and early 1980s Norton III found it "virtually impossible to oversee his diversified agribusiness using commercial airlines." In addition to the convenience of flying virtually when and where necessary, the aircraft's role in Norton III's diversification capability was its strongest selling point. "Any single element in diversification could be reason enough to control your own company air transportation," Norton III told a reporter for United's magazine, Outlook, "whether for reasons of prudent agricultural management of the crops, climate, geography, or season influences. All these things play a part in successful farming and the aircraft contributes the quick management response we need to stay ahead." See "United's Chairman uses company plane for 'close management,'"Outlook, 33, 34.

[31]The Packer, March 3, 1983. The article detailed also Norton III''s successful diplomatic move concerning the newly formed Physical Distribution Center. At the Trucking Division meeting at the United Convention in Honolulu in 1982, truck brokers were upset because of a lack of representation on the board of directors and the transportation committee. Animosity between truckers and rail transportation workers was evident. Thus Norton III helped form the Physical Distribution Council to discuss common problems among transportation purveyors. He also established a new position at United—chairman-elect—in order to groom a person to be chairman the following year.

[32]The Perishable Agricultural Commodities Act, USC 499 (c) ("PACA") was passed, according to various accounts, "to encourage fair trading practices in the marketing of perishable agricultural commodities, and in 1984 Congress, thanks to the efforts of John Norton III, amended PACA to impose a statutory trust in favor of any unpaid supplier of Produce (the Produce Supplier) held by a Commodity Seller."

[33]The Packer, June 1983.

[34]Norton III oral history interview with August, August 9, 2009.

Chapter 10: The Deputy Secretary of Agriculture

[1]Norton III oral history interview with Dewhirst, August 7, 2001; Norton III oral history interview with McCroskey, January 27, 2004; Norton III oral history interview with August, July 22, 2011. Through December 1984, rumors abounded about who would replace Richard Lyng as Deputy Secretary of Agriculture though Norton III, who was not even mentioned among several names during that month, emerged as President Reagan's first choice in early January 1985. In January 1985, the Nortons engaged a housekeeper to move in and watch their Paradise Valley home. They returned to Washington, DC prior to January 20, 1985 in time to attend Reagan's second inauguration and attended all the events surrounding it. There had been a terrible blizzard that forced the inaugural ceremony indoors due to the foul weather.

²Catherine Merlos, "Arizona Heavyweight: John Norton, possibly the next Deputy Secretary of Agriculture, could move from a big spot in western agriculture, and Calcot's chairmanship. To the spotlight in Washington, DC," *Outlook* (January-February 1985).

³"The Nomination of John R. Norton III, of Arizona, To Be Deputy Secretary of Agriculture," March 27, 1985, *Hearing Before the Committee on Agriculture, Nutrition, and Forestry*, 99 Cong. 1 Sess. (Washington, DC Government Printing Office: 1985), 14. Norton III told committee members that he also experienced losses due to the ravages of insects and plant diseases.

⁴Doris recalled that "It was kind of a whirlwind. We had built this house and the children were able to be on their own at this time but it was a little bit of a problem for me getting someone to live here and take care of the house and still keep an eye on teenaged Melanie. In Washington during the first months they gave us some wonderful receptions and a beautiful send off and we lived in this hotel in downtown Washington and they allowed us to stay on one of the top floors. So we had a base there that we lived in while we looked for an apartment. . . . So we found a completely furnished apartment . . . and that was very handy because it was close to Cathedral Avenue and close to everything. I loved DC." See Doris Norton oral history interview with August, June 6, 2011.

⁵See "Statement of Senator Barry Goldwater, A US Senator from Arizona to the Senate Committee on Agriculture, Nutrition, and Forestry, and, Statement of Honorable Dennis DeConcini, A US Senator from from Arizona, and Statement of Honorable Morris Udall, A US Representative from the First District of Arizona," March 13, 1985, 3. See also, Jack L. August Jr. and Dennis DeConcini, *Senator Dennis DeConcini: From the Center of the Aisle* (Tucson: University of Arizona Press, 2006).

⁶Dennis DeConcini, oral history interview with Jack L .August Jr., March 15, 2011, Phoenix, Arizona, author's files; Norton III oral history interview with August, July 22, 2011; "The Nomination of John R. Norton III, of Arizona, To Be Deputy Secretary of Agriculture," March 27, 1985, *Hearing Before the Committee on Agriculture, Nutrition, and Forestry*, 99 Cong. 1 Sess. (Washington, DC Government Printing Office: 1985), 2. Senator Jesse Helms (R-North Carolina) was Chairman of the Committee. Robert Dole (R-Kansas), Richard Lugar (R-Indiana), Thad Cochran (R-Mississippi), Rudy Boschwitz (R-Minnesota), Paula Hawkins (R-Florida), Mark Andrews (R-North Dakota), Pete Wilson (R-California), and Mitch McConnell (R-Kentucky) rounded out the majority. The Democrats were composed of Ranking Member Edward Zorinsky (D-Nebraska), Patrick Leahy (D-Vermont), John Melcher (D-Montana), David Pryor (D-Arkansas), David Boren (D-Oklahoma), Alan Dixon (D-Illinois), and Tom Harkin (D-Iowa).

⁷At the time of his nomination, Norton III listed on his honors and awards statements an abridged version of his numerous recognitions: "Distinguished Citizen Award by University of Arizona Alumni Association, 1981; Produce Man of the Year, 1983; and Entrepreneurial Fellow at the Karl Eller Center for the Study of Private Market Economy, University of Arizona, 1984."

⁸"Statement of Honorable Pete Wilson, A US Senator from California," to the Senate Committee on Agriculture, Nutrition, and Forestry, March 25, 1985, 5, 6.

⁹Norton added in the appendix of his confirmation hearings some biographical data that underscored his holdings and his business background: "Founded own company in Blythe in 1955 and engaged in agricultural production. The company now operates in California, Arizona, and Nevada. My home office is in Phoenix, Arizona with branches in

Blythe, Brawley, and Salina, California. J. R. Norton Company is involved in diversified crop production including year-round production in lettuce and seasonal production in strawberries, cantaloupes, honeydews, citrus, broccoli, celery, cauliflower, carrots, asparagus, cotton, wheat, alfalfa, and others." He also provided the committee with his previous memberships and affiliations: "Member, Salt-Gila Flood Control Advisory Commission; Agricultural Employment Relations Board; Board of Trustees, Palo Verde Irrigation District, 1957-1972; Board of Directors, Western Farm Management Company; State Chairman, Farmers for Ford Committee, 1976; Board of Directors, AZL Resources, Inc.; Board of Associates, Smithsonian Institution; Member, US Meat Export Federation; Board of Directors, California Cattle Feeders Association; Member and Chapter Chairman, Young Presidents' Organization; Director, Mountain States Legal Foundation; Director, Arizona Museum of Science and Technology; and Trustee, Heard Museum."

[10]Norton III oral history interview with August, July 22, 2011.

[11]Norton III oral history interview with McCroskey, January 27, 2004; Norton III oral history interview with August, July 22, 2011. The Spence and Norton partnership occupied no federal land and maintained no ownership interest in any feedlot or meatpacking operations. In contrast to his loss of compensation in the various enterprises, Norton III was not required to forego land owned and leased to and operated by the J. R. Norton Company as well as land leased to the Spence and Norton partnership. During his tenure he received his proportionate share of the annual cash rentals for these lands located in Merced, Imperial, and Riverside Counties in California.

[12]These counties in both states included Fresno, Imperial, Merced, Monterey, and Riverside Counties in California and La Paz, Mohave, Maricopa, and Pinal Counties in Arizona. A portion of the fruits and vegetable crops produced by the J. R. Norton Company and its subsidiaries was graded by employees of USDA's Agricultural Marketing Service (AMS) under the voluntary grading provisions of the Agricultural Marketing Act of 1946. Moreover, the J. R. Norton Company and its subsidiary, the Garin Company, were licensed as brokers, dealers, and commission merchants under AMS. In order to avoid actual or apparent conflict of interest between Norton III's financial interests and his duties as Deputy Secretary, he recused himself in any and all matters that may have had an effect either under the fruit and vegetable grading programs administered by AMS under PACA.

[13]In addition to these required sacrifices to serve in government Norton III had to make adjustments with his three Norton family trusts. They included the John R. Norton Jr. Living Trust dated July 9, 1982, the John R. Norton Jr. Irrevocable Trust dated July 9, 1982 and the Earlene P. Norton Testamentary Trust. The trusts held interests in many of the same lands, agricultural enterprises, business interests, and marketable securities in which Norton III also held interests, but he was neither the principal nor a residual beneficiary under the terms of any of the three trusts. He agreed to forego receipt of the Trustee's fee, which he had been receiving annually in return for services as Trustee of the John R. Norton Jr. Living Trust dated July 9, 1982, but determined to continue serving as Trustee during his service as Deputy Secretary of Agriculture under each of the above named trusts. Finally, he stated his intention to continue his service of Co-Trustee under the Retirement Trust for the employees of the J. R. Norton Company, and agreed to resign his position as Co-Trustee upon his confirmation as Deputy Secretary.

[14]At the time of his nomination, Norton III held the following business and civic memberships, some of which are noted in the related narrative above: "Director and Past

Chairman, United Fresh Fruit and Vegetables Association; Director and member of the executive committee, Ramada Inns, Inc., Director, Arizona Public Service; Director, United Bancorp of Arizona; Director and Vice President Turf Paradise, Inc.; Director and Vice President Calcot Ltd.; Director, Modern Ginning Col; Director and member of the Executive Committee of the Central Arizona Project Association; Director and Past Chairman, Western Growers Association; Member and Director, Chief Executives Association; Chairman, Secure Arizona's Future Economy; Member, board of governors, The Plaza Club; Member, Phoenix 40; Cochairman, Citizens for America; Director, Phoenix Metropolitan Chamber of Commerce; and Member of the Board, Council for National Policy.

[15]Norton oral history with August, July 22, 2011. Norton III, in pungent terms, recalled, "This has inconvenienced a lot of people over the years. . . . So this Senator Melcher, from Montana, he was a third-rate jerk, and he put a hold on my nomination. . . . I can't even remember what it was."

[16]The first, "informal" hearing, was held on Wednesday, March 20, 1985 due to Senator John Melcher's need to be on the floor of the Senate for various bills that required his attention.

[17]See Sally Clark, *Regulation and the Revolution in a United States Farm Productivity* (New York: Cambridge University Press, 1994); Ronald Knutson, J. B. Penn, Barry Flinchbaugh, and Joe L. Outlaw, *Agricultural and Food Policy*. 6th ed. (Upper Saddle River, NJ: Prentice Hall, 2007); Don Paarlberg, *Farm and Food Policy: Issues of the 1980s* (Lincoln: University of Nebraska Press, 1980); Luther Tweeten, *Farm Policy Analysis* (Boulder, Colorado: Westview Press, 1989); US Congress. House, Committee on Agriculture, Nutrition, and Forestry, *Payment-in-Kind Program*, 97 Cong. 2 Sess., December 16, 1982; US Congress. House, Committee on Appropriations. Subcommittee on Agriculture, Rural Development, and Related Agencies, *Agriculture, Rural Development, and Related Agencies Appropriations for 1985*, 98 Cong. 2 Sess., February 21, 1984; US Congress, Senate, Committee on Small Business. Subcommittee on Small Business: Family Farm. *Impact of the Payment-in-Kind Program on Agricultural Support Industries*, 98 Cong., 1 Sess., Senate hearing 98-348 (Washington, DC: US Government Printing Office, 1983); US General Accounting Office. *1983 Payment-in-Kind Program Overview: Its Design, Impact, and Cost* (Washington, DC: US Government Printing Office, 1985).

[18]Norton III Nomination Hearing, Senate Committee on Agriculture, Nutrition, and Forestry, March 13, 1985, 13-14.

[19]Ibid., see expecially 50-100.

[20]*Arizona Republic*, September 22, 1985.

[21]*Arizona Republic*, May 16, 1985; *Washington Post*, May 16, 1985. Norton III's recollection of his swearing in ceremony: Norton oral history interview with August, July 22, 2011.

[22]"Statement of John R. Norton III, Deputy Secretary of Agriculture-Designate," March 27, 1985, 12.

[23]*Washington Post*, September 20, 1985; *Arizona Republic*, September 22, 1985.

[24]The Food Security Act of 1985 (PL 99-198), was a five-year omnibus farm bill that allowed lower commodity price and income supports and established a dairy herd buyout program. This bill changed other USDA programs. Several conservation programs were created, including Sodbuster, Swampbuster, and the Conservation Reserve Program. Shortly thereafter the Technical Corrections to the Food Security Act of 1985 Amendments Technical Corrections to Food Security Act of 1985 Amendments (PL 99-253) enabled USDA discretionary power to require cross-compliance for wheat and feed grains instead

of mandating them. Moreover, it changed acreage base calculations, and specified election procedures for local Agricultural Stabilization and Conservation committees. Technical changes and other modifications were enacted by the Food Security Improvements Act of 1986 (PL 99-260), including limiting the nonprogram crops. That same year the Omnibus Budget Reconciliation Act of 1986 (PL 99-509) made changes in the 1985 Act requiring advance deficiency payments to be made to producers of 1987 wheat, feed grains, upland cotton, and rice crops at a minimum of 40 percent for wheat and feed grains and 30 percent for rice and upland cotton. See also, Norton III oral history interview with August, July 22, 2011.

[25]Doris Norton oral history interview with Jack L. August Jr., June 6, 2011.

[26]Norton III oral history interview with August, July 22, 2011.

[27]Norton III oral history interview with McCroskey, January 27, 2004; Norton oral history interview with August, August 8, 2009.

[28]John R. Norton to Jack L. August Jr., September 21, 2010, author's files.

[29]Congressman Jamie L. Whitten served from November 4, 1941 to January 3, 1995. His record of service was eclipsed only by Senator Robert Byrd, Senator Carl Hayden, and Congressman John Dingell. Originally a conservative segregationist as were most of his colleagues from Mississippi, Whitten opposed the US Supreme Court decision *Brown v. Board of Education* and signed the Southern Manifesto which condemned the landmark decision. Like most of the Mississippi congressional delegation, he voted against the Civil Rights Acts of 1957, 1960, 1964, 1965, and 1968. Although Whitten represented a district that grew increasingly suburban and Republican from the 1970s onward, his opposition to Reagan's programs had no impact on his electoral success. He knew well that upon his retirement a Republican would win election to his long-held seat. Finally, his role in US agricultural policy during the late twentieth century was so profound that the US Department of Agriculture Administration Building was named the Jamie L. Whitten Building.

[30]Norton III oral history interview with McCroskey, January 27, 2004.

[31]*Fresno Bee*, March 9, 1986.

[32]*Chicago Tribune*, January 8, 1986. "It's like farming, you're never done," Block said of his government service. "My judgment is that I've done a great deal. I've made a difference and I've accomplished the final objectives that I've had in my mind a little over a year ago and I think it's a good time."

[33]Lyng had been in Reagan's gubernatorial cabinet in California prior to the Reagan presidency. It appeared also that the former Deputy Secretary of Agriculture for four years prior to Norton III's short stint, had been passed over for the top slot in 1980 when Kansas Senator Robert Dole "imposed" John Block on Reagan in 1980. Many speculated that President Reagan owed Lyng for his loyal service in California and during his first term.

[34]*Houston Chronicle*, February 20, 1986; *Arizona Republic*, February 20, 1986; *Fresno Bee*, March 9, 1986. As the *Houston Chronicle* reported the change at USDA, "Richard Lyng, who was deputy secretary during President Reagan's first term, has been named by Reagan to be Block's successor. Lyng, a native of California, is awaiting Senate confirmation. There had been speculation that Norton, who holds extensive farming interests in California and Arizona, would leave because he and Lyng would make USDA top-heavy with Westerners."

[35]Norton III oral history interview with McCroskey, January 27, 2004; Norton III oral history with August, July 22, 2011.

[36]*Fresno Bee*, March 9, 1986.
[37]*Arizona Business Gazette* (Phoenix), November 11, 1986; *Fresno Bee*, March 9, 1986.
[38]*Arizona Business Gazette*, November 10, 1986.

Chapter 11: An Accurate Vision

[1]The J. R. Norton Company offices in 1986 were located at 24th Street and Culver, in Phoenix, three blocks south of the main east-west artery of McDowell Street, because it was reasonably close to Sky Harbor Airport and they could reach the farming enterprises in Salinas, Imperial Valley, and Palo Verde Valley, California as well as Willcox in Cochise County, Arizona. Indeed, over the years the operation grew so large and was so involved that he hired Art Carroll to run the lettuce side of the business. Norton III continued to fly, but in the early 1980s he hired a pilot for Carroll because Carroll took on responsibilities delegated by Norton III, who was a pilot. He flew over forty years. And he had a Beechcraft Bonanza, two Beechcraft Barons, Cessna 414, Cessna 421, Aero Commander, and Turbo Commander 1000, a very sophisticated high performance turbo prop plane. When he left the lettuce business in 1989, Norton III sold the Aero Commander, his last plane.

[2]At this time the CAP canal, the Hayden-Rhodes Aqueduct, had been completed to Phoenix and by 1991 it made it to the terminus outside Tucson. See August, *Vision in the Desert;* August, *Dividing Western Waters.*

[3]*Arizona Republic*, May 19, 1987. According to the obituary, Norton Jr. was a founding director of the Turf Paradise, the board of which his son, Norton III served on, and he was survived at the time by his third wife, Bonita, a stepdaughter, Kathryn Kent, a stepson, Dr. Tyler Kent, his son, and seven grandchildren.

[4]Norton oral history interview with Dewhirst, August 7, 2001.

[5]John R. Norton III oral history interview with Mona McCroskey, March 16, 2004, Phoenix, Arizona, Sharlot Hall Museum Oral History Project, Prescott, Arizona; Norton oral history interview with August, July 22, 2011.

[6]*Salinas Californian*, July 29, 1989. Nearly all of the four hundred employees, who lived or spent their summer months in Salinas harvested their last head of lettuce there in October. Also, the local offices at 271 Rianda Circle, and three farmland properties, Firestone Ranch, Thompson Ranch, and Martin Ranch, were set for sale. The remaining title lands in Salinas was rented to Duda & Sons, a diversified crop growing company out of Florida.

[7]Norton III interview with August, July 22, 2011.

[8]In Phoenix, Norton III served on the Heard Museum and the Phoenix Art Museum, where he donated $1.4 million to start a photo gallery.

[9]Ripley served as Secretary of the Smithsonian from 1964 to 1984 and oversaw tremendous expansion of the institution. Norton oral history interview with August, July 22, 2011.

[10]Norton oral history interview with McCroskey, March 16, 2004; Norton oral history interview with August, July 22, 2011. In addition to the boards already mentioned in the narrative, Norton III served as a Director of America West Airlines, Shamrock Foods, Pinnacle West Capital Corporation, Suncor Development Company, Arizona Business Leaders for Education, Ready Pac Produce, Terra Industries, Inc., Turf Paradise, AZL Resources, Arizona Agricultural Employment Relations Board, Western Farm Management Company, US Meat Export Federation, Mountain States Legal Foundation, Arizona Museum of Science and Technology, Heard Museum, Colorado River Association, Salt-Gila Flood Control Advisory Commission, Greater Phoenix Leadership, Secure Arizona's Future

Economy, Water for Arizona's Growing Economy, Metropolitan Chamber of Commerce, and Council for National Policy.

[11]Longtime Phoenix residents knew that the seeds of the Phoenix 40 were planted two generations ago, when Frank Snell, founding partner of the powerhouse law firm, Snell & Wilmer, and two other private sector titans, Walter Bimson and Eugene Pulliam, virtually shaped Phoenix's economic environment from the 1940s to the 1970s during the Norton III's father's era of influence. By 1974 they had grown old; Pulliam was eighty-three, Bimson, a close friend of President Franklin Delano Roosevelt, was eighty-five; Snell was seventy-six. A few younger lions approached them and in December 1974, the "Big Three" convened a group of business and civic leaders to "consider major community problems." In 1975 the Phoenix 40 was established. Over its first decades, chairmen like Herman Chanen, Richard Snell, Norton III, and Phoenix Suns managing partner, Jerry Colangelo served. Though derided as "elitist" on occasion, its membership and body of work embraced the enlightened philosophy that businessmen must be concerned not only with economics, but also with ethics and responsibility to their communities' needs. See Jack L. August Jr. "Origins of the Phoenix 40: Lessons in Civic Stewardship," *Arizona Food Industry Journal* (July 2009), 8,9.

[12]See, for example, *Chronicle of Higher Education*, July 10, 2009; *Washington Post*, October 26, 2004. Prior to becoming a for-profit education entrepreneur and political activist, Sperling was a professor at San Jose State University. There he promoted several liberal causes including helping to organize a faculty union and ran headlong into California education authorities, especially with his ideas about night school adult continuing education learners. Specifically, Sperling was founder and Executive Chairman of the Board of Apollo Group. He served as President of Apollo Group until February 1998, Chief Executive Officer of Apollo Group until August 2001 and Chairman of the Board until June 2004. He was Acting Executive Chairman of the Board from January 2006 to September 2008 and has served as Executive Chairman of the Board since September 2008. Prior to his involvement with Apollo Group, from 1961 to 1973, Dr. Sperling was a professor of Humanities at San Jose State University where he was the Director of the Right to Read Project and the Director of the National Science Foundation Cooperative College-School Science Program in Economics. He received his Doctor of Philosophy from Cambridge University, a Master of Arts from the University of California, Berkeley, and a Bachelor of Arts from Reed College. Norton III offered that the Apollo Group during his ten-year service, "The stock just went up and up. And Sperling had gone from a college professor at San Jose State to a billionaire.... And that's only half the story, because he gave half his stock to his son, Peter. So, really, the company has created more than $3 billion in wealth in just over twenty years." See Norton oral history with McCroskey, March 27, 2004.

[13]See www.ypo.org; Norton oral history interview with August, July 22, 2011.

[14]Doris added her personal invitation to the letter: "On behalf of the San Francisco Forum Committee, John and I extend our personal invitation to each of you to join us next October 15 to 20 for a 'San Francisco Romance,'—a delightful blend of old views and new visions. Our forum will be held at the famous St. Francis Hotel, the Grand Dame of San Francisco!"

[15]*San Francisco Chronicle*, October 18, 19, 1989; *San Jose Mercury News*, October 18, 19, 1989; *Sacramento Bee*, October 18, 19, 1989.

[16]Norton III oral history interview with August, July 22, 2011; Doris Norton oral history interview with August, June 6, 2011; Mortimer R. Feinburg to John R. Norton III, October 30, 1989, Norton Family Papers, Phoenix, Arizona. Feinburg described further the chaos

at the hotel: "There were also signs of confusion in the lobby. The hotel staff was intent on being polite, however serious the emergency.... The electricity and regular elevators were out in the entire building, plumbing was only available on the lower floors."

[17]John R. Norton III to Jack L. August Jr., May 14, 2010, author's files. See also www.goldwaterinstitute.org. Norton III described the Goldwater Institute: "The Goldwater Institute probably has four or five hundred people that support it financially.... But there are about fifteen or twenty really active people who are involved in the executive committee and the board runs the thing. We confine our activities to Arizona. Now we may touch an issue that's being debated in other states also, but we're dealing with it in respect to its Arizona impact. In other words, if we talk about public transportation, we might put out a paper, and the principles might apply as well in Michigan as they do in Arizona, but we're talking about this impact on Arizona and what our views are."

[18]John R. Norton III to Jack L. August, June 10, 2011, author's files.

[19]August, "John R. Norton III: Agribusiness Visionary and Arizona Philanthropist," AFIJ, 8; George Seitts, oral history interview with Jack L. August Jr., July 2, 2008; Soyean Shim oral history interview with Jack L. August Jr., October 8, 2008; *Norton News*, 1, 1 (October, 2008).

[20]Norton oral history interview with August, July 22, 2011.

[21]Norton oral history with August, July 22, 2011; August, *Vision in the Desert*, 156-201; Jack L. August Jr. and Grady Gammage Jr., "Shaped by Water: An Arizona Historical Perspective," in Bonnie G. Colby and Katharine L. Jacons, *Arizona Water Policy: Management Innovations in an Urbanizing, Arid Region* (Resources for the Future Press: Washington, DC, 2007), 18.

[22]Norton III oral history interview with August, July 22, 2011; August and Gammage, "Shaped by Water," in *Arizona Water Policy*, 18.

[23]The counties that comprised the new agency were Maricopa, Pinal, and Pima counties. See August, *Vision in the Desert*, 205-211; August, *Dividing Western Waters*, 95-115.

[24]John R. Norton III to Jack L. August Jr., July 27, 2011, author's files; August and Gammage, "Shaped by Water," *Arizona Water Policy*, 19.

[25]Additionally, it recast Arizona water law in substantive ways that resulted in the creation of a new agency of government, the Arizona Department of Water Resources (ADWR), which replaced the governor-appointed Arizona Water Commission. Indeed, after three decades the Groundwater Management Act stands as one of the towering innovations in Arizona's water management history. The AMA-based regulations, moreover, further strengthened the role of water as a "binding agent" in Arizona's urban growth by discouraging groundwater-dependent development within AMA.

[26]Don Campbell, "An Interim Report: Central Arizona Project," *Arizona Highways* (November 1968), 9.

[27]August, *Vision in the Desert*, 161.

[28]See August and Gammage, "Shaped by Water," in *Arizona Water Policy*, 20.

[29]Actually prior to passage of the legislation it was known as S 437. The measure encompassed three settlements, two with Indian communities and one with the federal government. Through this settlement the Gila River Nation will receive approximately 155,700 acre-feet of water while the Tohono O'odahm will receive approximately 37,800 acre-feet. The Gila River tribe will also lease another 40,000 acre-feet of their allotment as per the settlement agreement. Some 67,000 acre-feet of water that had been previously unclaimed will be distributed among Arizona cities, with an additional 96,000 acre-feet held in reserve for future allocation. Many considered the intricate settlement as a "giveaway"

to a handful of Indian communities which would remarket the water to Arizona cities at a huge profit. However the reality was more complex. For almost a century courts have upheld claims of Indian communities to federally reserved water rights. Because of the senior status of Indian claims to waters from rivers like the Gila, the reserved rights water doctrine could result in water being taken away from SRP and the cities of central Arizona. Realizing this, Arizona Senator Jon Kyl (R-Arizona), himself a water rights attorney, took the lead in helping structure a settlement of claims of the Gila River and other tribes. This resulted in the Arizona Water Settlement Act, which President George W. Bush signed into law on December 10, 2004. The agreement is one of the most complex and far-reaching in US history. See "Arizona Water Settlements Act," (PL 108-451) December 10, 2004. See Patti Jo King, "Bush Signs Water Settlement Act," *Indian Country Today*, January 17, 2005.

[30]Norton oral history interview with August, July 22, 2011.

[31]*Arizona Daily Star* (Tucson), May 4, 2001. In 1995 a well-financed "anti-CAP/anti-Tucson water" campaign spearheaded by auto dealer Bob Beaudry, used images of dying women in wheelchairs linked to "CAP Chemical Soup" to create fear in the community. Independent experts and community leaders who privately supported CAP water were unwilling to expose themselves as targets. The campaign resulted in the passage of the 1995 Water Consumer Protection Act, which banned the delivery of CAP water unless it is treated to a quality far beyond the Safe Drinking and Water Act standards. To this day Tucson deliveries are a mix of Colorado River water that has been recharged through the aquifer and recovered through wells and groundwater from Avra Valley to the west.

[32]August, and Gammage, "Shaped by Water," *Arizona Water Policy*, 23.

[33]Frank Herbert, *Dune* (New York: Ace Publishers, 1965); Norton oral history interview with August, July 22, 2011.

Index

Block, John "Jack", 202, 212-213, 215-217
Blythe, Thomas, 161-162
Blythe, California, 103, 136-137, 139, 160-167, 179, 184-191, 196, 206, 220-221, 230, 232, 236, 243
Boulder Canyon, 112, 114, 117, 119, 194
Boulder Canyon Project Act of 1928, 3, 120-121, 165-169, 174-176, 179
Boykin, J. C., 82
Bracero Program, 187-188
Brandeis, Louis, 166
Brawley, California, 186, 206
Breakenridge Survey of 1889, 8-10, 43-44, 47, 49, 241
Breakenridge, William N., 8-10, 47
Bridge Canyon Dam Plan, 172-173
Brooklyn, New York, 152
Buckeye Canal, 56-65
Buckeye Canal Company, 56-65, 75
Buckeye Irrigation Company, 56-65
Buckeye Valley, 56-65, 72, 76, 82-85, 91, 105
Buckeye, Arizona, 56-65, 75, 83, 91, 99, 125
Buckley, John D., 26
Buenos Aires Ranch, 138, 191-193
Bush, George H. W., 214

CAGRD (See Central Arizona Groundwater Replenishment)
Calcot, Ltd., 208
California Limitation Act of 1929, 179
California Cattle Feeders, 191
California Lettuce Marketing Board, 191
California State Automobile Association, 83
California Wright Act of 1887, 51-52
Calloway, Oliver P., 162
Camelback Mountain, 31
Cameron, Ralph, 114, 118-119
Camp McDowell, 19
Campbell Ranch, 128
Campbell, Thomas, 114
CAP (See Central Arizona Project)
CAPA (See Central Arizona Project Association)

Capitol Saloon, 20
Cardon, Dr. Bradley Pratt, 144-145, 192
Carey Act of 1894, 47, 68
Carpenter, Delpheus "Delph", 113-115, 119
Carter, Jimmy, 232
Cashion, Arizona, 125
Cashion, James Angus, 36, 94, 125
Cave Creek, 61
CAWCD (See Central Arizona Water Conservation District)
Central Arizona Groundwater Replenishment District (CAGRD), 236-237
Central Arizona Project (CAP), 69, 71, 161, 172-178, 181-182, 193-195, 219-220, 230-238, 239
Central Arizona Project Association (CAPA), 172, 175, 233
Central Arizona Water Conservation District (CAWCD), 232, 234-237
Central Valley Project, 178
Centralia, Illinois, 11
Chamberlin, A.W., 62
Chandler, Alexander J., 70
Chandler, Arizona, 127, 134
Chandler, Harry, 109
Chapman, Oscar, 180
Chief Executives Organization, 208, 224-225
Chihuahua, Mexico, 16
Chino Valley Ranch, 135
Chittenden Report, 53
Chittenden, Hiram, 52-53
Christy Road, 32, 127
Christy, I. M., 57
Christy, Loyd B., 95
Christy, William, 32-33
Churchill, Clark, 23-27, 34-35, 37-38, 40
Churchill, H. P., 62
Churchill, Margaretha, 26
citrus, 32, 36, 41-43, 98, 100, 150-151, 221, 239
Civil War, 11, 44, 65
Clanton, M. E., 60

Fuerte River, 16
Fulwiler, Julia L., 26, 40
Fulwiler, William Dunlap, 26, 34, 40

G. E. "Blondie" Hall Ranch, 135
Gadsden Purchase, 18, 191
Gambrel, Mary Nell, 134
Gardner, John, 20, 22
Garin Company, 186, 206-207
Garvey, Dan, 147, 149, 175
Gibson City, Illinois, 7, 11
Gila River, 14, 16-19, 24, 34-35, 59-64,
    67-68, 107-108, 115-116, 121, 167, 174,
    230, 234-235
Gila River Valley, 17, 35, 172
Glen Canyon Dam, 195
Glendale, Arizona, 100, 125, 134, 137, 192
Goetzmann, William H., 44
Goldwater Institute, 227-228
Goldwater, Barry M., 95, 180-181, 183-184,
    204, 227
Goodyear, Arizona, 98-100
Gosper, John, 60
Grand Canal, 48, 77-79
Grand Canal Co., 41
Gray, John Pinkham, 166
Great San Francisco Earthquake of 1989,
    225-227
Groundwater Management Act of 1980,
    232-233, 237
Gulf of California, 19, 107
Gulf of Mexico, 17

Hackberry, Arizona, 12-13, 26, 130, 142
Hakes, C. R., 8
Hancock, William H., 24
Hantranft, William, 132
Harding, Warren G., 114
Harlan, John, 195
Harer, Evan, 129
Harrison, Benjamin, 63, 96
Hassayampa River, 61, 64
Hatch, F. C., 24, 37-38
Hatch, New Mexico, 137, 186

Hatch, Orrin, 205
Hayden, Carl, 57, 74, 81, 95, 103, 111, 114,
    117-121, 123, 167, 170-181, 184, 194-
    195, 230, 241
Hays, Samuel P., 72
Heard Land and Cattle Company, 80
Heard, Dwight B., 70, 78-80, 83, 96-97
Helms, Jesse, 203, 209
Hendershott, Wells, 49
Henderson Livestock Commission
    Company, 151
Henderson, Stockton, 101
Heyman, Benjamin, 68-69
Hill, Louis C., 85, 88
Hill, Wesley A., 87
Hinton, Richard, 46-47, 53
Hirst, C. T., 58-59
Hitchcock, Ethan, 72, 74, 77, 79
Hobbie, Robert, 155-158
Hobson, W. A., 162
Hohokam, 14-15, 17-21, 25
Holt, W. F., 162
Homestead Act of 1862, 22, 74
Hoover Dam, 108, 120, 166-167, 170, 172,
    176
Hoover, Herbert, 106, 114-116, 119, 124,
    176
Hualapai Indian, 12-13
Hudson Reservoir and Canal Company, 49,
    67-69, 77, 85
Hudspeth Ranch, 128-130
Hughes, Charles Evans, 133
Hughes, Louis C., 56
Hundley Jr., Norris, 107, 240
Hunt, George W. P., 57, 106, 117-118, 141,
    166
Hurley v Abbott (1910), 89-90

Ickes, Harold, 168
IID (See Imperial Irrigation District)
Imperial Irrigation District (IID), 109-112,
    162, 164, 240
Imperial Valley, 108-113, 126, 137, 167,
    186, 198, 221

Pacific Fruit Express, 82
Packers and Stockyards Act, 199-200
Palo Verde Drainage District, 162-163
Palo Verde Irrigation District (PVID), 137, 161, 163-164, 185, 189, 191, 196-197, 230, 240, 242
Palo Verde Joint Levee District, 162-163
Palo Verde Land and Water Company, 162-163
Palo Verde Nuclear Generating Station, 223
Palo Verde Valley, 137, 160-164, 186, 190, 197, 221
Palo Verde Valley Times, 190
Paradise Valley, 139, 172, 185, 228
Parker Dam, 168-170
Parker Pump Plan, 173
Parker, Arizona, 168-170
Parker, Frank, 78
Parker, William, 21
Patten, Harold "Porky", 176-177, 181
Payment-in-Kind Program (PIK), 208-210
Peach Springs, Arizona, 12-13
Perishable Agricultural Commodities Act of 1930 (PACA), 200-202
Peterson, K. Berry, 166
Phelps, Early Clay, 101
Phoenix 40, 223-224
Phoenix and Maricopa County Board of Trade, 68, 70-71, 80, 93
Phoenix Art Museum, 228
Phoenix Chamber of Commerce, 8, 71, 104
Phoenix Daily Herald, 9-10
Phoenix Ditch Company, 19
Phoenix Gazette, 42, 176, 181
Phoenix Hay and Grain Company, 57
Phoenix Herald, 23, 37
Phoenix Railway, 43, 95
Phoenix Union High School, 97
Phoenix, Arizona, 8-10, 14, 19-29, 31-35, 38, 43, 49, 56-57, 59, 61, 64, 66, 67, 71, 75-76, 83, 89, 90, 91-98, 101-107, 124, 176, 206, 219, 223-224, 240
Pierce, Samuel, 216

PIK (See Payment-in-Kind Program)
Pilot Knob Project, 170
Pima County, 173
Pima Indians, 14, 34
Pinchot, Gifford, 71-73
Pistor, W. J., 144
Pittman Amendment, 121
Pittman, Key, 121
Polled Hereford Association of Arizona, 136
Poppe, David, 218
Poulson, Norris, 175, 177-178
Powell, John Wesley, 44-47, 52-53, 66-67, 108
Prescott, Arizona, 14, 19, 24-26, 32, 48, 96, 130, 162
Prior appropriation, 16, 18, 40, 74, 79, 89, 108, 112, 115, 120
Pryor, David, 209
Pulliam, Eugene, 182
PVID (See Palo Verde Irrigation District)
Pyle, Howard, 177, 179, 181, 184

Quayle, Dan, 225

Ramada Inns, Inc., 208, 223
Randall, Charles, 111
Rau, John, 20
Reagan, Nancy, 213, 227
Reagan, Ronald, 202-203, 208, 213-217, 223, 241, 243
Recker, Charles "Charlie", 129, 130, 149
Reid, F. A., 95
Report on the Arid Regions of the United States, 44-45, 108
Reserve Officers' Training Corps (ROTC), 145, 147, 149, 151, 242
Rhodes, John J., 180, 183, 184
Rice, G. W., 162
Rieke, Betsy, 235
Rio Grande, 16, 107
Ripley, S. Dillon, 222-223
Roberts, Eli G., 48
Roberts, Oscar, 82

Robson, Edward, 197
Rockaway Beach, New York, 153, 155
Rockwood, Charles, 108-109
Roosevelt Dam, 84, 86-90, 92-94, 97, 106
Roosevelt, Franklin D., 117, 133, 171, 187
Roosevelt, Theodore, 44, 55, 72-74, 76, 78, 87-90, 110, 241
Rose, Mark, 109
ROTC (See Reserve Officers' Training Corps)
Rough Riders, 87, 90
Rudd, Eldon, 204

S&N Farms, 197
Salinas Californian, 221
Salinas, California, 186, 206, 221
Salt Lake City, Utah, 4
Salt River, 8-11, 14, 16-19, 21, 23-25, 36-38, 41, 48-50, 52, 61-62, 67, 77, 84-86, 89, 97, 107-108, 115, 235
Salt River Project (SRP), 2-3, 26, 69, 71, 230, 234, 236, 239-240
Salt River Valley, 10-11, 14-15, 19, 21, 23-25, 29, 30, 34-36, 40-44, 46-48, 50, 55, 57, 59-61, 65-66, 68-72, 74-78, 82, 85, 87-94, 96-100, 103-106, 117, 123-128, 130, 135-136, 170, 219, 241
Salt River Valley Canal, 19-20, 42, 78-79
Salt River Valley Water Users' Association (SRVWUA), 55, 76-80, 90
Salton Sea, 107, 109
San Diego, California, 4, 110, 114, 132, 161-162
San Diego Gas & Electric (SDG&E), 196
San Joaquin Valley, 186-187
San Pedro River, 19
San Tan Ranches, 186, 206, 210
Santa Cruz County, 191
Santa Cruz River, 18
Santa Cruz Valley, 17
Santa Fe Railroad, 43, 82, 92-93, 128, 136
Saylor, John, 177-178
Scharf-Norton Institute for Constitutional Litigation, 227

Schultz, George, 213
SDG&E (See San Diego Gas and Electric)
Segregated Reservoir Sites Act of 1888, 63
Seligman, Arizona, 13, 97, 130, 151
Senate Committee of Agriculture, Nutrition, and Forestry, 203
Sharon, Frederick, 43
Sherman, Moses Hazeltine, 24, 34, 95-96, 103
Shim, Soyean, 228
Shoemaker, Willie, 135
Shoup, Oliver, 113
Sierra Estrella, 19
Simmons, James, 223
Simms, J. T., 34
Sinclair, Ward, 211-212
Skunk Creek, 24-25
Sky Harbor Airport, 103, 202
Slankard, Charlie, 58
Sloan, Richard E., 29-30, 88-89, 95, 241
Smith, Anson, 107
Smith, J. Y. T., 24
Smith, James "Bud", 197
Smith, Marcus Aurelius, 57
Smithsonian Institution, 222-223
Smythe, William Ellsworth, 50-51, 93
Sonoroa, Mexico, 17, 61, 108, 191
Southern Pacific Railroad, 43, 82, 93, 109, 125
Spain, Joshua L., 59-60
Spanish-American War, 64, 90
Sperling, John, 224
SRP (See Salt River Project)
SRVWUA (See Salt River Valley Water Users' Association)
St. Francis Hotel, 225-227
St. Joseph's Hospital, 97, 105, 123, 160, 187, 228
Stanford University, 101, 143, 176, 242
Stanford, Raleigh C., 166
Stanley, E. B., 143
Starer, Andrew, 19
Starer, Jacob, 19
Stauffer, Charles, 95
Stelzreide, 37
Stewart, William M., 8, 47

Welch, Richard, 176
Wells, Richard, 99
West Valley, Arizona, 76
Western Growers Association (WGA),
    198, 202, 208
WGA (See Western Growers Association)
White Tanks Canal Company, 75
Whitten, Jamie, 215
Wickenburg, Arizona, 18, 56, 137, 186
Wilbur, Ray Lyman, 168, 176
Wilkie, Wendall, 132-133, 184, 241
Williams, Ed, 162
Wilmer, Mark, 180, 193-194
Wilson, Fred O., 179
Wilson, Paul N., 235
Wilson, Pete, 205-206
Wilton, Edward H., 61
Winters v. United States (1908), 234
Wolfley, Lewis, 63
Wood, Joseph A., 139
World War I, 2, 91, 97-98, 100, 143, 199
World War II, 4, 126-127, 129-130, 134-
    135, 139, 143, 148-149, 155, 158-159,
    171, 179, 188, 242
Wozencraft, Oliver, 108-109
Wray, Gay Firestone, 193, 223
Wray, Peter, 139, 193
Wright, J.C., 48
Wyoming v. Colorado (1922), 113-116

X Bar One Ranch, 129, 141-142, 191

Yaqui River, 16
Yavapai County, 19, 57, 96, 128, 135, 207
Yavapai Indians, 85
Yorty, Sam, 177
Young Presidents Organization (YPO), 224
Young, John W., 12
Young, W.G., 12
YPO (See Young Presidents Organization)
Yuma, Arizona, 10, 48, 59, 83, 103, 107,
    112, 125, 160, 172

Zanjero, 15, 42